2nd Edition: Frontiers in Health Care for Older Adults

2nd Edition: Frontiers in Health Care for Older Adults

Editors

Francisco José Tarazona-Santabalbina
Sebastià Josep Santaeugènia González
José Augusto García Navarro
José Viña Ribes

MDPI

Basel • Beijing • Wuhan • Barcelona • Belgrade • Novi Sad • Cluj • Manchester

Editors
Francisco José Tarazona-Santabalbina
Universitat Catòlica de Valéncia Sant Vicent Màrtir
Valéncia
Spain

Sebastià Josep Santaeugènia González
Generalitat de Catalunya
Barcelona
Spain

José Augusto García Navarro
Consorci de Salut i Social de Catalunya
Barcelona
Spain

José Viña Ribes
University of Valencia
Valéncia
Spain

Editorial Office
MDPI
St. Alban-Anlage 66
4052 Basel, Switzerland

This is a reprint of articles from the Special Issue published online in the open access journal *International Journal of Environmental Research and Public Health* (ISSN 1660-4601) (available at: https://www.mdpi.com/journal/ijerph/special_issues/health_care_older_adults).

For citation purposes, cite each article independently as indicated on the article page online and as indicated below:

Lastname, A.A.; Lastname, B.B. Article Title. *Journal Name* **Year**, *Volume Number*, Page Range.

ISBN 978-3-7258-0981-3 (Hbk)
ISBN 978-3-7258-0982-0 (PDF)
doi.org/10.3390/books978-3-7258-0982-0

Contents

About the Editors

Francisco José Tarazona-Santabalbina

Francisco José Tarazona-Santabalbina, MD, PhD. Specialist in Geriatric Medicine. He is deputy director of the Hospital Universitario de la Ribera, Alzira, Spain, as well as he is member of CIBERFES (Network Center of Biomedic Investigaion in Frailty and Healthy Aging. He is professor in Geriatric in the Universitat Catòlica de Valéncia Sant Vicent Màrtir and member of the Executive Board of the Spanish Society of Geriatrics and Gerontology. He has published 69 papers on geriatrics and gerontology indexed journals.

Sebastià Josep Santaeugènia González

Sebastià Josep Santaeugènia González, , MD, PhD, MHA. He is Specialist in Internal Medicine and member of the Central Catalonia Chronicity Research Group (C3RG), Centre for Health and Social Care Research (CESS), Universitat de Vic–University of Vic-Central University of Catalonia (UVIC-UCC). Currently, he is General Director of the Parc Sanitari Sant Joan de Déu (Sant Boi del Llobregat) and he has published more than 50 papers on geriatric and palliative care indexed journals.

José Augusto García Navarro

José Augusto García Navarro is a geriatrician that has developed most of hiscareer in the planning and management of health services, with special attention to geriatric services and care for disabled elderly people. He is currently the director of the Health and Social Consortium of Catalonia, member of the Board of Directors of the Public Health System of Catalonia, professor of health management at the University of Barcelona and the International University of Catalonia. He is currently leading an important study about deinstitutionalization of disabled elderly people to improve the delivery of long-term care in Spain. He is the president of the Spanish Society of Geriatrics and Gerontology.

José Viña Ribes

José Viña studied Medicine at the University of Valencia, Spain, where he obtained his PhD after doing research work with Hans Krebs in Oxford, UK. Dr. Viña is Professor of Physiology at the University of Valencia, where he combines teaching with research, the latter in two main lines, aging and exercise. José Viña leads a successful research group called FRESHAGE that works on different aspects of aging, including healthy aging, exercise, and Alzheimer's disease. His h-index is 78. He has over 350 publications in international journals) on glutathione, mitochondria, oxidative stress, free radicals, nutrition, and exercise. Prof. Viña has received numerous awards and distinctions, including the Albert Struyvenberg Medal, awarded by the European Society for Clinical Research (ESCI) 2017, and the Research Trajectory Award from the Council of the University of Valencia 2019. Dr. Viña gave a lecture at the 2018 Gerontological Society of America presidential address. Dr. Viña is a member of the Royal Academy of Medicine of the Valencian Community. He was awarded honorary doctorates, from University of Buenos Aires (Argentina) and University of Rennes (France). Dr. Vina was President of the Society for Free Radical Research International (SFRR-I) 2020–21.

Dr. Viña's work has been recognized by his colleagues. Dr. Viña's articles (including book chapters) have been cited more than 21,100 times, resulting in an h-index of 78 and an extended h-index (including book chapter citations) of 85.

Dr. Viña was the author of the most cited article in the Biochemical Journal 1978, also in the Free Radical Biology and Medicine in 2003 and in the Journal of Gerontology in 2013. He was the most cited author in Free Radical Research 2008. He was the second most cited author most cited author the Biochemical Journal in 1980 and in Free Radical Biology and Medicine in 2019.

Preface

Frontiers in Health Care for Older Adults is the second, but not the last, edition of a challenge that began in 2021 with the first edition of this monograph focusing on health care for older adults. In this volume of 14 chapters, we address various concerns and new developments, focusing on highly prevalent pathologies such as anaemia and stroke; the relationship between successful longevity and the presence of geriatric syndromes such as frailty and sarcopenia; the benefits of prehabilitation protocols in reducing adverse events in programmed surgery of oncological aetiology; the consequences of falls in the older adults, and its relationship with mobility problems and pain associated with degenerative joint disease, the benefits of intensive home care for the health of the population with complex co-morbidity profiles and the need to adapt these programmes to rural care needs; the importance of frailty in the evolution of COPD older patients; not forgetting the role of pharmacists in detecting polypharmacy and inappropriate prescribing in this population.

In short, a collection of interesting scientific contributions that can help us to improve the quality of life of a group whose percentage of the total population is increasing every day.

Francisco José Tarazona-Santabalbina, Sebastià Josep Santaeugènia González, José Augusto García Navarro, and José Viña Ribes
Editors

International Journal of
Environmental Research and Public Health

MDPI

Article

Association between Anemia Severity and Ischemic Stroke Incidence: A Retrospective Cohort Study

Hui-Fen Chen [1], Tsing-Fen Ho [2,*], Yu-Hung Kuo [3] and Ju-Huei Chien [2,3,*]

[1] Department of Nephrology, Taichung Tzu-Chi Hospital, Buddhist Tzu-Chi Medical Foundation, No. 88 Fong-Shing Rd., Taichung City 42743, Taiwan
[2] Department of Medical Laboratory Science and Biotechnology, Central Taiwan University of Science and Technology, No. 666 Buzih Rd., Taichung City 40601, Taiwan
[3] Department of Research, Taichung Tzu-Chi Hospital, Buddhist Tzu-Chi Medical Foundation, No. 88 Fong-Shing Rd., Taichung City 42743, Taiwan
* Correspondence: tfho@ctust.edu.tw (T.-F.H.); tinachien0312@gmail.com (J.-H.C.)

Abstract: Stroke patients presenting with anemia at the time of stroke onset had a higher risk of mortality and development of other cardiovascular diseases and comorbidities. The association between the severity of anemia and the risk of developing a stroke is still uncertain. This retrospective study aimed to evaluate the association between stroke incidence and anemia severity (by WHO criteria). A total of 71,787 patients were included, of whom 16,708 (23.27%) were identified as anemic and 55,079 patients were anemia-free. Female patients (62.98%) were more likely to have anemia than males (37.02%). The likelihood of having a stroke within eight years after anemia diagnosis was calculated using Cox proportional hazard regression. Patients with moderate anemia had a significant increase in stroke risk compared to the non-anemia group in univariate analyses (hazard ratios [HR] = 2.31, 95% confidence interval [CI], 1.97–2.71, $p < 0.001$) and in adjusted HRs (adj-HR = 1.20, 95% CI, 1.02–1.43, $p = 0.032$). The data reveal that patients with severe anemia received more anemia treatment, such as blood transfusion and nutritional supplementation, and maintaining blood homeostasis may be important to preventing stroke. Anemia is an important risk factor, but other risk factors, including diabetes and hyperlipidemia, also affect stroke development. There is a heightened awareness of anemia's severity and the increasing risk of stroke development.

Keywords: anemia; ischemic stroke; elderly; hemoglobin; retrospective cohort study

Citation: Chen, H.-F.; Ho, T.-F.; Kuo, Y.-H.; Chien, J.-H. Association between Anemia Severity and Ischemic Stroke Incidence: A Retrospective Cohort Study. *Int. J. Environ. Res. Public Health* **2023**, *20*, 3849. https://doi.org/10.3390/ijerph20053849

Academic Editors: Francisco José Tarazona Santabalbina, Sebastià Josep Santaeugènia Gonzàlez, José Augusto García Navarro and José Viña

Received: 17 November 2022
Revised: 17 February 2023
Accepted: 17 February 2023
Published: 21 February 2023

1. Introduction

Stroke is a leading cause of death and disability worldwide [1]. Stroke survivors suffer from various impairments and complications affecting motor, sensory, visual, language, and cognitive functions [2,3]. Therefore, a stroke imposes a great burden on patients as well as their caregivers and family members. Stroke patients may be hospitalized or may frequently visit the emergency department owing to their long-term sequelae and disability, which not only dramatically increases the burden on caregivers and their family's finances, but also severely affects their quality of life. There are numerous recognized risk factors for stroke, such as hypertension, hyperlipidemia, diabetes mellitus, cigarette use, obesity, age, and physical activity [1,4,5]. Increases in the elderly population and life expectancy are also key reasons for the increase in number of stroke patients.

Anemia affects 15–32% of the world's population, is usually present in stroke patients, and can worsen with aging [6,7]. In 2019, the age groups of 15 to 19 and 95 and older, for both males and females, had the highest global point prevalence of anemia. The mean (range) global prevalence rates of mild, moderate, and severe anemia were approximately 54.1%, (53.8–54.4%), 42.5% (42.2–42.7%), and 3.4% (3.3–3.5%), respectively [8]. Elderly individuals may experience malnutrition and dyspepsia as their physical condition deteriorates with age, and this may affect their hematopoiesis functions, thereby causing anemia or pancytopenia. Anemia is also a risk factor for ischemic stroke and is related to high post-stroke mortality [9,10].

Nevertheless, previous research has suggested that anemia may raise the risk of stroke. However, the new stroke guidelines from the American Stroke Association (ASA) do not list anemia as a major stroke risk factor [11]. Here, we conducted a retrospective cohort study to investigate the association between the severity of anemia and stroke incidence. Owing to Taiwan's National Health Insurance (NHI) policy, anemia is rarely listed as a primary condition and may not be documented on patient medical records on the basis of *International Classification of Diseases, Tenth Revision (ICD-10)* codes. The laboratory data of anemia status were not available in Taiwan's NHI system, and the prevalence of anemia could be underestimated. Moreover, the data of association between anemia and comorbidities in the Taiwanese population are scarce. An evaluation of the stroke risk factors, especially anemia severity, could provide important information that may enhance medical care or even national healthcare planning. This study retrospectively evaluated the prevalence and characteristics of anemia in hospitalized patients and analyzed whether anemia severity based on the hemoglobin (Hb) level was associated with stroke development.

2. Materials and Methods

2.1. Study Cohort

This retrospective cohort study included 454,424 patients aged ≥20 years who had visited or were hospitalized at Taichung Tzu-Chi Hospital, Taiwan, from 2013 to 2019. A total of 71,787 patients underwent at least 1 blood Hb measurement performed using a Sysmex XE-5000 hematology analyzer (Sysmex Co., Kobe, Japan) within 1 year to confirm their anemia status. This study was approved by the Research Ethics Committee of Taichung Tzu-Chi Hospital (REC 111-02). The need for informed consent was waived owing to the retrospective nature of the study and the use of anonymous medical records.

2.2. Definition of Anemia and ICD Codes

Adult patients older than the age of 20 were included in this study. All participants in this study completed at least one Hb measurement, and persons who did not fulfill the predetermined criteria were not included. The date of laboratory Hb measurement was defined as the index date, and the anemia severity was classified according to the World Health Organization (WHO) criteria [12]. We categorized the patients into different groups according to their anemia severity. Anemia is defined as an Hb level of <13.0 g/dL for men and <12.0 g/dL for women. The cutoff for Hb in mild anemia was 11.0–11.9 g/dL for women and 11.0–12.9 g/dL for men, whereas the cutoffs for moderate and severe anemia were 8.0–10.9 and <8.0 g/dL, respectively, for both men and women. As shown in Figure 1, the exclusion criteria were as follows: (1) patients without Hb measurements; (2) receiving a diagnosis that might affect the Hb status, including gastric intestinal bleeding (*ICD-10* code K92.2), bleeding (*ICD-10* code R58), trauma (*ICD-10* code T79.2), excessive bleeding associated with menopause onset (*ICD-10* code N92.4), intraoperative and postprocedural complications of spleen, endocrine, and nervous system (*ICD-10* code D78, E36, G97), excessive bleeding with onset of menstrual bleeding (*ICD-10* code N92.2), traumatic hemorrhage of the cerebrum (*ICD-10* code S06.360A), hemorrhage from respiratory passages (*ICD-10* code R04.9), nontraumatic intracerebral hemorrhage (*ICD-10* code I61.9), spleen diseases (*ICD-10* code D73), pulmonary vessels diseases (*ICD-10* code I28), stomach and duodenum diseases (*ICD-10* code K31), acute myocardial infarction (*ICD-10* code I21), injury to an unspecified body region (*ICD-10* code T14), or absent, scanty, or rare menstruation (*ICD-10* code N91), before their index date until anemia diagnosis; (3) receiving a stroke diagnosis before the index date on the basis of the *ICD-10* codes I63; (4) not visiting our out-patient clinic or being hospitalized within the last 2 years; and (5) death or leaving against medical advice (DAMA) less than 1 month after the index date.

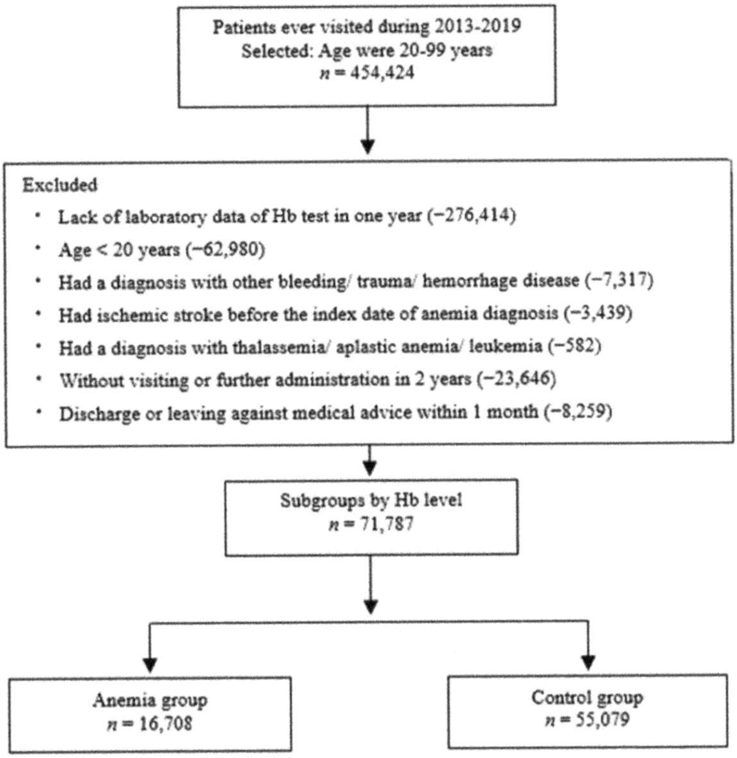

Figure 1. Flowchart of the patient enrollment process in this study. A total of 71,787 patients were included in this study. In total, 16,708 patients were subgrouped into the anemia group and 55,079 patients were subgrouped into the normal group.

A flowchart of the patient enrollment process is illustrated in Figure 1. All patients were grouped by sex and age (20–30, 31–40, 41–50, 51–60, 61–70, 71–80, and >80 years). The Hb status confirmation date was identified as the index date for the case and control groups, and stroke events were followed subsequently.

2.3. Outcome and Associated Factors

The eligibility of all patients was retrospectively determined in this cohort study. The severity of anemia was then subgrouped based on Hb level, and the stroke patients were those who had at least two *ICD-10* admission claims for clinic OPD visits or stroke-related hospitalization in our hospital during the study period. During the monitoring period, the occurrence of subsequent disease was examined. The occurrence of subsequent disease was analyzed during the observation period. Patients were individually tracked for 2–8 years, beginning on the index date, and followed thereafter. In this study, the outcome of stroke was defined as admission claims of *ICD-10* codeI63, cerebral infarction. The accuracy of diagnoses from claims data was verified in a previous study showing that the PPV and sensitivity of *ICD-10-CM* code I63 as a primary diagnosis of acute ischemic stroke were 92.7% and 99.4%, respectively [13]. We also analyzed the hazard ratio for comorbidities that were potentially linked to stroke: hypertension (I10–I13, I15), diabetes (E08–E11, E13), chronic kidney disease (CKD; N17–N19, I12, I13), chronic heart failure (I50), chronic obstructive pulmonary disease (J44, J60–70), hyperlipidemia (E78.0-E78.5), and atrial fibrillation (I48). The comorbidities were defined as the presence or absence of accompanying disease within one year before the index date of anemia. The national health insurance program (NHI) in Taiwan is mandatory for all citizens, and various medications and medical procedures were coded with unique code. In this study, six frequently prescribed drugs were included to investigate the efficacy of various anemia therapies for patients within six months after the hemoglobin measurement index date. These medications included iron (hydroxide-polymaltose complex, Yuanchou Chemical and Pharmaceutical Co., Ltd., Taiwan, NHI code AC46166100), ferric hydroxide sucrose complex (TCM Biotech international Corp. Taiwan, NHI code AC57884221), sodium ferrous citrate (Guang Heng Enterprise Co., Ltd. Taiwan, NHI code BC22097100), hydroxocobalamin (Shinlin Sinseng Pharmaceutical Co., Ltd. Taiwan, ACETATE, NHI code AC09754209), mecobalmin (Eisai Taiwan Inc., NHI code AC296301G0), folic acid (Johnson Chemical Pharmaceutical works Co., Ltd. Taiwan, NHI code AC346701G0), and blood transfusion (NHI code 94001C).

2.4. Statistical Analysis

Statistical analyses were conducted using the SAS statistical package (Version 9.4) and SPSS (version 28.0, SPSS Inc., Chicago, IL, USA) to examine the prevalence and clinical trends of anemia among the different age groups, sexes, and comorbidities. The categorical variables were assessed by applying a Chi-square test. The continuous variables were assessed by applying a t test. Furthermore, different predictors were used to estimate relative risks [14]. To examine the stroke risk associations with anemia, the deaths as competing risks of stroke were analyzed by using a Cox proportional cause-specific hazard model to calculate hazard ratios (HR), 95% confidence intervals (CIs), and two-sided p values. A two-sided p value of <0.05 was considered statistically significant. A multivariate Cox proportional cause-specific hazard regression model was adjusted for age, sex, and comorbidities. A proportional hazard assumption was evaluated by the Kolmogorov-type Supremum test; that was not violated.

3. Results

As shown in Figure 1, only 71,787 of the 454,424 patients who visited our facility qualified for the retrospective cohort research. The baseline characteristics of the case and control groups are summarized in Table 1. The mean Hb level was 14.2 ± 1.3 g/dL in the normal group and 10.7 ± 1.6 g/dL in the anemia group.

Table 1. Baseline characteristics of study cohort.

Characteristic n = 71,787	Normal n = 55,079 (76.73%)		All Anemia n = 16,708 (23.27%)		p	Mild Anemia n = 9065 (53.99%)		Moderate Anemia n = 6532 (39.27%)		Severe Anemia n = 1111 (6.75%)	
Hb (g/dL)	14.2 ± 1.3		10.7 ± 1.6		<0.001	11.8 ± 0.5		9.9 ± 0.8		6.7 ± 1.1	
Gender											
Male	26,184	47.54%	6185	37.02%		3933	43.39%	1880	28.78%	372	33.48%
Female	28,895	52.46%	10, 523	62.98%		5132	56.61%	4652	71.22%	739	66.52%
Age (years)	50.6 ± 16.3		59.1 ± 18.5		<0.001	58.9 ± 18.5		59.7 ± 18.6		57.9 ± 17.5	
20–30	7298	13.25%	1123	6.72%	<0.001	704	8.1%	705	7.78%	369	5.65%
31–40	9139	16.59%	1961	11.74%		1069	12.3%	1077	11.88%	756	11.57%
41–50	10,423	18.92%	2883	17.26%		1287	14.9%	1316	14.52%	1275	19.52%
51–60	12,475	22.65%	2586	15.48%		1451	16.8%	1489	16.43%	919	14.07%
61–70	9168	16.65%	2754	16.48%		1512	17.5%	1610	17.76%	999	15.29%
71–80	4732	8.59%	2990	17.90%		1536	17.7%	1671	18.43%	1147	17.56%
81 above	1844	3.35%	2411	14.43%		1100	12.7%	1197	13.20%	1067	16.33%
Comorbidities											
Hypertension	6497	11.80%	3408	20.40%	<0.001	1797	19.82%	1400	21.43%	211	18.99%
Diabetes	3713	6.74%	2467	14.77%	<0.001	1254	13.83%	1067	16.33%	146	13.14%
Chronic kidney disease	496	0.90%	1015	6.07%	<0.001	300	3.31%	580	8.88%	135	12.15%
Chronic heart failure disease	500	0.91%	446	2.67%	<0.001	206	2.27%	208	3.18%	32	2.88%
Chronic obstructive pulmonary disease	1159	2.10%	495	2.96%	<0.001	293	3.23%	179	2.74%	23	2.07%
Hyperlipidemia	2177	3.95%	643	3.85%	0.544	395	4.36%	222	3.40%	26	2.34%
Atrial fibrillation	270	0.49%	169	1.01%	<0.001	86	0.95%	73	1.12%	10	0.90%
Treatment											
Blood transfusion	1795	3.26%	3273	19.59%	<0.001	1017	11.22%	1577	24.14%	679	61.12%
Iron therapy	72	0.13%	662	3.96%	<0.001	67	0.74%	388	5.94%	207	18.63%
Folic acid supplement	344	0.62%	469	2.81%	<0.001	135	1.49%	226	3.46%	108	9.72%
Vitamin B12 supplement	68	0.12%	94	0.56%	<0.001	37	0.41%	38	0.58%	19	1.71%

Of the 16,708 anemia patients, 6185 (37.02%) were men and 10,523 (62.98%) were women. The mean age of the case group was 59.1 ± 18.5 years, and that of the control group was 50.6 ± 16.3 years. The case group had a higher incidence of comorbidities, including hypertension (11.80% versus 20.40%, $p < 0.001$), diabetes (6.74% versus 14.77%, $p < 0.001$), CKD (0.90% versus 6.07%, $p < 0.001$), chronic heart failure (0.91% versus 2.67%, $p < 0.001$), chronic obstructive pulmonary disease (2.10% versus 2.96%, $p < 0.001$), and atrial fibrillation (0.49% versus 1.01%, $p < 0.001$), than did the control group.

Table 2 presents the anemia severity and subsequent cases of stroke. The patients with anemia were further divided into three subgroups according to anemia severity, determined on the basis of Hb levels by WHO criteria [12]. Thus, of the 16,708 patients with anemia, 9065 (54.25%) had mild anemia, 6532 (39.09%) had moderate anemia, and 1111 (6.65%) had severe anemia. During follow-up, a total of 447 anemia patients (2.68%, 447/16,708) and 744 controls (1.35%, 744/55,079) were diagnosed as having stroke. Moreover, there were 740 non-anemia patient deaths and 1229 anemia patient deaths throughout the 8-year follow-up period (1.34% and 7.63%, respectively).

Table 2. Severity of anemia classified according to WHO criteria and subsequent stroke events.

Severity of Anemia	WHO Criteria (Hb, g/dL)		Study Cohort *n* = 71,787		Stroke Events *n* = 1191		Death Events *n* = 1969		Average Follow Up
	Male	**Female**	*n*	%	*n*	%	*n*	%	**Years**
Normal	>13.0	>12.0	55,079	—	744	1.35	740	1.34	2.32 ± 2.04
Anemia			16,708		447	2.68	1229	7.63	
Mild	11.0–12.9	11.0–11.9	9065	54.25	229	2.53	496	5.47	2.09 ± 1.95
Moderate	8.0–10.9	8.0–10.9	6532	39.09	193	2.95	599	9.17	1.98 ± 1.95
Severe	<8.0	<8.0	1111	6.65	25	2.25	134	12.06	1.95 ± 1.94

We observed a positive association between the severity of anemia, determined based on Hb measurements, and the risk of stroke. Figure 2 illustrates the cumulative incidence of stroke in the three subgroups of anemia severity during the 8-year follow-up. A higher incidence of stroke events was noted in the patients with moderate anemia after their diagnosis during the 8-year follow-up (log-rank test, $p < 0.001$).

Table 3 illustrates the univariate and adjusted associations between the risk of stroke and the severity of anemia, sex, age, and comorbidities. The risk of stroke was higher in the case group than in the control group. In univariate regression analysis, we found moderate anemia (HR = 2.31; 95% CI, 1.97–2.71) had a significant increase in stroke risk compared to the non-anemia group. After adjusting, we found the risk of stroke was higher in the patients with moderate anemia (adj-HR, 1.20; 95% CI, 1.02–1.43, $p = 0.032$) than in the controls. The same results were obtained for gender and age by both univariate analysis (HR = 1.66, 95% CI = 1.48–1.87, $p < 0.001$; HR = 1.07, 95% CI = 1.07–1.08, $p < 0.001$, respectively) and adjusted HRs (adj-HR = 1.64, 95% CI = 1.46–1.85, $p < 0.001$; adj-HR = 1.07, 95% CI = 1.065–1.074, $p < 0.001$, respectively). Furthermore, the case group had a higher prevalence of comorbidities than did the control group. However, only the comorbidities diabetes mellitus and hyperlipidemia, by both univariate analysis (HR = 2.86, 95% CI = 2.50–3.28, $p < 0.001$; HR = 1.89, 95% CI = 1.54–2.31, $p < 0.001$, respectively) and adjusted HRs (adj-HR, 1.48; 95% CI, 1.27–1.71; $p < 0.001$), (adj-HR, 1.13; 95% CI, 0.91–1.39; $p = 0.280$), were associated with a higher risk of stroke in the case group compared to the control group.

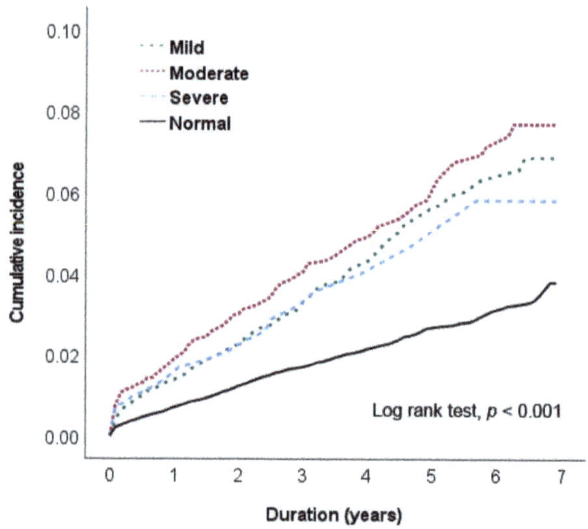

Figure 2. Cumulative risk of stroke based on the anemia severity during the 8-year follow-up: mild anemia (green curve), moderate anemia (red curve), severe anemia (blue curve), and non-anemia (controls; black curve).

Table 3. Risk associations between stroke and anemia, sex, age, and comorbidities.

Predictors	HR (95% CI)			
	Univariate	***p***	**Adjusted**	***p***
Severity of anemia				
Normal	—		—	
Mild	1.96 (1.69–2.27)	<0.001	0.98 (0.84–1.15)	0.795
Moderate	2.31 (1.97–2.71)	<0.001	1.20 (1.02–1.43)	0.032
Severe	1.73 (1.16–2.58)	0.007	0.99 (0.66–1.48)	0.943
Gender (male)	1.66 (1.48–1.87)	<0.001	1.64 (1.46–1.85)	<0.001
Age (years)	1.07 (1.07–1.08)	<0.001	1.07 (1.065–1.074)	<0.001
Comorbidity				
Hypertension	3.11 (2.76–3.50)	<0.001	1.26 (1.10–1.45)	0.001
Diabetes mellitus	2.86 (2.50–3.28)	<0.001	1.48 (1.27–1.71)	<0.001
Chronic kidney disease	2.65 (2.10–3.35)	<0.001	1.02 (0.80–1.31)	0.869
Chronic heart failure disease	3.18 (2.41–4.19)	<0.001	1.00 (0.74–1.35)	0.988
Chronic obstructive pulmonary disease	1.88 (144–2.46)	<0.001	0.76 (0.58–1.00)	0.053
Hyperlipidemia	1.89 (1.54–2.31)	<0.001	1.13 (0.91–1.39)	0.280
Atrial fibrillation	5.50 (3.98–7.59)	<0.001	1.84 (1.31–2.60)	<0.001

HR: hazard ratio; CI: confidence interval. $p < 0.05$ was considered statistically significant.

4. Discussion

This retrospective study evaluated the prevalence and characteristics of anemia and the risk of stroke. The strength of this study is that it identified the association between anemia and the risk of stroke by using a hospital-based database, from which the laboratory data were retrieved to classify the severity of anemia. In contrast to previous studies, which have estimated the risk of stroke associated with anemia by using data from Taiwan's NHI databases based on *ICD* codes and lacked conclusive laboratory Hb measurements [15,16], our study analyzed laboratory data and classified the patients into subgroups according to the severity of anemia to assess the associations between anemia severity and the risk of stroke. We also excluded patients with diseases that might interfere with our results, including those with a tendency of bleeding, other hemorrhagic disease, and persons who did not fulfill the predetermined criteria were also excluded. All participants in this study completed at least one Hb measurement, and persons who did not fulfill the predetermined criteria were then excluded. Our findings indicate that patients with moderate anemia showed an increased likelihood of stroke development.

In this retrospective analysis, there were more female anemic patients than male anemic patients. In the initial stage, the primary signs of mild anemia include fatigue, light skin, dizziness, debility, and headaches. Patients in the early stage of anemia or mild anemia may not seek medical care or consultations with physicians, particularly middle-aged men. Many male patients did not meet the criteria for hospital visits in 2 years. On the other hand, most women experience menopause at the age of 40–50 years; thus, some anemia symptoms, such as dizziness, fatigue, or paleness, may be overlooked or misdiagnosed as menopausal symptoms. Even when individuals visit a hospital or clinic, medical personnel tend to focus more on other maladies rather than anemia. However, if anemic condition is left untreated for a longer period, the consequences and complications can become more severe, causing shortness of breath, low blood pressure, arrhythmia, and even chronic heart failure. Results from this research demonstrated an increased risk of stroke occurrence in moderate anemia patients compared with the non-anemia control group. Additionally, the mortality rate in the severe anemia group was 12%, much higher than that of other patients with anemia in this study. Patients suffering from severe anemia might die from other illnesses caused by their feeble condition prior to having a stroke. As a result, the risk of stroke in the severe anemia group was observed to be lower than in the moderate anemia group.

According to statistical data from Taiwan's Ministry of the Interior, the population aged >65 years increased from 11.15% in 2012 to 16.68% in 2021. In the past two decades, the average life expectancy also increased from 76.75 to 81.30 years. The Council for Economic Planning and Development estimated that Taiwan will become a super-aged society by as early as 2025; moreover, the population aged ≥65 years is expected to account for >20% of all individuals [17]. This accelerated speed of aging has become a burden to the healthcare system and society. In this study, there is an upward trend in the prevalence of anemia with age (from 6.72% in the 20–30 age range to over 15% in the elderly age groups; Table 1). Our results are consistent with the global prevalence of anemia, indicating that the trend of anemia burden increases with age [18–20]. We observed that the anemia prevalence peaked at 17.3% in the 71–80 age group and at 14.4% in the >80 age group. Anemia rates in the 71–80 age range in this study cohort were 4.2% (2990/71,787) and 3.4% (2411/71,787), respectively. In this study, the prevalence of anemia in people over 60 is approximately 11%, which is lower than it is in other Asian countries, such as Korea, where it is 13.8% for people over 65 [20]. Anemia is a common condition in older adults and can be caused by various factors such as poor nutrition, chronic diseases, medication, and healthcare. Taiwan has a relatively high standard of living, and the population has access to a variety of nutritious foods, which helps to prevent nutrient deficiencies, including iron deficiency. Moreover, Taiwan has a well-developed healthcare and medical insurance system. The Ministry of Health and Welfare also promotes and encourages all citizens above the age of 45 to participate in adult health checkup programs. These programs enable the early detection of diseases such as cancer and other chronic disease, as well as delivery of comprehensive healthcare prior to the disease worsening [17]. Therefore, all those factors may contribute to reduce the overall prevalence of anemia in the population. In elderly people, anemia has been reported to be associated with cardiovascular disease [21], stroke [6], dementia [22], frailty [23], and high morbidity as well as mortality [24]. Because of Taiwan's NHI policy, however, anemia has rarely been listed as a primary condition in elderly people. According to the WHO recommendation, an anemia prevalence of >5% is considered to be of public health significance [12] and may require public health attention and intervention. The increased prevalence of anemia in the elderly should be considered an important public issue in Taiwan.

In this study, we also observed a higher prevalence of pre-existing comorbidities among the anemia group compared to the non-anemia population. The moderate to severe anemia patients had higher all-cause mortality compared to the non-anemia group; this trend was mentioned in previous studies [9,25]. Other unreported comorbidities may interfere with the association between anemia and stroke. Severe anemia might be corrected well, but mild to moderate anemia might become a chronic condition which eventually becomes associated with stroke. In this study, we observed that patients with severe anemia required blood transfusions more frequently than the group with moderate anemia and the control group. However, a study by Dr. Ren that was published in Nature Communications raises the possibility that blood transfusions might be advantageous to health even up to seven hours after a stroke in a mouse model. Their team discovered that replenishing 20% of the mouse's blood was sufficient to significantly lessen brain damage [26]. However, there are few studies focusing on maintaining hemodynamic condition in severe anemia patients to prevent stroke. Therefore, more studies might help to clarify the benefit from blood transfusions on this issue in the future. Furthermore, the different therapeutic strategies may explain why severe anemia portends lower stroke risk than other anemia severities.

Studies assessing the association between anemia and comorbidities in the Taiwanese population are rare. Anemia, a direct consequence of decreases in Hb and red blood cell (RBC) levels in circulation, is a multifactorial condition; lack of iron, folate, and vitamin B12 are well-known causes of anemia. The most common type of anemia is iron deficiency anemia, which may account for as much as 50% of all explained anemia cases [27]. Other diseases such as diabetes, chronic infections, inflammation, and CKD also affect RBC proliferation, erythropoietin production, androgen secretion, and myelodysplasia [28]. Anemia is also positively associated with impaired renal function. Taiwan has one of the highest number of cases of CKD and end-stage renal disease in the world; CKD is the most frequent cause of anemia [8,20,21,29]. The severity of anemia is directly related to the degree of renal dysfunction. CKD causes reduction in erythropoietin synthesis, subsequently resulting in decreased cell proliferation. At least one-third of anemia patients aged >65 years have CKD or autoimmune diseases/chronic infection [30]. Patients with CKD are also at a significant risk for stroke, including the ischemic and hemorrhagic subtypes. The mechanisms linked to higher risk of stroke in CKD patients include alterations in cardiac output, platelet function, regional cerebral perfusion, accelerated systemic atherosclerosis, altered blood brain barrier, and disordered neurovascular coupling [31]. Additionally, Dr. Poznyak also identified the atherosclerosis-specific features in chronic kidney disease (CKD) in a recent study [32]. The major symptoms of anemia may range from mild fatigue to severe systemic illnesses. In addition, accumulating evidence indicates that anemia engenders outcomes such as increased stroke [9], heart failure [33], hospitalization [25], and mortality [34], all of which impose a severe burden on healthcare systems. Furthermore, anemia is associated with increased iron overload, increased chances of viral infection [35], and increased risks of myocardial infarction [36]. We also analyzed other known conventional risk factors, such as hyperlipidemia and atrial fibrillation, that affect the development of stroke; the hazard ratio was slightly different to other investigations [9]. Hyperlipidemia is an important risk factor for stroke [4,37]. Atrial fibrillation (AF) is a frequent cardiac rhythm disease associated with various significant negative health outcomes, such as heart failure and stroke. Particularly in women, atrial fibrillation is linked to an increased long-term risk of stroke, heart failure, and all-cause death [38,39]. Many investigations also revealed that anemia is a frequently observed comorbidity in patients with AF and is associated with cardiovascular, stroke, and gastrointestinal bleeding [40].

In medical practice, those experiencing moderate to severe anemia are more likely to receive medical attention than those with mild anemia. This means that patients with moderate to severe anemia with signs of illness symptoms would be given blood transfusions, iron supplements, and vitamin B12, while mild anemia would more likely be overlooked [41–43]. Regarding the management of anemic patients, blood transfusions are often seen as an effective way to increase hemoglobin levels and improve their overall health. In this study, we examined patients who received transfusions and pharmacological therapy within six months of the diagnosis index date. According to our results, patients with moderate and severe anemia received a greater proportion of blood transfusion than those with mild anemia (24.14%, 61.12% vs. 11.22%, Table 1). Blood transfusions can maintain in the body's hemodynamics and alter the viscosity of the blood. Keeping the blood in balance in the body's circulation and offering better care may be a strategy to prevent stroke. However, blood transfusion is influenced by a number of circumstances and the decision of the healthcare professionals. Patients who receive frequent transfusions may also be exposed to an increased risk of stroke. To ascertain the beneficial effects of anemia therapies such as transfusion and other medication on reducing the chance of stroke, further research must be conducted.

Despite its strengths, our study has some limitations that should be noted. First, the different types of anemia, such as iron deficiency anemia or folic acid anemia, were not correctly defined in this study. Second, we could not analyze data regarding lifestyles or socioeconomic status, such as smoking, alcohol habits, obesity, education, or financial condition. Third, in order to confirm the validity of the diagnosis for anemia, we only included the patients with one Hb measurement, which could cause a potential selection bias in a retrospective study. The medical service of our hospital serves a population of approximately 2.8 million in the center area of Taiwan, and more than 700,000 clinical visits are made each year. Finally, we did not retrieve clinical data on atherosclerosis, nutrition, pregnancy, or endogenous hormones, which might be predisposing factors for stroke and the retrospective data from the hospital might still miss a few stroke patients who were diagnosed in other hospitals or died at home.

5. Conclusions

This study assessed the association between anemia and the risk of stroke. The prevalence of anemia was found to increase with age. A high prevalence of anemia is expected to impose a major medical burden in countries becoming super-aged societies. In this study, the risk of stroke was found to be associated with age, regardless of sex. Our study reveals that moderate anemia should be considered an increased risk factor associated with stroke incidence, and monitoring anemia severity as well as other risk factors and biomarkers is crucial in clinical practice.

Author Contributions: H.-F.C.: resources, methodology, data curation; T.-F.H.: conceptualization, formal analysis, writing—review and editing; Y.-H.K.: investigation; J.-H.C.: conceptualization, data curation, writing—original draft preparation. All authors have read and agreed to the published version of the manuscript.

Funding: This research was supported by the Ministry of Science and Technology of Taiwan (MOST 110-2637-H-166-001), and the Taichung Tzu-Chi Hospital, Taichung, Taiwan (TCCRD112-27). The funders had no role in study design, data collection and analysis, decision to publish, or preparation of the manuscript.

Institutional Review Board Statement: The study has been approved by the Ethics Committee of the Taichung Tzu-Chi Hospital (REC 111-02).

Informed Consent Statement: For this project, the IRB approved waiving documentation of informed consent.

Data Availability Statement: The datasets generated and analysed in this study are not publicly available due to Tzu Chi Hospital regulations.

Conflicts of Interest: The authors declare no conflict of interest.

References

1. Feigin, V.L.; Stark, B.A.; Johnson, C.O.; Roth, G.A.; Bisignano, C.; Abady, G.G.; Abbasifard, M.; Abbasi-Kangevari, M.; Abd-Allah, F.; Abedi, V.; et al. Global, regional, and national burden of stroke and its risk factors, 1990–2019: A systematic analysis for the Global Burden of Disease Study 2019. *Lancet Neurol.* **2021**, *20*, 795–820. [CrossRef] [PubMed]
2. Fisher, R.J.; Riley-Bennett, F.; Russell, L.; Lee, C.; Sturt, R.; Walker, M.; Sackley, C. Nominal group technique to establish the core components of home-based rehabilitation for survivors of stroke with severe disability. *BMJ Open* **2021**, *11*, e052593. [CrossRef]

3. Gottesman, R.F.; Seshadri, S. Risk Factors, Lifestyle Behaviors, and Vascular Brain Health. *Stroke* **2022**, *53*, 394–403. [CrossRef]
4. Guzik, A.; Bushnell, C. Stroke epidemiology and risk factor management. *Continuum Lifelong Learn. Neurol.* **2017**, *23*, 15–39. [CrossRef] [PubMed]
5. Bang, O.Y.; Ovbiagele, B.; Kim, J.S. Nontraditional risk factors for ischemic stroke: An update. *Stroke* **2015**, *46*, 3571–3578. [CrossRef] [PubMed]
6. Kaiafa, G.; Savopoulos, C.; Kanellos, I.; Mylonas, K.S.; Tsikalakis, G.; Tegos, T.; Kakaletsis, N.; Hatzitolios, A.I. Anemia and stroke: Where do we stand? *Acta Neurol. Scand.* **2017**, *135*, 596–602. [CrossRef]
7. Kassebaum, N.J.; Jasrasaria, R.; Naghavi, M.; Wulf, S.K.; Johns, N.; Lozano, R.; Regan, M.; Weatherall, D.; Chou, D.P.; Eisele, T.P.; et al. A systematic analysis of global anemia burden from 1990 to 2010. *Blood* **2014**, *123*, 615–624. [CrossRef]
8. Safiri, S.; Kolahi, A.A.; Noori, M.; Nejadghaderi, S.A.; Karamzad, N.; Bragazzi, N.L.; Sullman, M.J.M.; Abdollahi, M.; Collins, G.S.; Kaufman, J.S.; et al. Burden of anemia and its underlying causes in 204 countries and territories, 1990-2019: Results from the Global Burden of Disease Study 2019. *J. Hematol. Oncol.* **2021**, *14*, 185. [CrossRef]
9. Heo, J.; Youk, T.M.; Seo, K.D. Anemia is a risk factor for the development of ischemic stroke and post-stroke mortality. *J. Clin. Med.* **2021**, *10*, 2556. [CrossRef]
10. Kaiafa, G.; Kanellos, I.; Savopoulos, C.; Kakaletsis, N.; Giannakoulas, G.; Hatzitolios, A.I. Is anemia a new cardiovascular risk factor? *Int. J. Cardiol.* **2015**, *186*, 117–124. [CrossRef]
11. Kleindorfer, D.O.; Towfighi, A.; Chaturvedi, S.; Cockroft, K.M.; Gutierrez, J.; Lombardi-Hill, D.; Kamel, H.; Kernan, W.N.; Kittner, S.J.; Leira, E.C.; et al. 2021 Guideline for the prevention of stroke in patients with stroke and transient ischemic attack: A guideline from the American heart association/American stroke association. *Stroke* **2021**, *52*, e364–e467. [CrossRef] [PubMed]
12. WHO. Haemoglobin Concentrations for the Diagnosis of Anaemia and Assessment of Severity. 2011. Available online: https://apps.who.int/iris/bitstream/handle/10665/85839/WHO_NMH_NHD_MNM_11.1_eng.pdf?ua=1 (accessed on 16 February 2023).
13. Hsieh, M.T.; Hsieh, C.Y.; Tsai, T.T.; Wang, Y.C.; Sung, S.F. Performance of ICD-10-CM diagnosis codes for identifying acute ischemic stroke in a national health insurance claims database. *Clin. Epidemiol.* **2020**, *12*, 1007–1013. [CrossRef] [PubMed]
14. Yao, C.Y.; Chien, J.H.; Chuang, H.Y.; Ho, T.F. Associated factors with acute transfusion reaction from hospital online reporting events: A retrospective cohort study. *J. Patient Saf.* **2020**, *16*, e303–e309. [CrossRef] [PubMed]
15. Chang, Y.L.; Hung, S.H.; Ling, W.; Lin, H.C.; Li, H.C.; Chung, S.D. Association between ischemic stroke and iron-deficiency anemia: A population-based study. *PLoS ONE* **2013**, *8*, e82952. [CrossRef]
16. Sui, Y.; Hong, C.T.; Chien, L.N.; Liu, H.Y.; Chiou, H.Y.; Hsieh, Y.C. Association between anemia and stroke in females: Anationwide, population-based cohort study in Taiwan. *Int. J. Environ. Res. Public Health* **2020**, *17*, 7440. [CrossRef]
17. Lin, Y.-Y.; Huang, C.-S. Aging in Taiwan: Building a society for active aging and aging in place. *Gerontologist* **2015**, *56*, 176–183. [CrossRef]
18. Halawi, R.; Moukhadder, H.; Taher, A. Anemia in the elderly: A consequence of aging? *Expert Rev. Hematol.* **2017**, *10*, 327–335. [CrossRef]
19. Goodnough, L.T.; Schrier, S.L. Evaluation and management of anemia in the elderly. *Am. J. Hematol.* **2014**, *89*, 88–96. [CrossRef]
20. Chueh, H.W.; Jung, H.L.; Shim, Y.J.; Choi, H.S.; Han, J.Y. High anemia prevalence in Korean older adults, an advent healthcare problem: 2007–2016 KNHANES. *BMC Geriatr.* **2020**, *20*, 509. [CrossRef]
21. Kılıçgedik, A.; Dündar, C.; Tigen, M.K. Anemia in heart failure. *Anadolu Kardiyol. Derg.* **2012**, *12*, 65–70.
22. Hong, C.H.; Falvey, C.; Harris, T.B.; Simonsick, E.M.; Satterfield, S.; Ferrucci, L.; Metti, A.L.; Patel, K.V.; Yaffe, K. Anemia and risk of dementia in older adults: Findings from the Health ABC study. *Neurology* **2013**, *81*, 528–533. [CrossRef]
23. Ruan, Y.; Guo, Y.; Kowal, P.; Lu, Y.; Liu, C.; Sun, S.; Huang, Z.; Zheng, Y.; Wang, W.; Li, G.; et al. Association between anemia and frailty in 13,175 community-dwelling adults aged 50 years and older in China. *BMC Geriatr.* **2019**, *19*, 327. [CrossRef]
24. Culleton, B.F.; Manns, B.J.; Zhang, J.; Tonelli, M.; Klarenbach, S.; Hemmelgarn, B.R. Impact of anemia on hospitalization and mortality in older adults. *Blood* **2006**, *107*, 3841–3846. [CrossRef]
25. Michalak, S.S.; Rupa-Matysek, J.; Gil, L. Comorbidities, repeated hospitalizations, and age $\geq$ 80 years as indicators of anemia development in the older population. *Ann. Hematol.* **2018**, *97*, 1337–1347. [CrossRef] [PubMed]
26. Ren, X.; Hu, H.; Farooqi, I.; Simpkins, J.W. Blood substitution therapy rescues the brain of mice from ischemic damage. *Nat. Commun.* **2020**, *11*, 4078. [CrossRef] [PubMed]
27. Petry, N.; Olofin, I.; Hurrell, R.F.; Boy, E.; Wirth, J.P.; Moursi, M.; Donahue Angel, M.; Rohner, F. The proportion of anemia associated with iron deficiency in low, medium, and high human development index countries: A systematic analysis of national surveys. *Nutrients* **2016**, *8*, 693. [CrossRef] [PubMed]
28. Makipour, S.; Kanapuru, B.; Ershler, W.B. Unexplained anemia in the elderly. *Semin. Hematol.* **2008**, *45*, 250–254. [CrossRef] [PubMed]
29. Smith, D.L. Anemia in the elderly. *Am. Fam. Physician* **2000**, *62*, 1565–1572.
30. Stauder, R.; Valent, P.; Theurl, I. Anemia at older age: Etiologies, clinical implications, and management. *Blood* **2018**, *131*, 505–514. [CrossRef]
31. Ghoshal, S. Freedman BI. Mechanisms of Stroke in Patients with Chronic Kidney Disease. *Am. J. Nephrol.* **2019**, *50*, 229–239. [CrossRef]

32. Poznyak, A.V.; Sadykhov, N.K.; Kartuesov, A.G.; Borisov, E.E.; Sukhorukov, V.N.; Orekhov, A.N. Atherosclerosis specific features in chronic kidney disease (CKD). *Biomedicines* **2022**, *10*, 2094. [CrossRef] [PubMed]
33. Morita, S.; Malik, A.H.; Kuno, T.; Ando, T.; Kaul, R.; Yandrapalli, S.; Briasoulis, A. Analysis of outcome of 6-month readmissions after percutaneous left atrial appendage occlusion. *Heart* **2022**, *108*, 606–612. [CrossRef] [PubMed]
34. Sim, Y.E.; Wee, H.E.; Ang, A.L.; Ranjakunalan, N.; Ong, B.C.; Abdullah, H.R. Prevalence of preoperative anemia, abnormal mean corpuscular volume and red cell distribution width among surgical patients in Singapore, and their influence on one year mortality. *PLoS ONE* **2017**, *12*, e0182543. [CrossRef] [PubMed]
35. Cho, H.J.; Koo, J.W.; Roh, S.K.; Kim, Y.K.; Suh, J.S.; Moon, J.H.; Sohn, S.K.; Baek, D.W. COVID-19 transmission and blood transfusion: A case report. *J. Infect. Public Health* **2020**, *13*, 1678–1679. [CrossRef] [PubMed]
36. Yin, Z.; Yu, B.; Liu, W.; Lan, K. Blood transfusion and mortality in myocardial infarction: An updated meta-analysis. *Oncotarget* **2017**, *8*, 102254–102262. [CrossRef]
37. Parnianfard, N.; Sadat Ebrahimi, S.R.; Hoseini, P. Hyperlipidemia as a stroke risk factor: Study of patients from northwest of Iran from 2008 to 2013. *Atherosclerosis* **2020**, *315*, e196. [CrossRef]
38. Stewart, S.; Hart, C.L.; Hole, D.J.; McMurray, J.J. A population-based study of the long-term risks associated with atrial fibrillation: 20-year follow-up of the Renfrew/Paisley study. *Am. J. Med.* **2002**, *113*, 359–364. [CrossRef]
39. Raghunath, S.; Pfeifer, J.M.; Ulloa-Cerna, A.E.; Nemani, A.; Carbonati, T.; Jing, L.; van Maanen, D.P.; Hartzel, D.N.; Ruhl, J.A.; Lagerman, B.F.; et al. Deep neural networks can predict new-onset atrial fibrillation from the 12-lead ECG and help identify those at risk of atrial fibrillation-related stroke. *Circulation* **2021**, *143*, 1287–1298. [CrossRef]
40. Tu, S.J.; Hanna-Rivero, N.; Elliott, A.D.; Clarke, N.; Huang, S.; Pitman, B.M.; Gallagher, C.; Linz, D.; Mahajan, R.; Lau, D.H.; et al. Associations of anemia with stroke, bleeding, and mortality in atrial fibrillation: A systematic review and meta-analysis. *J. Cardiovasc. Electrophysiol.* **2021**, *32*, 686–694. [CrossRef]
41. Krishnasivam, D.; Trentino, K.M.; Burrows, S.; Farmer, S.L.; Picardo, S.; Leahy, M.F.; Halder, A.; Chamberlain, J.; Swain, S.; Muthucumarana, K.; et al. Anemia in hospitalized patients: An overlooked risk in medical care. *Transfusion* **2018**, *58*, 2522–2528. [CrossRef]
42. Lucca, U.; Tettamanti, M.; Mosconi, P.; Apolone, G.; Gandini, F.; Nobili, A.; Tallone, M.V.; Detoma, P.; Giacomin, A.; Clerico, M.; et al. Association of mild anemia with cognitive, functional, mood and quality of life outcomes in the elderly: The "Health and Anemia" study. *PLoS ONE* **2008**, *3*, e1920. [CrossRef] [PubMed]
43. Yotsueda, R.; Tanaka, S.; Taniguchi, M.; Fujisaki, K.; Torisu, K.; Masutani, K.; Hirakata, H.; Kitazono, T.; Tsuruya, K. Hemoglobin concentration and the risk of hemorrhagic and ischemic stroke in patients undergoing hemodialysis: The Q-cohort study. *Nephrol. Dial. Transplant.* **2018**, *33*, 856–864. [CrossRef] [PubMed]

International Journal of
Environmental Research and Public Health

MDPI

Article

Effect of Familial Longevity on Frailty and Sarcopenia: A Case–Control Study

Angel Belenguer-Varea [1,2,*], Juan Antonio Avellana-Zaragoza [1,2], Marta Inglés [3], Cristina Cunha-Pérez [2], David Cuesta-Peredo [4], Consuelo Borrás [5], José Viña [5] and Francisco José Tarazona-Santabalbina [1,2,6]

1 Division of Geriatrics, Hospital Universitario de la Ribera, 46600 Valencia, Spain
2 School of Doctorate, Universidad Católica de Valencia San Vicente Martir, 46001 Valencia, Spain
3 Freshage Research Group, Department of Physiotherapy, Faculty of Physiotherapy, University of Valencia, CIBERFES-ISCIII, INCLIVA, 46010 Valencia, Spain
4 Department of Quality Management, Hospital Universitario de la Ribera, 46600 Valencia, Spain
5 Freshage Research Group, Department of Physiology, Faculty of Medicine, University of Valencia, CIBERFES-ISCIII, INCLIVA, 46010 Valencia, Spain
6 Centro de Investigación Biomédica en Red Fragilidad y Envejecimiento Saludable (CIBERFES), 46010 Valencia, Spain
* Correspondence: abelenguervareal@gmail.com

Abstract: Familial longevity confers advantages in terms of health, functionality, and longevity. We sought to assess potential differences in frailty and sarcopenia in older adults according to a parental history of extraordinary longevity. A total of 176 community-dwelling subjects aged 65–80 years were recruited in this observational case–control study, pair-matched 1:1 for gender, age, and place of birth and residence: 88 centenarians' offspring (case group) and 88 non-centenarians' offspring (control group). The main variables were frailty and sarcopenia based on Fried's phenotype and the European Working Group on Sarcopenia in Older People (EWGSOP) definitions, respectively. Sociodemographics, comorbidities, clinical and functional variables, the presence of geriatric syndromes, and laboratory parameters were also collected. Related sample tests were applied, and conditional logistic regression was performed. Cases had a higher percentage of robust patients (31.8% vs. 15.9%), lower percentages of frailty (9.1% vs. 21.6%) and pre-frailty (59.1% vs. 62.5%) ($p = 0.001$), and lower levels of IL-6 ($p = 0.044$) than controls. The robust adjusted OR for cases was 3.00 (95% CI = 1.06–8.47, $p = 0.038$). No significant differences in muscle mass were found. Familial longevity was also associated with less obesity, insomnia, pain, and polypharmacy and a higher education level and total and low-density lipoprotein cholesterol. The results suggest an inherited genetic component in the frailty phenotype, while the sarcopenia association with familial longevity remains challenging.

Keywords: aging; function; muscle; interleukin-6; heredity; frailty; sarcopenia; longevity

Citation: Belenguer-Varea, A.; Avellana-Zaragoza, J.A.; Inglés, M.; Cunha-Pérez, C.; Cuesta-Peredo, D.; Borrás, C.; Viña, J.; Tarazona-Santabalbina, F.J. Effect of Familial Longevity on Frailty and Sarcopenia: A Case–Control Study. *Int. J. Environ. Res. Public Health* **2023**, *20*, 1534. https://doi.org/10.3390/ijerph20021534

Academic Editor: Paul B. Tchounwou

Received: 28 November 2022
Revised: 10 January 2023
Accepted: 12 January 2023
Published: 14 January 2023

1. Introduction

The aging of the world population has become a reality, and the proportion of people aged over 60 years is growing faster than any other age group. In 2018, the population aged 65 years and over exceeded the number of children under 5 years worldwide for the first time in history, and by 2050, the number of people aged 80 years and over will have tripled, from 143 million in 2019 to 426 million (United Nations, Department of Economic and Social Affairs, Population Division (2019). World Population Prospects 2019: Highlights. ST/ESA/SER.A/423). This demographic change is particularly fast in the European Union (EU) as a result of both a longer life expectancy and low fertility rates. The Spanish population aged 65 or over on 1 January 2022 was 9,620,055 people (20.2% of the total population), of whom 16.8% were 85 or older, and 6.4% were 90 years old or more [1]. Spain, with the rapid growth of people over 80 and 90 years old in the last decade, is among the countries with the oldest populations and could have the highest life expectancy in the world by 2040 [2]. Such circumstances could lead to an uncontrollable increase in morbidity and dependency that would compromise the sustainability of the health system. Therefore, increasing research and knowledge on factors related to comorbidity, geriatric syndromes, and disability (to prevent and avoid them), as well as longevity and successful aging (to promote them) should be encouraged.

Frailty and sarcopenia have emerged as two key geriatric syndromes in the aging process and have been associated with adverse events, including falls, fractures, functional impairment, disability, cognitive impairment, hospital admissions, and institutionalization [3–5]. These events are likely to limit the quality of life and longevity and increase healthcare costs [6]. Longevity, on the other hand, has been positively associated with a lower frequency of sarcopenia and a delay in the onset of physical frailty and cognitive impairment [7]. Their prevalence may vary depending on the characteristics of different studies, the geographical area, and the gender and age of the participants. In Europe, the prevalence could range between 6% and 27% for frailty [8] and between 1% and 29% for sarcopenia [9].

Frailty and sarcopenia, both characterized by the limited capacity of the organism to cope with stressors, are favored by common underlying illnesses and lifestyle and environmental factors [10]. Furthermore, they are dynamic processes, with the possibility of transitions from lower to higher levels and, more difficult, from higher to lower states in the case of frailty [11,12]. The presence of sarcopenia, on the other hand, also appears to modulate transitions in frailty status [13]. Regrettably, both syndromes are frequently detected only when one or both are well established and, often after a seemingly minor event, the individual is already suffering a health crisis with significant functional loss and dependency.

Previous studies have suggested the existence of inheritance patterns in frailty- and sarcopenia-related variables, such as strength, gait speed, and overall physical fitness [14–16]. However, the question of how heritability contributes to the onset of these two syndromes remains unanswered.

The delay in functional decline observed in individuals with extraordinary longevity [17] suggests better homeostasis and capacity to adapt and recover from stressors, which might be related to genetic and epigenetic factors, including greater genomic integrity, more preserved methylation, and a characteristic microRNA profile [18–20]. The offspring of individuals with favorable genetic characteristics also seem to inherit some advantages from their parents regarding overall health, functionality, and longevity [21–30]. However, the link between genetic and clinical characteristics in these lineages is poorly understood, thus highlighting the need to investigate heritability in aging-related conditions.

In this study, we used a pair-matched case–control approach to investigate frailty and sarcopenia in community-dwelling persons 65 years and older who are descendants of long-lived individuals and compare them with a population-based control group of age- and gender-matched individuals without a parental history of extraordinary longevity.

2. Materials and Methods

2.1. Study Design and Participants

This was a cross-sectional matched-control analysis of community-dwelling individuals aged 65 to 80 years who had been born in the area and were usual residents (i.e., >6 months per year) in the Health Department of La Ribera (249,063 people, 71% in small urban areas, Valencian Community, Spain). The study was conducted between 9 March 2015 and 6 February 2017.

Inclusion criteria for cases were as follows: Cases must be 65 to 80 years old, have been born in the study area and lived at least 6 months a year in it, and reside at home (community-dwelling). Finally, they must have at least one parent alive of 97 years or older.

Inclusion criteria for controls were as follows: The control had to be the same age (±5 years) as the case he or she was matched with, have been born in the study area and lived at least 6 months a year in it, and reside at home (community-dwelling). Finally, his or her parents must have died before 90 years old.

Exclusion criteria for both groups were as follows: being diagnosed with a terminal illness, having a life expectancy of fewer than 6 months, or not signing the informed consent form.

The study was conducted according to the principles of the Declaration of Helsinki, and the study protocol was approved by the Research Ethics Committee of the Hospital Universitario de La Ribera de Alzira (Valencia, Spain). Results were reported following the Strengthening the Reporting of Observational Studies in Epidemiology (STROBE) guidelines [31].

2.2. Procedures

Three nurses were hired as field researchers. They were specifically trained by the research team in all activities in which they were involved: selecting and recruiting patients; informing patients, relatives, or caregivers and requesting informed consent; organizing and carrying out home visits; handing out questionnaires for medical histories; performing comprehensive geriatric assessments; performing physical and functional tests; measuring bioelectrical impedance; taking blood samples and performing their initial processing; and transferring the samples to the laboratory. After the training process, it was verified that the field researchers had a high degree of agreement.

Study candidates were selected by the field researchers following a three-stage procedure. First, the entire population database of the Health Department was screened for individuals aged 97 years or more. Candidates were contacted (either directly or via their relatives/caregivers) and informed about the study. After obtaining oral informed consent, a home visit was scheduled in which candidates were asked about living offspring aged 65 to 80 years and permission to contact them. Second, offspring candidates were contacted, informed about the study, and invited to enroll. Candidates who provided written informed consent were included in the study as cases unless they met any exclusion criteria. Together with the inclusion of cases, a population-based matched control group was established by pairing 1:1 for gender, age (±5 years), and place of birth and residence. Controls were screened among individuals in the population registry of the Valencian Community. The same eligibility criteria, except that their parents must have deceased before 90 years old, and the same procedure for obtaining informed consent were used. During home visits, members of both groups underwent the same physical evaluation, questionnaires, functional and cognitive tests, bioelectrical impedance measurement, and blood draws.

The initial protocol can be freely consulted on the internet at https://www.educacion.gob.es/teseo/mostrarRef.do?ref=1943613# (accessed on 30 December 2022).

2.3. Primary Outcomes

The two co-primary outcomes were frailty and sarcopenia. Frailty was assessed based on Linda Fried's phenotypic criteria [32]: unintentional weight loss (>4.5 kg in the last year), exhaustion (the subject reported that any activity was too strenuous or he/she was unable to continue carrying it out at least 3 days in the previous week), low physical activity (weekly time walked was 2 h/week in women and 2.5 h/week in men [33]), weakness (grip strength, measured with a JAMAR© analog dynamometer, below the cut-off point, stratified by body mass index (BMI) and gender), and slowness (15-foot (4.572 m) gait speed above the cut-off point, stratified by gender and height, or inability to complete the test). Individuals were considered frail if they met 3 or more criteria; pre-frail if they met 1 or 2; and robust if none of the criteria were met.

The presence of sarcopenia was assessed according to the 2018 European Working Group on Sarcopenia in Older People (EWGSOP) definition [34]. Muscle mass was assessed by electrical bioimpedance analysis (BIA) using a four-sensor measurement device (OMRON BF500© manufactured by OMRON. 's-Hertogenbosch, The Nederlands). For low muscle quantity, a cut-off point of 7 kg/m^2 in men and 6 kg/m^2 in women was adopted. Muscle strength was determined by measuring the grip strength with JAMAR©, and cut-off points of 27 kg in men and 16 kg in women were used. Muscle performance was measured by gait speed in 4.572 m; the cut-off point was 0.8 m/s in both genders.

2.4. Secondary Variables

Sociodemographic and clinical variables were also collected to better characterize the sample, such as educational level, smoking, alcohol intake, medications, physical exercise, nutritional risk (DETERMINE scale), the performance of instrumental (Lawton and Brody index) or basic activities of daily living (Barthel index), limitations in physical activity or disability (modified Rankin scale), mobility ability (Holden's Functional Ambulation Classification (FAC)), the presence of comorbidities (age-adjusted Charlson comorbidity index), cognitive status (Spanish adaptation of the Mini-Mental State Examination, MEC-Lobo), depression and anxiety (Goldberg depression and anxiety scale), falls in the last 3–6 and 12 months, pain (visual analog scale), quality of life (Spitzer Quality of Life Index), social assessment (OARS scale), resource consumption, blood pressure, weight in kilograms, and height in centimeters.

Laboratory parameters included fasting glucose, lipid profile, renal and hepatic profile, C-reactive protein (CRP), interleukin-6 (IL-6), testosterone, sex-hormone-binding globulin (SHBG), thyroid-stimulating hormone (TSH), 25-hydroxycholecalciferol (25-OH D), iron profile, vitamin B12, folic acid, and complete blood count.

2.5. Statistical Analysis

Variables were assessed for normality using the Kolmogorov–Smirnov test. Quantitative variables were described using the mean and standard deviation (SD) and the median and interquartile range (IQR, defined as the difference between 75th and 25th percentiles) for normally and non-normally distributed variables, respectively. Categorical variables were described as frequency and percentage (%). Tests for paired data were used in order to investigate differences between groups. McNemar's test was applied for categorical variables with 2 × 2 contingency tables. In the case of quantitative variables with a normal distribution, Student's *t*-test for related samples was applied to compare means between the two groups. The Wilcoxon signed-rank test for matched samples was applied when a quantitative variable did not follow a normal distribution, and also in the case of ordinal categorical variables with more than two categories [35,36]. For the study of the whole sample, encompassing both groups together (i.e., some tables in Supplementary Materials), tests for independent samples were applied. Specifically, Chi-squared for categorical variables, Pearson's correlation for quantitative variables, and the Kruskal–Wallis test for quantitative variables that did not meet the normality assumption. The adjusted odds ratio (OR) for robustness was assessed using a conditional logistic regression (CLR) that included all variables potentially confounding the effect. We considered a variable to be a candidate for CLR if it was associated with familial longevity and with robustness in controls with a p-value less than 0.2 and caused a change in the OR equal to or greater than 10% [37]. A sensitivity analysis was performed for the serum levels of IL-6, CRP, ferritin, leukocytes, and lymphocytes. We took into consideration that some of the individuals included in the case group could be siblings. Thus, significant differences between siblings and nonsiblings in the case group were assessed for age, gender, BMI, comorbidity, frailty, muscle mass, gait speed, handgrip strength, sarcopenia, inflammatory parameters, and lipids. Data processing and statistical analysis were performed with SPSS Statistics for Macintosh v21 (Armonk, NY, USA: IBM Corp) and Stata 16.0 for PC.

3. Results

3.1. Participant Disposition and Characteristics

Sixty-three long-lived people were interviewed, resulting in 96 offspring being contacted for eligibility. After considering the inclusion and exclusion criteria, 88 cases and

88 controls were analyzed (Figure 1). Sixty percent of the recruited samples were female, and the median age was 70 years (IQR = 7). The most prevalent comorbidity was hypertension (58%), followed by dyslipidemia (43.2%) and fractures in any location (36.4%) (Table S1: Comorbidities). The most frequent geriatric syndromes were polypharmacy (38.1%), followed by insomnia (34.9%), pain (34.7%), depression (21.6%), and falls within the last year (21.1%) (Table S2: Geriatric Syndromes).

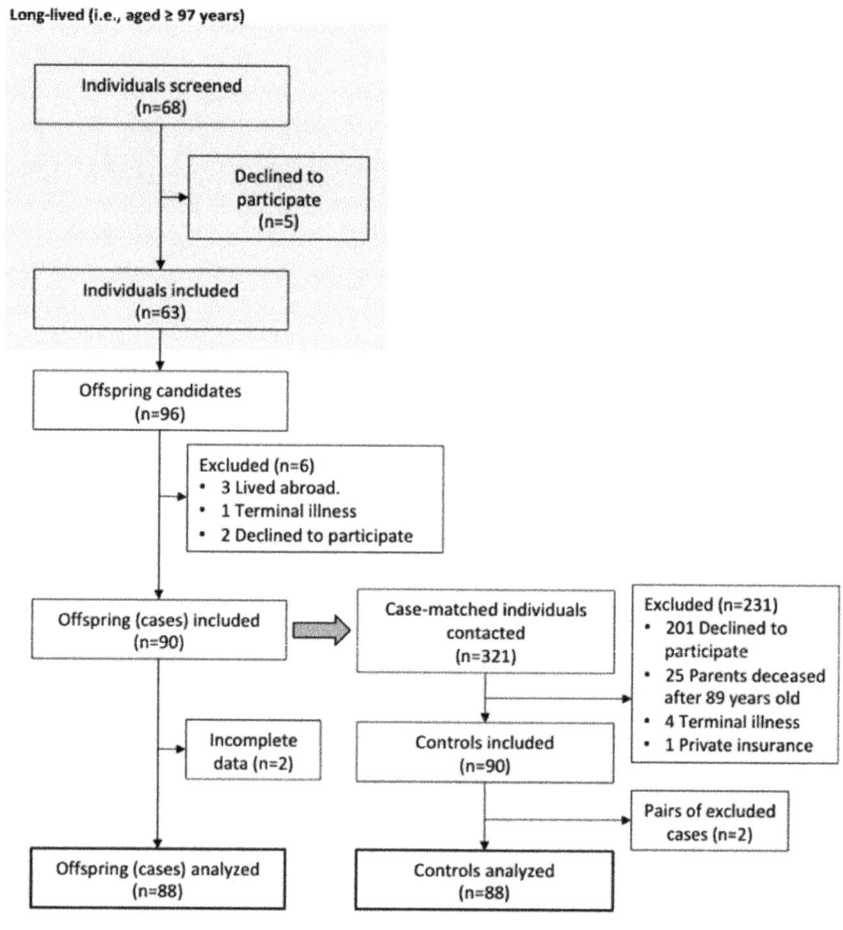

Figure 1. Flow chart of subjects' selection: shaded rectangle = selection process of long-lived; arrow pointing downward = progress flow; arrow pointing to the right = exit flow; gray arrow = start of control selection.

Table 1 shows the main demographic, clinical, and functional characteristics. Compared with cases, individuals in the control group had a significantly higher weight (with a higher obesity prevalence), had a higher frequency of insomnia and pain, and were prescribed more drugs, with a higher percentage of polypharmacy. The two groups also differed in their education levels, with a lower percentage of university degrees and a higher percentage of the non-completion of primary education in controls than in cases. No significant differences were observed between the groups in smoking, alcohol intake, exercise/physical activity, nutritional risk, scores on the different functionality scales, falls, depression, anxiety, cognitive status, quality of life, social resources, and health service utilization. The age-adjusted Charlson index did not reveal significant differences in the overall comorbidity burden; however, in individuals over 70 years, the score was significantly lower in the case group (median = 3; IQR = 1) than in controls (median = 4; IQR = 1) ($p = 0.022$) (Figure S1: Charlson index).

Table 1. Demographic, health, and functional characteristics.

	Overall n = 176	Controls n = 88	Cases n = 88	p-Value [a]
Age, years, median (IQR)	70 (7)	69 (7)	70 (6)	0.69
Female, n (%)	106 (60)	53 (60)	53 (60)	1
Weight, kg, mean ± SD	74.1 ± 14.4	76.9 ± 13.6	71.2 ± 14.6	0.008 *
BMI, kg/m^2, median (IQR)	28.1 (5.6)	29.2 (6.1)	27.6 (4.5)	0.007 *
Underweight, n (%)	5 (2.9)	1 (1.1)	4 (4.6)	0.004 *
Normal weight, n (%)	30 (17.1)	12 (13.6)	18 (20.7)	
Overweight, n (%)	79 (45.1)	35 (39.8)	44 (50.6)	
Obese, n (%)	61 (34.9)	40 (45.5)	21 (24.1)	
Current smoker, n (%)	23 (13.1)	12 (13.6)	11 (12.5)	0.424
Alcohol use, n (%)	46 (26.6)	26 (29.9)	20 (23.9)	0.472
Mild physical activity [b], n (%)	96 (56.1)	52 (59.1)	44 (53%)	0.417
Regular exercise [b], n (%)	123 (69.9)	64 (72.7)	59 (67)	0.372
Nutritional risk [c], n (%)	44 (25)	23 (26.1)	21 (23.8)	0.819
Sleeping hours				
Hours per day, median (IQR)	7 (2)	7 (2)	7.5 (1)	0.640
Insomnia, n (%)	59 (34.9)	37 (44.6)	22 (25.6)	0.030 *
Charlson, median (IQR)	3 (2)	3 (2)	3(2)	0.350
Lawton, median (IQR)	8 (0)	8 (0)	8 (0)	0.719
Barthel, median (IQR)	100 (0)	100 (0)	100 (0)	0.305
Rankin ≤1, n (%)	170 (97.7)	84 (96.6)	86 (98.8)	0.613
FAC ≥4, n (%)	133 (95.7)	69 (94.5)	64 (97)	0.461
QL Spitzer, median (IQR)	10 (1)	10 (0)	10 (1)	0.839
Falls ≥1/12 months, n (%)	34 (21.1)	18 (22.2)	16 (19.8)	0.148
Pain [d], n (%)	61 (34.7)	38 (43.2)	23 (26.1)	0.029 *
Goldberg scale				
Anxiety, n (%)	33 (18.8)	21 (23.9)	12 (13.6)	0.137
Depression, n (%)	38 (21.6)	18 (20.5)	20 (22.7)	0.850
MMSE-Lobo, median (IQR)	32 (4)	33 (3)	32 (5)	0.592
Polypharmacy [e], n (%)	67 (38.1)	43 (48.9)	24 (27.3)	0.003 *
N° of drugs, median (IQR)	3 (3)	3 (5)	2 (3)	0.001 *
OARS good or excellent, n (%)	172 (98.3)	85 (96.6)	88 (100)	0.554
Lives alone, n (%)	8 (4.5)	5 (5.7)	3 (3.4)	0.727
Education, n (%)				0.034 *
No primary	24 (13.8)	16 (18.2)	8 (9.3)	
Primary	108 (62.1)	52 (59.1)	56 (65.1)	
Secondary	28 (16.1)	17 (19.3)	11 (12.8)	
University	14 (8)	3 (3.4)	11 (12.8)	
Healthcare utilization, medical visits/year, median (IQR)				
Primary care	2 (3)	2 (3)	2 (2)	0.644
Specialized care	1 (2)	1 (2)	0 (2)	0.518

Notes: Mean and standard deviation (SD) for normal quantitative variables. Median and interquartile range (IQR) for non-normal quantitative variables; n = patients; % = percentage; BMI = body mass index; FAC = functional ambulation classification; QL = quality of life; MMSE = Mini-Mental State Examination; OARS = Older Americans Resources and Services; [a] = McNemar test for categorical variables, Student's t-test for related samples for quantitative variables with normal distribution, and Wilcoxon signed-rank test for matched samples for non-normal quantitative variables and ordinal categorical variables of more than 2 categories; [b] = 150 min per week or more; [c] = 3 points or more on Determine scale; [d] = 2 days per week or more; [e] = regular use of at least five medications; * $p < 0.05$.

Regarding analytical variable measurements (Table 2), cases presented significantly higher levels of total and LDL cholesterol. The percentage of individuals with an LDL/ApoB ratio lower than 1.3, which is associated with the presence of small and dense LDL particles, was significantly higher in the control group. Cases showed higher levels of SHBG and lower levels of IL-6 than controls. No significant differences were observed in the other variables analyzed.

Table 2. Laboratory results.

	Overall n = 176	Controls n = 88	Cases n = 88	*p*-Value [a]
Glucose, mg/dL, median (IQR)	95 (25)	96 (27)	93 (21)	0.113
<100, n (%)	108 (61.4)	49 (55.7)	59 (67)	0.084
100–125, n (%)	40 (22.7)	22 (25)	18 (20.5)	
≥126, n (%)	28 (15.9)	17 (19.3)	11 (12.5)	
Albumin, g/dL, median (IQR)	4.4 (0.3)	4.4 (0.3)	4.3 (0.3)	0.301
Transferrin, mg/dL, mean ± SD	277.3 ± 40.8	281.3 ± 46.2	273.3 ± 34.4	0.506
Ferritin, ng/dL, median (IQR)	89 (101)	85 (110.5)	89 (96)	0.274
Total cholesterol, mg/dL, mean ± SD	199.1 ± 39.5	191.7 ± 37.3	206.4 ± 40.4	0.015 *
LDL, mg/dL, mean ± SD	115.7 ± 35.3	110.1 ± 32.1	121.3 ± 37.5	0.043 *
HDL, mg/dL, median (IQR)	56.5 (20)	55.5 (18)	58 (22)	0.213
VLDL, mg/dL, median (IQR)	23 (14)	23 (13)	22.5 (14)	0.370
Triglycerides, mg/dL, median (IQR)	113 (70)	115 (68)	112 (72)	0.360
ApoB, mg/dL, mean ± SD	101.5 ± 20.6	98.9 ± 21.3	104 ± 19.7	0.141
LDL/ApoB Ratio < 1.3, n (%)	135 (77.1)	74 (84.1)	61 (70.1)	0.038 *
Vitamin B12, pg/mL, median (IQR)	372 (188)	376 (188)	371 (191)	0.549
Folic acid, ng/mL, median (IQR)	10.5 (6.7)	11.3 (7.1)	9.4 (6.7)	0.360
25-OHD, ng/mL, mean ± SD	19.6 ± 6.5	20.3 ± 6.5	18.8 ± 6.5	0.193
TSH, mcU/mL, median (IQR)	1.59 (1.20)	1.48 (1.12)	1.67 (1.07)	0.562
Total testosterone, ng/mL, median (IQR)	0.58 (3.47)	0.60 (3.13)	0.57 (3.67)	0.328
Free testosterone, ng/mL, median (IQR)	0.08 (0.05)	0.009 (0.05)	0.008 (0.05)	0.334
SHBG, nmol/L, median (IQR)	50.3 (26.2)	47.6 (23.1)	56.1 (33.1)	0.004 *
Hemoglobin, g/dL, mean ± SD	14.4 ± 1.2	14.3 ± 1.2	14.4 ± 1.3	0.502
Leukocytes, $\times 10^9$/L, mean ± SD	6.72 ± 1.79	6.84 ± 1.89	6.60 ± 1.70	0.475
Lymphocytes, $\times 10^9$/L, median (IQR)	1.80 (0.75)	1.90 (0.80)	1.80 (0.78)	0.733
CRP, mg/L, median (IQR)	1.49 (2.34)	1.92 (2.24)	1.18 (2.14)	0.155
IL-6, pg/mL, median (IQR)	1.20 (1.26)	1.45 (1.38)	1.03 (0.96)	0.044 *

Notes: Mean and standard deviation (SD) for normal quantitative variables. Median and interquartile range (IQR) for non-normal quantitative variables; n = patients; % = percentage; LDL = low-density lipoprotein cholesterol; HDL = high-density lipoprotein cholesterol; VLDL = very-low-density lipoprotein cholesterol; ApoB = apolipoprotein B; 25-OHD = 25-hydroxycholecalciferol; TSH = thyroid-stimulating hormone; SHBG = sex-hormone-binding globulin; CRP = C-reactive protein; IL-6; interleukin-6; [a] = McNemar test for categorical variables, Student's *t*-test for related samples for quantitative variables with normal distribution, and Wilcoxon signed-rank test for matched samples for non-normal quantitative variables and ordinal categorical variables of more than 2 categories; * $p < 0.05$.

Thirty-eight individuals from the case group were siblings (43.2%), and no statistically significant differences were found between sibling and nonsibling cases for age, gender, BMI, comorbidity, frailty, muscle mass, gait speed, handgrip strength, sarcopenia, inflammatory parameters, and lipids.

3.2. Frailty and Sarcopenia

Overall, 27 (15.3%) individuals in the study sample were frail, and 107 (60.8%) were pre-frail. The prevalence of frailty and pre-frailty was significantly higher in women: 17% and 65.1%, respectively, among women vs. 12.9% and 54.3% among men ($p = 0.023$). Reduced physical activity and weakness were the most prevalent frailty criteria (55.4% and 50.3%, respectively). According to EWGSOP criteria, 4.6% of the overall sample had sarcopenia, and 1.7% had severe sarcopenia. The overall prevalence of sarcopenia was higher in women (9.4%) than in men (1.4%) ($p = 0.009$).

The percentage of robust patients was significantly higher among cases (31.8%) than controls (15.9%) ($p = 0.001$). Likewise, lower percentages of frail and pre-frail individuals were observed in the cases (9.1% and 59.1%, respectively) than in the control group (21.6% and 62.5%, respectively) (Figure 2A). Of all frailty criteria, weakness, slowness, and exhaustion were significantly less prevalent in cases than in controls (Table 3). The crude OR of cases for robustness was 3.33 (95% IC = 1.38–8.06, $p = 0.009$). The attributable fraction in the case group was 70% (95% IC = 23–90%) and 22% in the population (95% = IC 7–35%). After the confounding assessment, the number of medications and blood levels of 25-hydroxycholecalciferol were selected. Both variables were associated with robustness in controls and with familial longevity, with a p-value < 0.2, and caused a change in the OR equal to or greater than 10%. After conditional logistic regression had been performed, the robust adjusted OR for cases was 3.00 (95% IC = 1.06–8.47, $p = 0.038$).

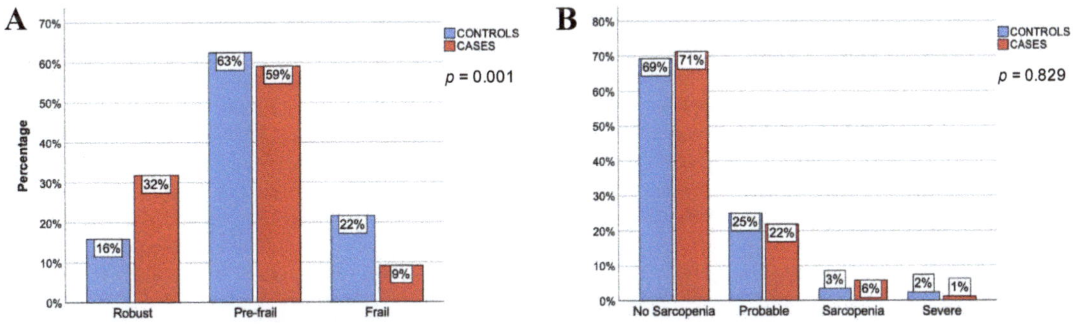

Figure 2. Results for frailty (panel **A**) and sarcopenia (panel **B**): Y-axis = percentage of subjects within group; X-axis = categories; blue columns = controls; red columns = cases; p = statistical significance.

Table 3. Frailty and sarcopenia components.

	Controls n = 88	Cases n = 88	p-value [a] n = 88
Frailty, n (%)			
Weight Loss	5 (5.7)	6 (6.8)	1
Exhaustion	15 (17)	4 (4.5)	0.013 *
Reduced activity	54 (61.4)	43 (49.4)	0.082
Weakness	50 (56.8)	38 (43.7)	0.045 *
Slowness	22 (25)	9 (10.3)	0.004 *
Sarcopenia, n (%)			
Low muscle mass	8 (9.4)	10 (11.8)	0.815
Weakness	27 (30.7)	25 (28.7)	0.038 *
Slowness	28 (32.6)	14 (16.3)	0.018 *

Notes: [a] = McNemar test; * $p < 0.05$.

When using the 2018 EWGSOP criteria, we found no significant differences in sarcopenia between groups (Figure 2B). Cases showed lower percentages of weakness and slowness; however, no significant differences were observed regarding muscle mass (Table 3).

4. Discussion

In this pair-matched case–control observational study, a familial history of parental longevity was associated with greater robustness and lower frailty in their 65–80-year-old offspring. However, the longevity history did not significantly contribute to the presence of sarcopenia due to the lack of differences in muscle mass. Besides being more robust, descendants of long-lived individuals had a higher education level, a lower prevalence of obesity, lower levels of IL-6, higher total and LDL cholesterol (with a lower percentage of subjects with an LDL/ApoB ratio of less than 1.3), and a lower prevalence of geriatric syndromes such as pain, insomnia, and polypharmacy.

The prevalences of frailty, pre-frailty, and robustness in the total sample, as well as the percentage of women, were within the ranges reported in the existing literature [8]. Previous research on the history of familial longevity and frailty has considered the latter either with a definition based on the accumulation of deficits or with a phenotypic definition.

In our opinion, the frailty index could be more appropriate for assessing more complex geriatric patients, while phenotypic categorization may be more adequate for younger, less complex older people or those at risk of frailty [38,39], such as those in our sample. Kim et al. and Arosio et al., with the first approach, found that family longevity was significantly associated with a lower rate of frailty [40,41]. Nevertheless, research using a phenotypic definition has not always presented similar results. In the Long-Life Family Study, with the Scale of Aging Vigor in Epidemiology (SAVE) [42] based on Fried's criteria, moderate heritability and less frailty in offspring were found. On the other hand, The LonGenity study concluded that, though the offspring of parents with exceptional longevity had better measures of physical function, there were no differences in frailty using Fried's criteria [43]. The results obtained in our study could be explained by a different design, with younger participants from a different geographic location and, presumably, with greater genetic heterogeneity than the aforementioned study. Furthermore, the cases we studied had significantly lower levels of IL-6, and this may suggest a lower degree of age-related inflammation, consistent with their lower degree of frailty [44,45].

Three other geriatric syndromes that were less prevalent in centenarians' offspring were insomnia, pain, and polypharmacy. Sleep is one of the processes affected by aging, with lower sleep efficiency and REM sleep [46], and older adults with more ADL and IADL limitations have a higher risk of experiencing a decline in sleep quality [47]. A certain degree of heritability for sleep–wake patterns and insomnia has been suggested [48,49], but the mechanisms involved are still under investigation, and its association with longevity is controversial. In fact, the empirical support for an increase in mortality risk with insomnia is inconsistent, although it might be higher with the use of hypnotic medication [50], and previous research did not show significant differences in sleep patterns between centenarians' offspring and their controls [51]. In contrast, our familial longevity case group had significantly less insomnia than their matched controls. The relationship between sleep disturbances and frailty status seems more evident [52]; however, the implications of these associations are not well established. Insomnia may be a marker for different conditions, such as poor health, comorbidities, disabilities, social impairment, or, interestingly, inflammatory cytokines. Both conditions, frailty and insomnia, might share some underlying mechanisms. Thus, we understand that the differences observed for insomnia support the consistency of the results found for frailty and familial longevity in the present study.

Although no significant differences were found between groups in anxiety, depression, cognition, falls, fractures, or osteoarthritis (Tables S1 and S2), our control group reported significantly more pain than offspring enriched for human longevity. Pain is a complex experience that involves sensory-discriminative, affective-motivational, and cognitive-evaluative dimensions. When investigating the relationships of parental longevity with the regional brain structure, some differences in zones involved in transferring and processing sensory and nociceptive information, which might support our results, have been found, but more research is needed on this issue [53].

Regarding pharmacological therapy, a lower use of medications has been observed previously in descendants of centenarians [28], which could be related to their lower comorbidity incidence, better physical and cognitive function, and health perception. Nevertheless, differences in the Charlson index were observed in our study, being statistically significant only in people over 70 years of age. Thus, this difference in comorbidity after the age of 70, in addition to the higher prevalence of pain and insomnia in the control group, and their probable need for treatment might have influenced differences in polypharmacy.

Some metabolomic and lipidomic studies have been published, trying to identify a specific profile for aging, longevity, and frailty [54–57]. However, this goal was beyond our scope. The metabolic profiles of our patients were evaluated with traditional analytical variables, such as the lipid profile, fasting glucose and others (Table 2), weight, and BMI. Except for total and LDL cholesterol and BMI, no significant differences were observed between the case group and the controls.

The relation between cholesterol, frailty, and longevity is complex. Higher levels of total and LDL cholesterol have been linked to cardiovascular disease (CVD), and it is also known that centenarians have lower levels than younger people. Nevertheless, in some studies [58,59], higher concentrations have been also associated with centenarians' offspring, theoretically enriched for longevity, and less frail elderly. We also found this apparently less favorable lipid profile (i.e., higher levels of total and LDL cholesterol) in our case group with familial longevity. However, in this regard, it is worth mentioning that small, dense low-density lipoproteins seem to be a better marker for cardiovascular disease outcomes [60]. We did not have access to any of the different laboratory techniques used to separate LDL fractions into subfractions. Notwithstanding the above, we found that the control group, despite having lower cholesterol levels, more frequently showed an LDL/ApoB ratio of less than 1.3, which suggests the greater presence of small and dense LDL particles and, therefore, a higher cardiovascular risk for controls [60,61]. This result is in line with a previous study [58], which suggested that the offspring of long-lived individuals, despite presenting higher levels of total and LDL cholesterol, could have a better profile in LDL particles, which might explain the lower incidence of cardiovascular disease reported in previous research [62]. Hypercholesterolemia is frequently associated with obesity and overweight, and a positive correlation of total and LDL cholesterol levels with age and BMI is known, especially up to age 65 [63]. In our sample, total and LDL cholesterol showed a slight (less than 0.3), but statistically significant, negative correlation with BMI (Table S3). However, because of the small magnitude of the coefficient and the loss of significance (except for LDL) when stratified by case group, we cannot draw any relevant conclusions from this result.

Differences in carbohydrate metabolism, associated with lower insulin resistance and a lower percentage of diabetes, with family longevity have been extensively reported in previous studies [22,23,25–27]. Along this line, we found that the case group showed a non-significant statistical tendency ($p = 0.08$) to present less diabetes and altered basal glucose, which could reflect an underpowered sample to detect differences. Notwithstanding the above, previous studies also found no significant differences in this regard [28,64]. Thus, we should also consider that different characteristics of the studied populations, in terms of environmental factors, habits, lifestyle, and diabetes prevalence, might have contributed to these results.

Previous research has suggested some degree of heritability for both muscle mass and function [15,65], lower morbidity and mortality, and slower functional loss in centenarians' offspring [24,64], raising the possibility that they may also have a lower prevalence of sarcopenia. Still, in the present study, no significant differences were detected, probably due to the low prevalence found using the EWGSOP criteria [66]. Likewise, it is worth mentioning that this result was due to the lack of differences in the muscle mass index between groups, while muscular performance, as assessed by handgrip strength and gait speed, was clearly better in the offspring of long-lived individuals. This lack of correlation might suggest that the relationship between mass and function could vary depending on factors related to the population studied [67].

Chronic inflammation and obesity have been linked to frailty in previous research. In the present study, familial longevity, in addition to higher robustness and better muscular performance, was also significantly associated with a lower prevalence of obesity and lower levels of inflammation, as measured by Il-6 levels. This raises the question of whether these differences might confound the results. It is worth mentioning in this regard that neither BMI nor IL-6 levels were significantly associated with robustness or frailty in the global sample (Table S4). Therefore, statistically, they do not meet the confounding criteria. Previous research suggests that both frailty and obesity might be influenced by genes and underlying mechanisms related to inflammation and energy metabolism [57,68–70]. Thus, we understand the lower inflammation, BMI, and frailty prevalence found in our case group are more likely a question of internal consistency than a source of a potential confounding effect.

In contrast to previous studies [28,43], no significant differences were observed in the cognitive and functional profiles of the two groups. Similarly, no differences were observed between the groups in nutritional status, toxic habits, exercise, and physical activity. Existing studies have found that centenarians' offspring have a lower incidence of comorbidities, later disease onset, lower polypharmacy, and a lower mortality risk than their peers without a history of family longevity [21,25,28]. In the present study, the offspring of long-lived individuals showed a lower polypharmacy prevalence and a lower Charlson index score, but we only observed this last result when the analysis was conducted in the 70+ age group (Figure S1), suggesting that the difference in comorbidity is likely to be more evident at older ages. This might be a consequence of a slower pace of aging, which would be in line with previous research [26,71].

The biological mechanisms that could explain the advantages for frailty and muscle performance observed in centenarians' offspring compared with their controls are still under investigation. Centenarians have specific characteristics that lead to the deceleration of the aging rate throughout their lives [72]. Along this line, in previous research, they showed the differential expression of mRNA and miRNA related to processes, such as cellular damage protection and the modulation of the immune response, associated with healthy aging and frailty [20,73]. Additionally, it has been observed that some special traits of centenarians might be inherited by their offspring, distinguishing them from non-centenarians' offspring [57]. The former could possess better-preserved metabolic patterns in order to face the increase in energy demand associated with the maintenance of homeostasis, health status, and functionality. Furthermore, in a subsample of subjects from the present study, we found that offspring overexpressed genes related to bone growth activation, muscle development, skeletal development, and cell differentiation [74], which could partly explain the differences found in frailty and muscle performance.

Some limitations in our study should be considered. Its cross-sectional nature increases its susceptibility to bias, as it is not conducive to the establishment of causal relationships. The study population was geographically restricted to our Health Department, and thus, the number of candidates for the study was limited. Assuming the robustness prevalence found and the matched case–control groups with a paired data design, our sample size would have been enough for 85% power. Its matched design allowed us to control for important variables, such as age, gender, birthplace, and residence, as well as increase efficiency in obtaining identical sample sizes [75].

We must consider the possibility that some variables may have influenced our results, confounding the association between familial longevity and frailty. In relation to this, it is worth mentioning that a study variable could have a confounding effect on the results if it satisfies the three following properties: there must be statistically significant differences between cases and controls regarding the variable studied, it must have a statistically significant association with frailty, and it must not be an effect of familial longevity (intermediate variable) [76]. The last criterion is usually the most complex to address, so the process normally begins with the verification of the first two, and if they are met, it proceeds to the verification of the third criterion.

Some lifestyle habits can influence the onset and progression of frailty [77]. Regarding the presence of habits such as smoking, alcohol intake, or regular physical exercise in our study, no significant differences were observed between the case group and the controls (Table 1), and no association was found between these variables and frailty (Table S5). The results for physical activity were somewhat different. Although no significant differences were found between the case group and the control group (Table 1), participants with moderate to vigorous physical activity were significantly more robust and less frail than those with mild activity in the whole sample (Table S5). This contrast between regular physical exercise and physical activity might be explained by better accuracy when we classify the level of physical activity according to the usual tables ([78]) with respect to the generic question of whether or not a person exercises regularly. After these considerations, we consider it unlikely that differences in these habits could have influenced our results in a decisive way.

The relation between body weight and frailty is still under investigation, with the probability that the underweight BMI category has a higher risk [79]. As previously mentioned in this discussion, although BMI was significantly higher in the control group, there was not a significant association between BMI and frailty in our overall sample (Table S4). On the other hand, despite the low prevalence of sarcopenia observed, we also explored the presence of sarcopenic obesity and its possible influence on our results. Only 11 participants met the EWGSOP criteria for sarcopenia; none of them were obese, and in both groups, the majority were in the normal-weight category (Table S5). Therefore, despite the important relation between BMI, sarcopenic obesity, and frailty demonstrated in previous research, it seems unlikely to us that differences in weight, or sarcopenic obesity, could have a relevant confounding effect on our results.

Additionally, no differences in other potentially confounding variables, such as nutrition and physical activity, were found in the bivariate analysis. Furthermore, two variables that met the criteria for confounding were identified, and the conditional logistic regression carried out supports our findings. Therefore, it seems probable that the extent of the genetic enrichment effect in the case group, besides the study design, might explain the ability to find significant differences in the frailty phenotype. We should also consider that some degree of nonresponse bias could be possible due to the unknown characteristics of the controls who refused to participate. This bias occurs when key characteristics of the study make a difference between respondents and nonrespondents. In previous research, respondents had higher education levels and reported better health and satisfaction levels than nonrespondents [80,81]. These aspects, if they occurred in our study, could have reduced our power to detect differences rather than the opposite, thus highlighting our results. Additionally, it should be noted that the use of BIA to assess muscle mass might have influenced the results for sarcopenia. BIA has some advantages, such as portability and ease of use, but on the other hand, it offers less validity than magnetic resonance imaging (MRI), computerized tomography (CT), or dual-energy X-ray absorptiometry (DXA). In addition, using this device requires precise instructions to minimize errors. Regarding this, it is worth mentioning that the field researchers were trained both in this technique and in how to conduct the tests and implement questionnaires in order to try to control potential sources of bias.

Finally, we have to consider that the population studied was clearly delimited and belonged to an area of the Spanish Mediterranean coast, which might limit the external validity of the results. Notwithstanding the above, and although long-lived families may also have healthier living habits or better socioeconomic conditions, the benefits associated with family longevity appear to be independent of such factors and may be largely attributable to a genetic influence [82]. In this way, we observed a 22% attributable fraction in the population, which is similar to the Robustness Index Ratio observed by Serena Dato et al. in the Longitudinal Study of Aging Danish Twins [6].

5. Conclusions

In summary, our findings indicate that the offspring of long-lived individuals have significantly lower odds of developing frailty within the age range of 65 to 80 years compared with an age- and gender-matched control group. This finding suggests an inherited component of the frailty phenotype, consistent with the lower levels of markers of underlying inflammatory processes and the better muscular performance. However, results regarding the prevalence of sarcopenia in this group of offspring are controversial, and the enhanced muscular performance conflicts with the lack of significant differences in muscular mass.

Supplementary Materials: The following supporting information can be downloaded at: https://www.mdpi.com/article/10.3390/ijerph20021534/s1. Figure S1: Charlson index; differences between matched case–control pairs; Figure S2: Charlson index bar chart with error bars; Table S1: Comorbidities; Table S2: Geriatric syndromes; Table S3: Pearson correlations among total and LDL cholesterol and body mass index; Table S4: Body mass index, IL-6, and frailty; Table S5: Lifestyle habits and frailty; Table S6: Sarcopenia patients according to their BMI; Table S7: Body mass index and LDL and total cholesterol.

Author Contributions: A.B.-V. thought out, designed, and directed the project. A.B.-V. and J.A.A.-Z. directed experimental work. A.B.-V. and F.J.T.-S. wrote the paper. C.C.-P. performed the translation. D.C.-P. made substantial contributions to the acquisition of data. M.I., C.B. and J.V. reviewed and improved the manuscript quality. All authors have read and agreed to the published version of the manuscript.

Funding: This research was funded by the Ministry for de Economy and Competitiveness, Carlos III Health Institute (Spain), "Red temática de investigación cooperativa en envejecimiento y fragilidad" (RD12/0043/0033. European Regional Development Funds), and the Health Department grants for research (2012–2013, 2014–2015) of the Hospital Universitario de La Ribera in Alzira (Valencia, Spain). The funders had no role in the design of the study; in the collection, analyses, or interpretation of data; in the writing of the manuscript; or in the decision to publish the results.

Institutional Review Board Statement: The study was conducted according to the guidelines of the Declaration of Helsinki and approved by the Institutional Ethics Committee of Hospital Universitario de la Ribera (Alzira, Valencia) (CEI-CI:20131216, 16 December 2013).

Informed Consent Statement: Informed consent was obtained from all subjects involved in the study.

Data Availability Statement: Data analyzed in the current study are available from the corresponding author upon reasonable request.

Acknowledgments: We acknowledge all the funders of this study; Juan Manuel De la Cámara de las Heras for the bibliographic help received; Mario Ortuño for the help received with the storage of biological samples and laboratory data; and Nadia Dolz Fernandez for her supervision in the final writing of the article.

Conflicts of Interest: The authors declare no conflict of interest.

References

1. Instituto Nacional de Estadística. Población Residente Por Fecha, Sexo y Grupo de Edad. Available online: https://www.ine.es/jaxiT3/Tabla.htm?t=9689 (accessed on 24 September 2022).
2. Foreman, K.J.; Marquez, N.; Dolgert, A.; Fukutaki, K.; Fullman, N.; McGaughey, M.; Pletcher, M.A.; Smith, A.E.; Tang, K.; Yuan, C.W.; et al. Forecasting Life Expectancy, Years of Life Lost, and All-Cause and Cause-Specific Mortality for 250 Causes of Death: Reference and Alternative Scenarios for 2016–40 for 195 Countries and Territories. *Lancet* **2018**, *392*, 2052–2090. [CrossRef] [PubMed]
3. Vermeiren, S.; Vella-Azzopardi, R.; Beckwée, D.; Habbig, A.K.; Scafoglieri, A.; Jansen, B.; Bautmans, I.; Bautmans, I.; Vert, D.; Beyer, I.; et al. Frailty and the Prediction of Negative Health Outcomes: A Meta-Analysis. *J. Am. Med. Dir. Assoc.* **2016**, *17*, 1163.e1–1163.e17. [CrossRef] [PubMed]
4. Zhang, Y.; Hao, Q.; Ge, M.; Dong, B. Association of Sarcopenia and Fractures in Community-Dwelling Older Adults: A Systematic Review and Meta-Analysis of Cohort Studies. *Osteoporos. Int.* **2018**, *29*, 1253–1262. [CrossRef]
5. Shimada, H.; Tsutsumimoto, K.; Doi, T.; Lee, S.; Bae, S.; Nakakubo, S.; Makino, K.; Arai, H. Effect of Sarcopenia Status on Disability Incidence Among Japanese Older Adults. *J. Am. Med. Dir. Assoc.* **2020**, *22*, 846–852. [CrossRef] [PubMed]
6. Álvarez-Bustos, A.; Rodríguez-Sánchez, B.; Carnicero-Carreño, J.A.; Sepúlveda-Loyola, W.; Garcia-Garcia, F.J.; Rodríguez-Mañas, L. Healthcare Cost Expenditures Associated to Frailty and Sarcopenia. *BMC Geriatr.* **2022**, *22*, 747. [CrossRef]
7. Landi, F.; Cruz-Jentoft, A.J.; Liperoti, R.; Russo, A.; Giovannini, S.; Tosato, M.; Capoluongo, E.; Bernabei, R.; Onder, G. Sarcopenia and Mortality Risk in Frail Older Persons Aged 80 Years and Older: Results from IlSIRENTE Study. *Age Ageing* **2013**, *42*, 203–209. [CrossRef]
8. Santos-Eggimann, B.; Cuénoud, P.; Spagnoli, J.; Junod, J. Prevalence of Frailty in Middle-Aged and Older Community-Dwelling Europeans Living in 10 Countries. *J. Gerontol. Ser. A Biol. Sci. Med. Sci.* **2009**, *64*, 675–681. [CrossRef]
9. Cruz-Jentoft, A.J.; Landi, F.; Schneider, S.M.; Zúñiga, C.; Arai, H.; Boirie, Y.; Chen, L.K.; Fielding, R.A.; Martin, F.C.; Michel, J.; et al. Prevalence of and Interventions for Sarcopenia in Ageing Adults: A Systematic Review. Report of the International Sarcopenia Initiative (EWGSOP and IWGS). *Age Ageing* **2014**, *43*, 48–759. [CrossRef]

10. Wilson, D.; Jackson, T.; Sapey, E.; Lord, J.M. Frailty and Sarcopenia: The Potential Role of an Aged Immune System. *Ageing Res. Rev.* **2017**, *36*, 1–10. [CrossRef]
11. Liu, Z.Y.; Wei, Y.Z.; Wei, L.Q.; Jiang, X.Y.; Wang, X.F.; Shi, Y.; Hai, H. Frailty Transitions and Types of Death in Chinese Older Adults: A Population-Based Cohort Study. *Clin. Interv. Aging* **2018**, *13*, 947–956. [CrossRef]
12. Trevisan, C.; Vetrano, D.L.; Calvani, R.; Picca, A.; Welmer, A.K. Twelve-Year Sarcopenia Trajectories in Older Adults: Results from a Population-Based Study. *J. Cachexia. Sarcopenia Muscle* **2022**, *13*, 254–263. [CrossRef] [PubMed]
13. Álvarez-Bustos, A.; Carnicero-Carreño, J.A.; Davies, B.; Garcia-Garcia, F.J.; Rodríguez-Artalejo, F.; Rodríguez-Mañas, L.; Alonso-Bouzón, C. Role of Sarcopenia in the Frailty Transitions in Older Adults: A Population-Based Cohort Study. *J. Cachexia. Sarcopenia Muscle* **2022**, *13*, 2352–2360. [CrossRef] [PubMed]
14. Dato, S.; Montesanto, A.; Lagani, V.; Jeune, B.; Christensen, K.; Passarino, G. Frailty Phenotypes in the Elderly Based on Cluster Analysis: A Longitudinal Study of Two Danish Cohorts. Evidence for a Genetic Influence on Frailty. *Age* **2012**, *34*, 571–582. [CrossRef]
15. Cournil, A.; Jeune, B.; Skytthe, A.; Gampe, J.; Passarino, G.; Robine, J.M. Handgrip Strength: Indications of Paternal Inheritance in Three European Regions. *J. Gerontol. Ser. A Biol. Sci. Med. Sci.* **2010**, *65A*, 1101–1106. [CrossRef] [PubMed]
16. Garatachea, N.; Lucia, A. Genes, Physical Fitness and Ageing. *Ageing Res. Rev.* **2013**, *12*, 90–102. [CrossRef]
17. Andersen, S.L.; Sebastiani, P.; Dworkis, D.A.; Feldman, L.; Perls, T.T. Health Span Approximates Life Span among Many Supercentenarians: Compression of Morbidity at the Approximate Limit of Life Span. *J. Gerontol. A. Biol. Sci. Med. Sci.* **2012**, *67*, 395–405. [CrossRef]
18. Engberg, H.; Oksuzyan, A.; Jeune, B.; Vaupel, J.W.; Christensen, K. Centenarians—A Useful Model for Healthy Aging? A 29-Year Follow-up of Hospitalizations among 40000 Danes Born in 1905. *Aging Cell* **2009**, *8*, 270–276. [CrossRef]
19. Gentilini, D.; Mari, D.; Castaldi, D.; Remondini, D.; Ogliari, G.; Ostan, R.; Bucci, L.; Sirchia, S.M.; Tabano, S.; Cavagnini, F.; et al. Role of Epigenetics in Human Aging and Longevity: Genome-Wide DNA Methylation Profile in Centenarians and Centenarians' Offspring. *Age* **2013**, *35*, 1961–1973. [CrossRef]
20. Serna, E.; Gambini, J.; Borras, C.; Abdelaziz, K.M.; Mohammed, K.; Belenguer, A.; Sanchis, P.; Avellana, J.A.; Rodriguez-Mañas, L.; Viña, J. Centenarians, but Not Octogenarians, up-Regulate the Expression of MicroRNAs. *Sci. Rep.* **2012**, *2*, 961. [CrossRef]
21. Terry, D.F.; Wilcox, M.; McCormick, M.A.; Lawler, E.; Perls, T.T. Cardiovascular Advantages Among the Offspring of Centenarians. *J. Gerontol. Ser. A Biol. Sci. Med. Sci.* **2003**, *58*, M425–M431. [CrossRef]
22. Atzmon, G.; Schechter, C.; Greiner, W.; Davidson, D.; Rennert, G.; Barzilai, N. Clinical Phenotype of Families with Longevity. *J. Am. Geriatr. Soc.* **2004**, *52*, 274–277. [CrossRef] [PubMed]
23. Terry, D.F.; Wilcox, M.A.; McCormick, M.A.; Pennington, J.M.Y.; Schoenhofen, E.A.; Andersen, S.L.; Perls, T.T. Lower All-Cause, Cardiovascular, and Cancer Mortality in Centenarians' Offspring. *J. Am. Geriatr. Soc.* **2004**, *52*, 2074–2076. [CrossRef] [PubMed]
24. Schoenmaker, M.; de Craen, A.J.M.; de Meijer, P.H.E.M.; Beekman, M.; Blauw, G.J.; Slagboom, P.E.; Westendorp, R.G.J. Evidence of Genetic Enrichment for Exceptional Survival Using a Family Approach: The Leiden Longevity Study. *Eur. J. Hum. Genet.* **2006**, *14*, 79–84. [CrossRef] [PubMed]
25. Adams, E.R.; Nolan, V.G.; Andersen, S.L.; Perls, T.T.; Terry, D.F. Centenarian Offspring: Start Healthier and Stay Healthier. *J. Am. Geriatr. Soc.* **2008**, *56*, 2089–2092. [CrossRef] [PubMed]
26. Westendorp, R.G.J.; Van Heemst, D.; Rozing, M.P.; Frölich, M.; Mooijaart, S.P.; Blauw, G.J.; Beekman, M.; Heijmans, B.T.; De Craen, A.J.M.; Slagboom, P.E. Nonagenarian Siblings and Their Offspring Display Lower Risk of Mortality and Morbidity than Sporadic Nonagenarians: The Leiden Longevity Study. *J. Am. Geriatr. Soc.* **2009**, *57*, 1634–1637. [CrossRef]
27. Stijntjes, M.; de Craen, A.J.M.; van Heemst, D.; Meskers, C.G.M.; van Buchem, M.A.; Westendorp, R.G.J.; Slagboom, P.E.; Maier, A.B. Familial Longevity Is Marked by Better Cognitive Performance at Middle Age: The Leiden Longevity Study. *PLoS ONE* **2013**, *8*, 1–8. [CrossRef]
28. Bucci, L.; Ostan, R.; Cevenini, E.; Pini, E.; Scurti, M.; Vitale, G.; Caruso, C.; Sansoni, P.; Fanelli, F.; Pasquali, R.; et al. Centenarians' Offspring as a Model of Healthy Aging: A Reappraisal of the Data on Italian Subjects and a Comprehensive Overview. *Aging* **2016**, *8*, 510–519. [CrossRef]
29. Barral, S.; Singh, J.; Fagan, E.; Cosentino, S.; Andersen-Toomey, S.L.; Wojczynski, M.K.; Feitosa, M.; Kammerer, C.M.; Schupf, N. Age-Related Biomarkers in LLFS Families with Exceptional Cognitive Abilities. *J. Gerontol.—Ser. A Biol. Sci. Med. Sci.* **2017**, *72*, 1683–1688. [CrossRef]
30. Shadyab, A.H.; Manson, J.E.; Wenjun, L.I.; Gass, M.; Brunner, R.L.; Naughton, M.J.; Cannell, B.; Howard, B.V.; Lacroix, A.Z. Parental Longevity Predicts Healthy Ageing among Women. *Age Ageing* **2018**, *47*, 853–860. [CrossRef]
31. Vandenbroucke, J.P.; Von Elm, E.; Altman, D.G.; Gøtzsche, P.C.; Mulrow, C.D.; Pocock, S.J.; Poole, C.; Schlesselman, J.J.; Egger, M. Strengthening the Reporting of Observational Studies in Epidemiology (STROBE): Explanation and Elaboration. *PLoS Med.* **2007**, *4*, 1628–1654. [CrossRef]
32. Fried, L.P.; Tangen, C.M.; Walston, J.; Newman, A.B.; Hirsch, C.; Gottdiener, J.; Seeman, T.; Tracy, R.; Kop, W.J.; Burke, G.; et al. Frailty in Older Adults: Evidence for a Phenotype. *J. Gerontol. Ser. Biological Sci. Med. Sci.* **2001**, *56*, M146–M156. [CrossRef] [PubMed]
33. Rodríguez-Mañas, L.; Bayer, A.J.; Kelly, M.; Zeyfang, A.; Izquierdo, M.; Laosa, O.; Hardman, T.C.; Sinclair, A.J. An Evaluation of the Effectiveness of a Multi-Modal Intervention in Frail and Pre-Frail Older People with Type 2 Diabetes—the MID-Frail Study: Study Protocol for a Randomised Controlled Trial. *Trials* **2014**, *15*, 34. [CrossRef] [PubMed]

34. Cruz-Jentoft, A.J.; Baeyens, J.P.; Bauer, J.M.; Boirie, Y.; Cederholm, T.; Landi, F.; Martin, F.C.; Michel, J.P.; Rolland, Y.; Schneider, S.M.; et al. Sarcopenia: European Consensus on Definition and Diagnosis. *Age Ageing* **2018**, *39*, 1–16. [CrossRef]
35. Tudor, G.; Koch, G.G. Review of Nonparametric Methods for the Analysis of Crossover Studies. *Stat. Methods Med. Res.* **1994**, *3*, 345–381. [CrossRef] [PubMed]
36. Scheff, S.W. Chapter 8—Nonparametric Statistics. In *Fundamental Statistical Principles for the Neurobiologist*; Scheff, S.W., Ed.; Academic Press: Cambridge, MA, USA, 2016; pp. 157–182. ISBN 978-0-12-804753-8.
37. Maldonado, G.; Greenland, S. Simulation Study of Confounder-Selection Strategies. *Am. J. Epidemiol.* **1993**, *138*, 923–936. [CrossRef]
38. Amblàs-Novellas, J.; Espaulella-Panicot, J.; Inzitari, M.; Rexach, L.; Fontecha, B.; Romero-Ortuno, R. The Challenge of Clinical Complexity in the 21st Century: Could Frailty Indexes Be the Answer? *Rev. Esp. Geriatr. Gerontol.* **2017**, *52*, 159–166. [CrossRef]
39. Kulminski, A.M.; Ukraintseva, S.V.; Kulminskaya, I.V.; Arbeev, K.G.; Land, K.; Yashin, A.I. Cumulative Deficits Better Characterize Susceptibility to Death in Elderly People than Phenotypic Frailty: Lessons from the Cardiovascular Health Study. *J. Am. Geriatr. Soc.* **2008**, *56*, 898–903. [CrossRef]
40. Kim, S.; Welsh, D.A.; Cherry, K.E.; Myers, L.; Jazwinski, S.M. Association of Healthy Aging with Parental Longevity. *Age* **2013**, *35*, 1975–1982. [CrossRef]
41. Arosio, B.; Ferri, E.; Casati, M.; Mari, D.; Vitale, G.; Cesari, M. The Frailty Index in Centenarians and Their Offspring. *Aging Clin. Exp. Res.* **2019**, *31*, 1685–1688. [CrossRef] [PubMed]
42. Sanders, J.L.; Singh, J.; Minster, R.L.; Walston, J.D.; Matteini, A.M.; Christensen, K.; Mayeux, R.; Borecki, I.B.; Perls, T.; Newman, A.B. Association Between Mortality and Heritability of the Scale of Aging Vigor in Epidemiology. *J. Am. Geriatr. Soc.* **2016**, *64*, 1679–1683. [CrossRef]
43. Ayers, E.; Barzilai, N.; Crandall, J.P.; Milman, S.; Verghese, J. Association of Exceptional Parental Longevity and Physical Function in Aging. *Age* **2014**, *36*, 9677. [CrossRef]
44. Soysal, P.; Stubbs, B.; Lucato, P.; Luchini, C.; Solmi, M.; Peluso, R.; Sergi, G.; Isik, A.T.; Manzato, E.; Maggi, S.; et al. Inflammation and Frailty in the Elderly: A Systematic Review and Meta-Analysis. *Ageing Res. Rev.* **2016**, *31*, 1–8. [CrossRef]
45. Franceschi, C.; Garagnani, P.; Parini, P.; Giuliani, C.; Santoro, A. Inflammaging: A New Immune–Metabolic Viewpoint for Age-Related Diseases. *Nat. Rev. Endocrinol.* **2018**, *14*, 576–590. [CrossRef]
46. Mazzotti, D.R.; Guindalini, C.; Moraes, W.A.D.S.; Andersen, M.L.; Cendoroglo, M.S.; Ramos, L.R.; Tufik, S. Human Longevity Is Associated with Regular Sleep Patterns, Maintenance of Slow Wave Sleep, and Favorable Lipid Profile. *Front. Aging Neurosci.* **2014**, *6*, 134. [CrossRef] [PubMed]
47. Lee, Y.H.; Kong, D.; Lee, Y.T.H.; Lin, C.H.; Liu, C.T.; Chang, Y.C. Functional Disabilities and Changes in Sleep Quality and Duration among Older Adults: Results from a Longitudinal Study in China, 2005–2014. *Eur. Geriatr. Med.* **2022**, *13*, 967–975. [CrossRef]
48. Kocevska, D.; Barclay, N.L.; Bramer, W.M.; Gehrman, P.R.; Van Someren, E.J.W. Heritability of Sleep Duration and Quality: A Systematic Review and Meta-Analysis. *Sleep Med. Rev.* **2021**, *59*, 101448. [CrossRef]
49. Madrid-Valero, J.J.; Rubio-Aparicio, M.; Gregory, A.M.; Sánchez-Meca, J.; Ordoñana, J.R. The Heritability of Insomnia: Systematic Review and Meta-Analysis of Twin Studies. *Sleep Med. Rev.* **2021**, *58*, 101437. [CrossRef]
50. Lovato, N.; Lack, L. Insomnia and Mortality: A Meta-Analysis. *Sleep Med. Rev.* **2019**, *43*, 71–83. [CrossRef] [PubMed]
51. Klein, L.; Gao, T.; Barzilai, N.; Milman, S. Association between Sleep Patterns and Health in Families with Exceptional Longevity. *Front. Med.* **2017**, *4*, 214. [CrossRef]
52. Ensrud, K.E.; Blackwell, T.L.; Redline, S.; Ancoli-Israel, S.; Paudel, M.L.; Cawthon, P.M.; Dam, T.T.L.; Barrett-Connor, E.; Leung, P.C.; Stone, K.L. Sleep Disturbances and Frailty Status in Older Community-Dwelling Men. *J. Am. Geriatr. Soc.* **2009**, *57*, 2085–2093. [CrossRef] [PubMed]
53. Tian, Q.; Pilling, L.C.; Atkins, J.L.; Melzer, D.; Ferrucci, L. The Relationship of Parental Longevity with the Aging Brain—Results from UK Biobank. *GeroScience* **2020**, *42*, 1377–1385. [CrossRef]
54. Collino, S.; Montoliu, I.; Martin, F.-P.J.; Scherer, M.; Mari, D.; Salvioli, S.; Bucci, L.; Ostan, R.; Monti, D.; Biagi, E.; et al. Metabolic Signatures of Extreme Longevity in Northern Italian Centenarians Reveal a Complex Remodeling of Lipids, Amino Acids, and Gut Microbiota Metabolism. *PLoS ONE* **2013**, *8*, e56564. [CrossRef]
55. Pradas, I.; Jové, M.; Huynh, K.; Puig, J.; Ingles, M.; Borras, C.; Viña, J.; Meikle, P.; Pamplona, R. Exceptional Human Longevity Is Associated with a Specific Plasma Phenotype of Ether Lipids. *Redox Biol.* **2019**, *21*, 101127. [CrossRef] [PubMed]
56. Picca, A.; Coelho-Junior, H.J.; Cesari, M.; Marini, F.; Miccheli, A.; Gervasoni, J.; Bossola, M.; Landi, F.; Bernabei, R.; Marzetti, E.; et al. The Metabolomics Side of Frailty: Toward Personalized Medicine for the Aged. *Exp. Gerontol.* **2019**, *126*, 110692. [CrossRef]
57. Li, G.H.; Han, F.; Xiao, F.H.; Gu, K.S.Y.; Shen, Q.; Xu, W.; Li, W.X.; Wang, Y.L.; Liang, B.; Huang, J.F.; et al. System-Level Metabolic Modeling Facilitates Unveiling Metabolic Signature in Exceptional Longevity. *Aging Cell* **2022**, *21*, e13595. [CrossRef] [PubMed]
58. Barzilai, N.; Atzmon, G.; Schechter, C.; Schaefer, E.J.; Cupples, A.L.; Lipton, R.; Cheng, S.; Shuldiner, A.R. Unique Lipoprotein Phenotype and Genotype Associated with Exceptional Longevity. *J. Am. Med. Assoc.* **2003**, *290*, 2030–2040. [CrossRef] [PubMed]
59. Gueresi, P.; Miglio, R.; Monti, D.; Mari, D.; Sansoni, P.; Caruso, C.; Bonafede, E.; Bucci, L.; Cevenini, E.; Ostan, R.; et al. Does the Longevity of One or Both Parents Influence the Health Status of Their Offspring? *Exp. Gerontol.* **2013**, *48*, 395–400. [CrossRef]
60. Liou, L.; Kaptoge, S. Association of Small, Dense LDL-Cholesterol Concentration and Lipoprotein Particle Characteristics with Coronary Heart Disease: A Systematic Review and Meta-Analysis. *PLoS ONE* **2020**, *15*, e0241993. [CrossRef] [PubMed]

61. Hirano, T.; Ito, Y.; Yoshino, G. Measurement of Small Dense Low-Density Lipoprotein Particles. *J. Atheroscler. Thromb.* **2005**, *12*, 67–72. [CrossRef]
62. Andersen, S.L.; Sweigart, B.; Sebastiani, P.; Drury, J.; Sidlowski, S.; Perls, T.T. Reduced Prevalence and Incidence of Cognitive Impairment among Centenarian Offspring. *J. Gerontol. Ser. A Biol. Sci. Med. Sci.* **2019**, *74*, 108–113. [CrossRef]
63. Gostynski, M.; Gutzwiller, F.; Kuulasmaa, K.; Döring, A.; Ferrario, M.; Grafnetter, D.; Pajak, A. Analysis of the Relationship between Total Cholesterol, Age, Body Mass Index among Males and Females in the WHO MONICA Project. *Int. J. Obes.* **2004**, *28*, 1082–1090. [CrossRef] [PubMed]
64. Ayers, E.; Barzilai, N.; Crandall, J.P.; Milman, S.; Verghese, J. Association of Family History of Exceptional Longevity with Decline in Physical Function in Aging. *J. Gerontol. Ser. A Biol. Sci. Med. Sci.* **2017**, *72*, 1649–1655. [CrossRef] [PubMed]
65. You, J.Y.; Kim, Y.J.; Shin, W.Y.; Kim, N.Y.; Cho, S.H.; Kim, J.H. Heritability of Muscle Mass in Korean Parent–Offspring Pairs in the Fifth Korean National Health and Nutrition Examination Survey (KNHANES V). *Maturitas* **2018**, *114*, 67–72. [CrossRef] [PubMed]
66. Masanés, F.; Rojano i Luque, X.; Salvà, A.; Serra-Rexach, J.A.; Artaza, I.; Formiga, F.; Cuesta, F.; López Soto, A.; Ruiz, D.; Cruz-Jentoft, A.J. Cut-off Points for Muscle Mass—Not Grip Strength or Gait Speed—Determine Variations in Sarcopenia Prevalence. *J. Nutr. Health Aging* **2017**, *21*, 825–829. [CrossRef] [PubMed]
67. Hayashida, I.; Tanimoto, Y.; Takahashi, Y.; Kusabiraki, T.; Tamaki, J. Correlation between Muscle Strength and Muscle Mass, and Their Association with Walking Speed, in Community-Dwelling Elderly Japanese Individuals. *PLoS ONE* **2014**, *9*, e111810. [CrossRef]
68. Inglés, M.; Gambini, J.; Carnicero, J.A.; García-García, F.J.; Rodríguez-Mañas, L.; Olaso-González, G.; Dromant, M.; Borrás, C.; Viña, J. Oxidative Stress Is Related to Frailty, Not to Age or Sex, in a Geriatric Population: Lipid and Protein Oxidation as Biomarkers of Frailty. *J. Am. Geriatr. Soc.* **2014**, *62*, 1324–1328. [CrossRef]
69. El Assar, M.; Angulo, J.; Carnicero, J.A.; Walter, S.; García-García, F.J.; López-Hernández, E.; Sánchez-Puelles, J.M.; Rodríguez-Mañas, L. Frailty Is Associated With Lower Expression of Genes Involved in Cellular Response to Stress: Results From the Toledo Study for Healthy Aging. *J. Am. Med. Dir. Assoc.* **2017**, *18*, 734.e1–734.e7. [CrossRef]
70. Rakib, A.; Kiran, S.; Mandal, M.; Singh, U.P. MicroRNAs: A Crossroad That Connects Obesity to Immunity and Aging. *Immun. Ageing* **2022**, *19*, 64. [CrossRef]
71. Terry, D.F.; Wilcox, M.A.; McCormick, M.A.; Perls, T.T. Cardiovascular Disease Delay in Centenarian Offspring. *J. Gerontol. A. Biol. Sci. Med. Sci.* **2004**, *59*, 385–389. [CrossRef]
72. Belenguer-Varea, Á.; Tarazona-Santabalbina, F.J.; Avellana-Zaragoza, J.A.; Martínez-Reig, M.; Mas-Bargues, C.; Inglés, M. Oxidative Stress and Exceptional Human Longevity: Systematic Review. *Free Radic. Biol. Med.* **2020**, *149*, 51–63. [CrossRef]
73. Borras, C.; Abdelaziz, K.M.; Gambini, J.; Serna, E.; Inglés, M.; de la Fuente, M.; Garcia, I.; Matheu, A.; Sanchís, P.; Belenguer, A.; et al. Human Exceptional Longevity: Transcriptome from Centenarians Is Distinct from Septuagenarians and Reveals a Role of Bcl-XL in Successful Aging. *Aging* **2016**, *8*, 3185–3208. [CrossRef] [PubMed]
74. Inglés, M.; Belenguer-Varea, A.; Serna, E.; Mas-Bargues, C.; Tarazona-Santabalbina, F.J.; Borrás, C.; Vina, J. Functional Transcriptomic Analysis of Centenarians' Offspring Reveals a Specific Genetic Footprint That May Explain That They Are Less Frail than Age-Matched Non-Centenarians' Offspring. *J. Gerontol. Ser. A* **2022**, *10*, 1931–1938. [CrossRef]
75. Kupper, L.L.; Karon, J.M.; Kleinbaum, D.G.; Morgenstern, H.; Lewis, D.K. Matching in Epidemiologic Studies: Validity and Efficiency Considerations. *Biometrics* **1981**, *37*, 271. [CrossRef] [PubMed]
76. Jager, K.J.; Zoccali, C.; MacLeod, A.; Dekker, F.W. Confounding: What It Is and How to Deal with It. *Kidney Int.* **2008**, *73*, 256–260. [CrossRef] [PubMed]
77. Hoogendijk, E.O.; Afilalo, J.; Ensrud, K.E.; Kowal, P.; Onder, G.; Fried, L.P. Frailty: Implications for Clinical Practice and Public Health. *Lancet* **2019**, *394*, 1365–1375. [CrossRef]
78. Ainsworth, B.E.; Haskell, W.L.; Whitt, M.C.; Irwin, M.L.; Swartz, A.M.; Strath, S.J.; O'Brien, W.L.; Bassett, J.; Schmitz, K.H.; Emplaincourt, P.O.; et al. Compendium of Physical Activities: An Update of Activity Codes and MET Intensities. *Med. Sci. Sports Exerc.* **2000**, *32*, S498–S504. [CrossRef]
79. Xu, L.; Zhang, J.; Shen, S.; Hong, X.; Zeng, X.; Yang, Y.; Liu, Z.; Chen, L.; Chen, X. Association between Body Composition and Frailty in Elder Inpatients. *Clin. Interv. Aging* **2020**, *15*, 313–320. [CrossRef]
80. Airoldi, C.; Ferrante, D.; Mirabelli, D.; Azzolina, D.; Magnani, C. Evaluation of Nonresponse Bias in a Case–Control Study of Pleural Mesothelioma. *Int. J. Environ. Res. Public Health* **2020**, *17*, 6146. [CrossRef]
81. Mazor, K.M.; Clauser, B.E.; Field, T.; Yood, R.A.; Gurwitz, J.H. A Demonstration of the Impact of Response Bias on the Results of Patient Satisfaction Surveys. *Health Serv. Res.* **2002**, *37*, 1403–1417. [CrossRef] [PubMed]
82. Gubbi, S.; Schwartz, E.; Crandall, J.; Verghese, J.; Holtzer, R.; Atzmon, G.; Braunstein, R.; Barzilai, N.; Milman, S. Effect of Exceptional Parental Longevity and Lifestyle Factors on Prevalence of Cardiovascular Disease in Offspring. *Am. J. Cardiol.* **2017**, *120*, 2170–2175. [CrossRef] [PubMed]

International Journal of
Environmental Research and Public Health

MDPI

Article

Before-and-After Study of the First Four Years of the Enhanced Recovery after Surgery (ERAS®) Programme in Older Adults Undergoing Elective Colorectal Cancer Surgery

Cristina Martínez-Escribano [1,2,*], Francisco Arteaga Moreno [2], David Cuesta Peredo [3], Francisco Javier Blanco Gonzalez [4], Juan Maria De la Cámara-de las Heras [5] and Francisco J. Tarazona Santabalbina [6,7]

1 Anaesthesiology, Resuscitation and Therapeutics of Pain, Hospital Universitario de la Ribera, Ctra de Corbera, km 1, 46600 Alzira, Valencia, Spain
2 Medical School, Universidad Católica de Valencia San Vicente Mártir, 46001 Valencia, Valencia, Spain
3 Quality of Care Department, Hospital Universitario de la Ribera, 46600 Alzira, Valencia, Spain
4 General Surgery Service, Hospital Universitario de la Ribera, 46600 Alzira, Valencia, Spain
5 Library Service, Hospital Universitario La Ribera, 46600 Alzira, Valencia, Spain
6 Geriatric Medicine Department, Hospital Universitario de la Ribera, 46600 Alzira, Valencia, Spain
7 Centro de Investigación Biomédica en Red Fragilidad y Envejecimiento Saludable (CIBERFES), 28029 Madrid, Spain
* Correspondence: martinez_criesc@gva.es

Citation: Martínez-Escribano, C.; Arteaga Moreno, F.; Cuesta Peredo, D.; Blanco Gonzalez, F.J.; De la Cámara-de las Heras, J.M.; Tarazona Santabalbina, F.J. Before-and-After Study of the First Four Years of the Enhanced Recovery after Surgery (ERAS®) Programme in Older Adults Undergoing Elective Colorectal Cancer Surgery. *Int. J. Environ. Res. Public Health* **2022**, *19*, 15299. https://doi.org/10.3390/ijerph192215299

Academic Editor: Stergios Boussios

Received: 28 August 2022
Accepted: 17 November 2022
Published: 19 November 2022

Abstract: Background: The aim of this study was to determine whether the inclusion of older patients undergoing elective colorectal cancer resection in the Enhanced Recovery After Surgery (ERAS®) programme could improve clinical outcomes during hospital admission. Methods: A before-and-after study in ≥70-year-old patients electively admitted for colorectal cancer resection was designed. In total, 213 patients were included in the ERAS® group, and 158 were included in the control group. Results: The average age was 77.9 years old (SD 5.31) and 57.14% of them were men, with a Charlson Index score of 3.42 (SD 3.32). The ERAS® group presented a lower transfusion rate of 42 (19.7%), compared to 75 (47.5%) in the control group ($p < 0.001$). The crude odds ratio (OR) for transfusion was 0.27 (95% CI 0.17–0.43; $p < 0.001$), and the adjusted odds ratio was 0.26 (95% CI 0.14–0.48; $p < 0.001$). The ERAS® group had a lower percentage of patients with moderate–severe malnutrition on admission, at 23.4% (37 patients) against 36.2% in the control group (42 patients) ($p = 0.023$), with an OR of 0.47 (95% CI 0.29–0.75; $p < 0.002$) and an adjusted OR of 0.48 (95% CI 0.29–0.78; $p = 0.003$). The number of patients who required admission to the intensive care unit (ICU) was also markedly lower: 54 from the ERAS® group (25.4%) versus 71 from the control group (44.9%) ($p < 0.001$). Conclusions: The inclusion of ≥70-year-old adults in the ERAS® programme resulted in a decrease in transfusions, number of erythrocyte concentrates transfused, and number of ICU admissions, along with improved nutritional status.

Keywords: colorectal surgery; geriatric assessment; ERAS; postoperative complications; older patients

1. Introduction

Cancer is the leading cause of death in adults in developed countries [1]. Globally, colorectal cancer (CRC) is the third most common cancer in men and the second most common in women. In Spain, CRC became the leading cause of death in women in 2020, surpassing breast cancer [2]. Seven out of ten patients diagnosed with CRC [3] and about 60% of patients undergoing elective or urgent surgery due to colorectal cancer are over 65 years of age [4].

Surgery plays a key role in CRC treatment [5] but is associated with a high complication rate that can range from 8% to 63% [6] and global perioperative mortality of between 1%

and 12% [7]. Old age adds further mortality and perioperative complications, hampers functional recovery, and increases costs [8]. Although some studies found no differences when age was the only factor taken into consideration [9], ageing-related factors such as frailty and the presence of geriatric syndromes were associated with an increase in mortality and morbidity [10,11]. Similarly, factors such as anaemia and malnutrition, which are common in gastrointestinal pathology, have been associated with worse postoperative outcomes.

Anaemia is an independent risk factor for complications, prolonged hospital stay, and increased mortality in any type of surgery [12,13]. In the case of CRC, it has been related to the advanced stage and proximal location of the tumour, and it has also been attributed prognostic value due to its relationship with overall survival and cancer-specific survival [14]. It is common in older adults [15], especially in the case of colorectal neoplasm, where blood loss is frequent [16]. Moreover, the transfusion rate itself increases complications and mortality [17,18]. Malnutrition is another strong predictor of morbidity, mortality, prolonged hospitalisation, and readmissions [19]. Consequently, the Spanish Multimodal Rehabilitation Group (GERM) and the European Society of Nutrition and Metabolism (ESPEN) emphasise screening for and correcting nutritional deficiencies before surgery [20,21] within enhanced recovery programmes like Fast-Track or ERAS® (Enhanced Recovery After Surgery) [22].

These programmes were developed in order to reduce surgical stress, accelerate recovery, and improve the postoperative outcome in patients undergoing colorectal surgery [23]. The implementation of ERAS® programmes has been shown to reduce complications and shorten hospital stays over the last few years, improving the cost-effectiveness of these processes as a result [24]. These improved results have been evinced for both scheduled and urgent surgeries [25,26]. Several studies have shown the safety of these early recovery programmes [27], but it remains to be seen to what extent these improve hospital outcomes in older adults.

The aim of this study was to assess whether the ERAS® programme implemented in our hospital for ≥70-year-old patients improved their nutritional status and reduced their transfusion rate, postoperative complications, hospital stay, and mortality.

2. Materials and Methods

A quasi-experimental before-and-after study in a hospital environment was designed to include all patients consecutively admitted for elective surgery due to colorectal neoplasm to the General Surgery Department of *Hospital Universitario La Ribera* (HULR) from 1 January 2011 to 31 December 2019. The sample consisted of 213 patients who were included in the ERAS® programme since its introduction in 2016 and a further 158 patients previously operated on and treated via traditional means who were included in the control group between 1 January 2011 to 31 December 2015. A bivariate analysis was completed between two periods in the control group (1 January 2011 to 30 June 2013 versus 1 July 2013 to 31 December 2015) to check the homogeneity in the clinical practice during these two periods. No statistical differences were found between groups. HULR is a tertiary care hospital and covers a population of 263,001 inhabitants, of which 19.4% are people aged 65 and over.

2.1. Eligibility Criteria

2.1.1. Inclusion Criteria

Patients aged 70 or over electively admitted to the General and Gastrointestinal Surgery Department to undergo curative surgery (which seeks to remove the entire tumour, nearby lymph nodes included) for colorectal cancer resection, stage I–III at diagnosis, were included.

2.1.2. Exclusion Criteria

The study excluded emergency surgery hospital admissions, patients with metastasis at the time of diagnosis, those who had relapsed or were receiving palliative surgery, and patients with an expectation of less than 6 months, according to the Palliative Prognosis

Score (PaP Score) [28]. In total, 9.2% of patients from the initial ERAS group and 8.7% from the initial control group did not fulfil the eligibility criteria and were excluded from the sample. The causes were basically progression of the disease at the time of surgery that led to a palliative approach, unresectability, or rapid deterioration of the patient that did not allow adherence to the programme.

2.2. Sample Size

The sample size was calculated considering a transfusion rate of 50% prior to the onset of the ERAS® protocols, with an estimated reduction in said rate of 20% by fixing alpha and beta error values of 5%. These data evidenced the need to include a minimum of 154 patients per group. Subsequently, a calculation of the power of the study with the drafted sample was carried out, obtaining a power of 99.5%.

2.3. Intervention

The ERAS® protocol developed for the preoperative period in HULR consists of the diagnosis and treatment of anaemia (haemoglobin levels < 13 g/dL in men and <12 g/dL in women) through dosing ferric carboxymaltose, depending on the Hb levels and according to the medication data sheet, 2 to 4 weeks prior to surgery [29]. An assessment of nutritional status through the Controlling Nutritional Status (CONUT) score was performed. Patients classified as risk-free or with a slight risk of malnutrition (CONUT 0–4) were given dietary advice. Those who were classified with moderate–severe risk (CONUT > 4) were prescribed enteral supplements [30]. All patients were referred to a specialised physiotherapist who instructed them in the management of a respiratory incentive to improve lung function and assigned them an exercise chart taking into account several characteristics of the patient. Throughout this time, a telephonic follow-up was carried out by the nursing staff of the surgery unit, who coordinated the process and stayed in contact with patients and their families. One of the fundamental objectives was that the time between diagnosis and surgery should not exceed 4–5 weeks.

From admission to discharge, patients were treated following the recommendations of the 2014 Enhanced Recovery in Abdominal Surgery (RICA) clinical pathway, elaborated by the Spanish Multimodal Rehabilitation Group (GERM) [31].

Patients assisted before the introduction of the ERAS® protocol in our hospital were treated as recommended via the clinical pathways of both anaesthesia and general surgery of that time, which involved no nutritional intervention, nor iron administration, nor physical activity guides.

2.4. Variables and Outcomes

Several variables were included: demographic variables (age and sex), anthropometric variables (weight, height, and body mass index (BMI)), frailty according to the Balducci Scale (validated for geriatric oncology because of its simplicity and agility) [32] and presence of geriatric syndromes, comorbidity and Charlson Index [33], tumour location and staging; laboratory data, proinflammatory state markers (C-reactive protein (PCR) and procalcitonin) and hospital process data, complications, intensive care unit admissions, hospital stay, number of reinterventions, readmissions, hospital mortality, and 1-year mortality.

The principal outcome was to determine whether the introduction of the ERAS® programme reduced the anaemia and malnutrition incidence at the time of surgery. It was considered that a patient suffered from anaemia if their haemoglobin levels were <13 g/dL in men and <12 g/dL in women, according to the WHO (World Health Organisation) classification [34]. The transfusion incidence (percentage of transfused patients out of the total number of operated patients) and the Total Transfusion Index (number of erythrocyte concentrates used per operated patient) were registered. The nutritional status was assessed at admission and discharge by scoring on the CONUT nutritional index, calculated from albumin serum levels, overall cholesterol, and lymphocyte count. Depending on the calculated value, patients were classified as risk-free or with slight risk of malnutrition if the total was CONUT ≤ 4 and as moderate–severe risk if CONUT > 4.

The effect of our intervention was also studied to determine whether it was reflected in the incidence of complications, both medical (delirium, heart failure or respiratory insufficiency, infections, etc.) and surgical (suture dehiscence, intestinal pseudo-obstruction (understood as a lack of gastrointestinal transit and oral tolerance set out 5 days post-surgery) and surgical wound infection), in the number of ICU admissions and reinterventions, hospital stay, readmissions rate, hospital mortality, and 1-year mortality.

2.5. Statistical Analysis

Data were analysed using version 22 of the statistical software program SPSS (SPSS Inc., Chicago, IL, USA).

A description of the qualitative variables (including dichotomous variables) through the use of absolute and relative frequencies was made. For quantitative variables, measures of central tendency (mean) were used, along with measures of dispersion (standard deviation (SD) or interquartile range (IQR)), depending on whether or not variables met normal distribution criteria as determined using the Kolmogorov–Smirnov test.

A bivariate calculation was performed for the variables mentioned in the main and secondary objectives. Student's *t*-test was used for quantitative variables with a normal distribution, and the Chi-Square test was used for qualitative variables. A binary logistic regression was made for the "transfused erythrocyte concentrates" variable by calculating the crude and adjusted odds ratio (OR) for the following variables: age, sex, Charlson Index, frailty, tumour stage, and CONUT score at hospital admission.

A multiple logistic regression model was built in order to study the need for transfusions, presence of malnutrition, need for ICU admissions, and hospital stays of $\leq$6 days. Survival was estimated using the Kaplan–Meier statistical method, and the difference between groups was estimated using the Mantel–Haenszel test.

Variables related to 1-year mortality were assessed using Cox's proportional hazards model and defined as deaths that occurred in the following 365 days.

Moreover, the Number Needed to Treat (NNT) calculation was performed using the NNT macro for SPSS [34], in order to determine the need for transfusion and the presence of malnutrition, ICU admissions, and stays longer than 6 days.

The significance threshold was set to a value of $p < 0.05$.

3. Results

In total, 371 patients were included in the study period (213 in the ERAS® group and 158 in the control group), of whom 212 (57.1%) were men. The average age was 77.9 years old (ranging from 70 to 96 years old), and the average Charlson Index was 3.4 (SD 3.3). The most frequent tumour location was the colon, with an incidence of 55%. No differences were found in relation to sex between groups in the bivariate analysis, but age differences were found: the average age was significantly higher in the ERAS® group, 78.5 years old (SD 5.14) compared to 77.0 (SD 5.41) in the control group ($p = 0.009$, Table 1). Likewise, the ERAS® group presented a higher incidence of myocardial ischaemia, heart failure, diabetes mellitus, and frailty, together with a significantly higher localisation rate of colon neoplasm—61.7% compared to 55.1% in the control group ($p = 0.018$). Table 1 presents the main features of each group at hospital admission.

The percentage of laparoscopies was substantially higher in the ERAS® group—66% compared to 45% in the control group ($p < 0.001$). The duration of surgery was also higher, with 197 min (SD 65.32) in the ERAS® group against 170 min (SD 64.0) in the control group ($p < 0.001$). The number of reoperated patients was higher in the ERAS® group, at 10.3% compared to 4.4% in the control group ($p = 0.049$, Table 2).

Table 1. Baseline characteristics of the ERAS and control group.

	ERAS (n = 213)	Non-ERAS (n = 158)	p
Age (years), mean (SD)	78.5 (5.14)	77.0 (5.41)	0.009
Sex n (%)			
Male	119 (55.9%)	93 (58.9%)	0.597
Female	94 (44.1%)	65 (41.1%)	
BMI, mean (SD)	28.7 (4.49)	28.1 (4.64)	0.681
Charlson Comorbidity Index, mean (SD)	3.37 (3.19)	3.50 (3.53)	0.168
Pathological history			
Frailty signs n (%)			
0	72 (33.8%)	73 (46.2%)	0.018
1 or more	141 (66.2%)	85 (53.8%)	
Dementia n (%)	5 (2.3%)	5 (3.2%)	0.749
Stroke n (%)	18 (8.5%)	12 (7.6%)	0.849
Heart failure n (%)	56 (26.3%)	19 (12.0%)	0.001
Myocardial ischaemia n (%)	36 (16.9%)	9 (5.7%)	0.001
Chronic pulmonary disease n (%)	35 (16.4%)	26 (16.5%)	1.000
Diabetes mellitus n (%)	70 (32.9%)	29 (18.4%)	0.002
Chronic renal insufficiency n (%)	14 (6.6%)	13 (8.2%)	0.551
ASA score n (%)			
I	85 (39.9%)	63 (39.9%)	0.912
II	121 (56.8%)	91 (57.6%)	
III	7 (3.3%)	4 (2.5%)	
Endovenous iron treatment before surgery	114 (53.5%)	0(0%)	<0.001
Tumour location n (%)			
Colon	143 (67.1%)	87 (55.1%)	0.018
Rectosigmoid	70 (32.9%)	71 (44.9%)	
Stage n (%)			
I	68 (31.9%)	36 (22.8%)	0.014
II	64 (30.0%)	55 (34.8%)	
III	69 (32.4%)	45 (28.5%)	
IV	12 (5.6%)	22 (13.9%)	

Legend: ERAS: Enhanced Recovery After Surgery; BMI: body mass index; ASA: American Society of Anaesthesiologists.

Table 2. Results of bivariate analysis of variables during hospital stay and hospital discharge.

	ERAS (n = 213)	Non-ERAS (n = 158)	p
Type of surgery n (%)			
Open	73 (34.3%)	76 (49.0%)	<0.001
Laparoscopy	140 (65.7%)	72 (46.5%)	
Surgery duration (minutes), mean (SD)	197 (65.32)	170 (63.99)	<0.001
Haemoglobin (g/dL) at admission, mean (SD)	12.3 (1.70)	12.2 (2.01)	0.492
Anaemia n (%)	114 (53.5%)	87 (55.1%)	0.833

Table 2. *Cont.*

	ERAS (n = 213)	Non-ERAS (n = 158)	p
Lymphocytes (1.10^9/L), mean (SD)	2.35 (0.97)	2.41 (1.00)	0.563
Cholesterol (mg/dL), mean (SD)	147.4 (43.09)	134.7 (43.71)	0.006
Albumin (g/dL), mean (SD)	3.64 (0.65)	3.38 (0.80)	0.001
C-reactive protein (mg/mL), mean (SD)	63.4 (68.92)	51.2 (63.90)	0.127
Procalcitonin (mg/mL), mean (SD)	0.46 (1.30)	0.37 (0.85)	0.409
CONUT score at admission, mean (SD)	2.70 (2.79)	3.64 (3.34)	0.009
CONUT > 4 at admission n (%)	50 (23.6%)	59 (40.2%)	0.010
Hospital stay (days), mean (DS)	11.5 (10.20)	11.4 (8.58)	0.926
Admissions of at most 6 days n (%)	57 (26.8%)	28 (17.7%)	0.046
ICU admissions n (%)	54 (25.4%)	71 (44.9%)	<0.001
Adverse events in the postoperative period n (%)	124 (53.7%)	103 (65.2%)	0.530
Medical complications n (%)	45 (21.1%)	29 (18.4%)	0.599
Delirium n (%)	15 (7.0%)	12 (7.6%)	0.843
Heart complications n (%)	19 (8.9%)	7 (4.4%)	0.104
Respiratory complications n (%)	12 (5.6%)	11 (7.0%)	0.666
Digestive complications n (%)	5 (2.3%)	0 (0.0%)	0.075
Urinary tract infection n (%)	8 (3.8%)	4 (2.5%)	0.568
Surgical infection n (%)	12 (5.6%)	5 (3.2%)	0.321
Surgical complications n (%)	57 (26.8%)	53 (33.5%)	0.169
Intestinal pseudo-obstruction n (%)	50 (23.5%)	46 (29.1%)	0.232
Suture dehiscence n (%)	9 (4.2%)	5 (3.2%)	0.784
Reintervention n (%)	22 (10.3%)	7 (4.4%)	0.049
Transfusion rate n (%)	42 (19.7%)	75 (47.5%)	<0.001
TTI (SD)	0.52 (1.24)	1.68 (2.75)	<0.001
Minimum haemoglobin (g/dL) at episode, mean (SD)	10.11 (1.38)	9.73 (1.53)	0.013
Haemoglobin at discharge, mean (SD)	11.03 (1.39)	10.76 (1.27)	0.056
Cholesterol at discharge (mg/dL), mean (SD)	132.2 (29.84)	124.6 (33.14)	0.023
Albumin at discharge (g/dL), mean (SD)	3.18 (0.39)	3.04 (0.40)	0.001
CONUT at discharge, mean (SD)	4.69 (2.40)	5.72 (2.86)	0.002
CONUT > 4 at discharge n (%)	83 (39.2%)	86 (54.8%)	0.002
Readmissions n (%)	5 (2.3%)	4 (2.5%)	1.000
Hospital mortality n (%)	9 (4.2%)	4 (2.5%)	0.570
1-year mortality n (%)	24 (11.3%)	12 (7.6%)	0.288

Legend: ERAS: Enhanced Recovery After Surgery; TTI: Total Transfusion Index; CONUT: Controlling Nutritional Status; n: total number; %: percentage; SD: standard deviation; g: grams; dL: decilitres; mg: milligrams; L: litres.

There were no differences in haemoglobin levels at admission or at discharge, but differences in the minimum levels of haemoglobin registered during the hospital stay were significant. A meaningful reduction in the transfusion rate was observed in the ERAS® group—19% against 47% in the control group ($p < 0.001$). The Total Transfusion Index was considerably lower in the ERAS® group, at 0.52 (SD 1.24) compared to 1.68 (SD 2.75) in the control group ($p < 0.001$, Table 2). There were no differences in the overall incidence of medical or surgical complications (Table 2).

It was observed in the multivariate analysis that patients included in the ERAS® group presented a crude OR of transfusion of 0.27 (95% confidence interval (95% CI) 0.17–0.43; $p < 0.001$), with an adjusted OR of 0.26 (CI 95% 0.14–0.48; $p < 0.001$) (Table 3). The Number Needed to Treat (NNT) in the ERAS® programme in order to avoid transfusion was 4 (Table 4).

Table 3. Multivariate logistic regression analysis of significant variables in association with participation in the ERAS® programme, including the following adjusted variables: age, sex, Charlson Index, frailty, tumour stage, and CONUT score at admission.

	Crude OR	95% CI	*p*	Adjusted OR	95% CI	*p*
Transfusion	0.27	0.17–0.43	<0.001	0.26	0.14–0.48	<0.001
CONUT > 4 at admission	0.46	0.29–0.76	0.002	0.48	0.29–0.78	0.003
CONUT > 4 at discharge	0.52	0.34–0.80	0.003	0.55	0.36–0.85	0.007
ICU admissions	0.34	0.21–0.57	<0.001	0.42	0.27–0.65	<0.001
LOS ≥ 6 days	1.41	0.73–2.75	0.309	1.39	0.75–2.68	0.311

Legend: CONUT: Controlling Nutritional Status; ICU: intensive care unit; LOS: length of stay; OR: odds ratio; CI: confidence interval.

Table 4. Number Needed to Treat in order to benefit from participation in the programme.

	NNT	95% CI
Transfusion	3.6	2.7–5.5
CONUT < 4 at admission	6.9	4.2–18.3
CONUT > 4 at discharge	6.4	3.9–18.7
ICU admissions	5.1	3.4–10.2
LOS ≤ 6 days	4.4	3.2–7.3

Legend: NNT: Number Needed to Treat; CONUT: Controlling Nutritional Status; ICU: intensive care unit; LOS: length of stay; CI: confidence interval.

In regard to nutritional assessment, the ERAS® group presented a statistically lower percentage of moderate or severe malnutrition (estimated using the CONUT score) at admission—23.6% against 40.1% in the control group ($p = 0.010$). In fact, the average score in the CONUT nutritional screening was also significantly lower upon admission in ERAS® group patients—2.7 (SD 2.8) against 3.64 (SD 3.6) in the control group ($p = 0.012$). The improvement in nutritional status was also reproduced at hospital discharge, at which point patients in the ERAS® group had a CONUT score of 4.69 (SD 2.40) compared to 5.72 (SD 2.86) in the control group ($p = 0.003$).

It was observed in the multivariate analysis that patients in the ERAS® group presented a lower risk of moderate or severe malnutrition at admission, with an OR of 0.47 (95% CI 0.29–0.75; $p < 0.002$) and an adjusted OR of 0.48 (95% CI 0.29–0.78; $p = 0.003$). The OR of presenting moderate–severe malnutrition at discharge was significantly lower in the ERAS® programme patients, at 0.52 (95% CI 0.34–0.80; $p < 0.003$), with an adjusted OR of 0.55 (95% CI 0.36–0.85; $p = 0.007$) (Table 3). The NNT calculated in a patient in order to avoid malnutrition at admission was 7 (Table 4).

This lower transfusion rate and improved nutritional status translated into a higher percentage of patients with a hospital stay of ≤6 days in the ERAS® group—26.8% against 17.7% of the control group ($p = 0.046$). There were no differences in the total hospital stay, the mean being 11.5 days (SD 10.2) in the ERAS® group compared to 11.4 days (SD 8.58) in the control group ($p = 0.926$). The increased complexity of the profile of patients in the ERAS® group did not lead to a longer stay, and the percentage of patients that required admission to the ICU was significantly lower—25.4% compared to 44.9% in the control group ($p \leq 0.001$). No differences were found in the overall incidence of surgical or medical complications (Table 3). Further, no differences in hospital mortality or 1-year mortality were found.

4. Discussion

The data from our study show that the participation of older patients in the ERAS® programme since its implementation in 2016 has resulted in a substantial decrease in the following variables: the number of patients requiring transfusion, number of erythrocyte concentrates used, number of patients with moderate or severe malnutrition at the moment of surgery and at discharge, and ICU admissions.

The progressive population dynamics are leading to an increase in older-age patients that will eventually need surgical procedures [34,35], and this can be seen in the higher age of the intervention group compared to the control group. The reduction in the number of patients requiring transfusion observed in this study after the implementation of the enhanced recovery programme was previously also described for patients undergoing thoracic and orthopaedic surgery [36,37].

In this study, the probability of receiving a transfusion was related to the laparoscopic approach and initial haemoglobin, but the probability of requiring transfusion decreased for all patients who participated in the ERAS® programme, regardless of preoperative anaemia levels.

This outcome can be related to the intravenous administration of ferric carboxymaltose before surgery [29,38], the application of more restrictive transfusion strategies, and less invasive surgical techniques. These proceedings are included in the ERAS® protocol [39,40].

Although nutritional deficiency is frequent in patients with neoplasm, especially in gastrointestinal neoplasm, the protocol managed to significantly reduce the percentage of patients with moderate or severe malnutrition at admission, thanks to the presurgical nutritional intervention that was carried out. Moreover, the risk of malnutrition is high in hospitalised patients and increases as hospital stays are prolonged [41]. In our case, the number of patients with malnutrition increased with the stay duration, but it was significantly lower in the ERAS® group than in the control group.

These results show the need for pre-surgical prehabilitation protocols like ERAS®, through which patients who undergo elective colorectal surgery can be optimised and have an early recovery. In this regard, the diagnosis and early control of anaemia and malnutrition are crucial for decreasing anaemia incidence at the moment of surgery, reducing the intake of blood-derivative drugs [36,39,42,43], and improving nutritional status [19,21,40,44].

No reduction was observed in the incidence of medical and surgical complications, hospital stay, or mortality, as reported in previous studies [38]. However, it is worth noting that patients in the ERAS® group presented higher prevalence of frailty, diabetes, and chronic heart disease, which may have influenced the absence of significant differences. In fact, the higher prevalence of frailty in the ERAS® group, as previously described, could be associated with an increase in complications, hospital stay, and readmissions and a reduction in survival [45]; these results correlate better with the presence of frailty than with age, morbidity, or even the severity of the surgical process [10,46]. Despite this more complex profile, the ERAS® group presented a significantly higher proportion of patients with a hospital stay of less than 6 days and a decrease in ICU admissions.

Patients in the ERAS® group underwent laparoscopy to a larger extent than the control group and presented a longer surgery duration, which was not associated with an increment in the incidence of intestinal pseudo-obstruction in the postoperative period. In line with previous studies, our intestinal pseudo-obstruction incidence was high (23.5%), despite the decreased transfusion rate, use of morphics, and guided management of fluid therapy, as all these interventions are included in ERAS® patient management and are also related to this complication [47].

Limitations

The study presents the typical biases of quasi-experimental studies, such as those of selection, especially the use of a historical cohort as a comparison group, and of confusion [48]. The main limitation of the study results from its nonrandomised design, which led to differences between groups in terms of the prevalence of frailty, diabetes, heart failure, and myocardial ischaemia. The bivariate analysis shows a more complex profile of patients in the intervention group than in the control, which confers greater internal validity to the results obtained by the intervention. A registration of patient adherence to the programme, which would have sustained the obtained outcomes, was not performed. Personal histories from the non-ERAS® group were retrospectively collected, with biases resulting from the loss of information this entails.

5. Conclusions

The ERAS® intervention reduced the need for transfusions and the number of transfused erythrocyte concentrates; furthermore, it improved the nutritional status of patients at admission and discharge, reduced ICU admissions, and increased the percentage of patients with hospital stays of less than 6 days. These results suggest that elderly patients can also benefit from participation. Further studies in patients in this age range are necessary to more exactly determine the true potential of the ERAS® programme in these patients.

Author Contributions: All the authors have made substantial contributions to the manuscript, are aware of the contents, and have approved the present version. Conceptualisation (C.M.-E., F.J.B.G. and F.J.T.S.); Data curation (C.M.-E., F.A.M., D.C.P., F.J.B.G., J.M.D.l.C.-d.l.H. and F.J.T.S.); Formal analysis (F.J.T.S.); Investigation (C.M.-E., F.J.B.G. and F.J.T.S.); Methodology (C.M.-E. and F.J.T.S.); Project administration (C.M.-E.); Resources (C.M.-E. and F.J.T.S.); Software (C.M.-E. and F.J.T.S.) Writing—original draft (C.M.-E. and F.J.T.S.); Writing—review and editing (C.M.-E. and F.J.T.S.). All authors have read and agreed to the published version of the manuscript.

Funding: This research received no external funding.

Institutional Review Board Statement: This study was accepted by the Clinical Research Ethics Committee (CEIC) of Hospital Universitario de la Ribera (registration code HULR06112019) and was carried out in conformance to the Declaration of Helsinki, updated by the World Medical Association in October 2008, and in accordance with the legislation in force at the time of its completion (Spanish Royal Decree-Law 223/2004 and Spanish Biomedical Research Act of 2007).

Informed Consent Statement: Patient consent was waived because data were collected from electronic health records (SIAS© Software, Alzira, Comunidad Valenciana, Spain). Since this was a descriptive study, it was not necessary to obtain informed consent from patients.

Data Availability Statement: Not applicable.

Acknowledgments: The authors would like to thank Nadia Dolz, medical translator from Hospital Universitario de la Ribera, for the translation and proofreading during the elaboration of this manuscript.

Conflicts of Interest: The authors declare no conflict of interest.

References

1. Yusuf, S.; Joseph, P.; Rangarajan, S.; Islam, S.; Mente, A.; Hystad, P.; Brauer, M.; Kutty, V.R.; Gupta, R.; Wielgosz, A.; et al. Modifiable risk factors, cardiovascular disease, and mortality in 155 722 individuals from 21 high-income, middle-income, and low-income countries (PURE): A prospective cohort study. *Lancet* **2020**, *395*, 795–808. [CrossRef]
2. Sung, H.; Ferlay, J.; Siegel, R.L.; Laversanne, M.; Soerjomataram, I.; Jemal, A.; Bray, F. Global Cancer Statistics 2020: GLOBOCAN Estimates of Incidence and Mortality Worldwide for 36 Cancers in 185 Countries. *CA A Cancer J. Clin.* **2021**, *71*, 209–249. [CrossRef] [PubMed]
3. Cancer Research, UK. 2021. Available online: http://www.cancerresearchuk.org/cancer-info/cancerstats/types/bowel/incidence/ (accessed on 2 February 2022).

4. Bruns, E.R.J.; van den Heuvel, B.; Buskens, C.J.; van Duijvendijk, P.; Festen, S.; Wassenaar, E.B.; van der Zaag, E.S.; Bemelman, W.A.; van Munster, B.C. The effects of physical prehabilitation in elderly patients undergoing colorectal surgery: A systematic review. *Color Dis.* **2016**, *18*, O267–O277. [CrossRef] [PubMed]
5. Pallis, A.G.; Papamichael, D.; Audisio, R.; Peeters, M.; Folprecht, G.; Lacombe, D.; Van Cutsem, E. EORTC Elderly Task Force experts' opinion for the treatment of colon cancer in older patients. *Cancer Treat. Rev.* **2010**, *36*, 83–90. [CrossRef] [PubMed]
6. Fagard, K.; Wolthuis, A.; Verhaegen, M.; Flamaing, J.; Deschodt, M. A retrospective observational study of enhanced recovery after surgery in older patients undergoing elective colorectal surgery. *PLoS ONE* **2020**, *15*, e0232857. [CrossRef] [PubMed]
7. Okabe, H.; Ohsaki, T.; Ogawa, K.; Ozaki, N.; Hayashi, H.; Akahoshi, S.; Ikuta, Y.; Ogata, K.; Baba, H.; Takamori, H. Frailty predicts severe postoperative complications after elective colorectal surgery. *Am. J. Surg.* **2019**, *217*, 677–681. [CrossRef] [PubMed]
8. Heriot, A.G.; Tekkis, P.P.; Smith, J.J.; Cohen, C.R.G.; Montgomery, A.; Audisio, R.A.; Thompson, M.R.; Stamatakis, J.D. Prediction of Postoperative Mortality in Elderly Patients with Colorectal Cancer. *Dis. Colon Rectum* **2006**, *49*, 816–824. [CrossRef]
9. Shalaby, M.; Di Lorenzo, N.; Franceschilli, L.; Perrone, F.; Angelucci, G.P.; Quareisma, S.; Gaspari, A.L.; Sileri, P. Outcome of colorectal surgery in elderly populations. *Ann. Coloproctol.* **2016**, *32*, 139–143. [CrossRef] [PubMed]
10. Fagard, K.; Leonard, S.; Deschodt, M.; Devriendt, E.; Wolthuis, A.; Prenen, H.; Flamaing, J.; Milisen, K.; Wildiers, H.; Kenis, C. The impact of frailty on postoperative outcomes in individuals aged 65 and over undergoing elective surgery for colorectal cancer: A systematic review. *J. Geriatr. Oncol.* **2016**, *7*, 479–491. [CrossRef]
11. Li, J.L.; Henderson, M.A.; Revenig, L.M.; Sweeney, J.F.; Kooby, D.A.; Maithel, S.K.; Master, V.A.; Ogan, K. Frailty and one-year mortality in major intra-abdominal operations. *J. Surg. Res.* **2016**, *6*, 507–512. [CrossRef] [PubMed]
12. Beattie, W.S.; Karkouti, K.; Wijeysundera, D.N.; Tait, G. Risk Associated with Preoperative Anemia in Noncardiac. *Anesthesiology* **2009**, *110*, 574–581. [CrossRef]
13. Feng, S.; MacHina, M.; Beattie, W.S. Influence of anaemia and red blood cell transfusion on mortality in high cardiac risk patients undergoing major non-cardiac surgery: A retrospective cohort study. *Br. J. Anaesth.* **2017**, *118*, 843–851. [CrossRef] [PubMed]
14. Väyrynen, J.P.; Tuomisto, A.; Väyrynen, S.A.; Klintrup, K.; Karhu, T.; Mäkelä, J.; Herzig, K.-H.; Karttunen, T.J.; Mäkinen, M.J. Preoperative anemia in colorectal cancer: Relationships with tumor characteristics, systemic inflammation, and survival. *Sci. Rep.* **2018**, *8*, 1126. [CrossRef] [PubMed]
15. Pang, W.W.; Schrier, S.L. Anemia in the elderly. *Curr. Opin. Hematol.* **2012**, *19*, 133–140. [CrossRef] [PubMed]
16. Girelli, D.; Marchi, G.; Camaschella, C. Anemia in the elderly. *HemaSphere* **2018**, *2*, e40. [CrossRef] [PubMed]
17. Acheson, A.G.; Brookes, M.J.; Spahn, D.R. Effects of Allogeneic Red Blood Cell Transfusions on Clinical Outcomes in Patients Undergoing Colorectal Cancer Surgery. *Ann. Surg.* **2012**, *256*, 235–244. [CrossRef]
18. Althoff, F.C.; Neb, H.; Herrmann, E.; Trentino, K.M.; Vernich, L.; Füllenbach, C.; Freedman, J.; Waters, J.H.; Farmer, S.; Leahy, M.F.; et al. Multimodal Patient Blood Management Program Based on a Three-pillar Strategy: A Systematic Review and Meta-analysis. *Ann. Surg.* **2019**, *269*, 794–804. [CrossRef] [PubMed]
19. Seretis, C.; Kaisari, P.; Wanigasooriya, K.; Shariff, U.; Youssef, H. Malnutrition is associated with adverse postoperative outcome in patients undergoing elective colorectal cancer resections. *J. Buon* **2018**, *23*, 36–41. [PubMed]
20. Chan, S.P.; Ip, K.Y.; Irwin, M.G. Peri-operative optimisation of elderly and frail patients: A narrative review. *Anaesthesia* **2019**, *74*, 80–89. [CrossRef]
21. Weimann, A.; Braga, M.; Carli, F.; Higashiguchi, T.; Hübner, M.; Klek, S.; Laviano, A.; Ljungqvist, O.; Lobo, D.N.; Martindale, R.; et al. ESPEN guideline: Clinical nutrition in surgery. *Clin. Nutr.* **2017**, *36*, 623–650. [CrossRef]
22. Kehlet, H. Fast-track surgery—An update on physiological care principles to enhance recovery. *Langenbeck's Arch. Surg.* **2011**, *396*, 585–590. Available online: https://link.springer.com/article/10.1007/s00423-011-0790-y (accessed on 5 March 2019). [CrossRef]
23. Kehlet, H.; Wilmore, D.W. Multimodal strategies to improve surgical outcome. *Am. J. Surg.* **2002**, *183*, 630–641. [CrossRef]
24. Bagnall, N.M.; Malietzis, G.; Kennedy, R.H.; Athanasiou, T.; Faiz, O.; Darzi, A. A systematic review of enhanced recovery care after colorectal surgery in elderly patients. *Color Dis.* **2014**, *16*, 947–956. [CrossRef] [PubMed]
25. Zhao, J.; Sun, J.; Gao, P.; Chen, X.; Song, Y.; Huang, X.; Xu, H.; Wang, Z. Fast-track surgery versus traditional perioperative care in laparoscopic colorectal cancer surgery: A meta-analysis. *BMC Cancer* **2014**, *14*, 607. [CrossRef] [PubMed]
26. Lohsiriwat, V.; Jitmungngan, R.; Chadbunchachai, W.; Ungprasert, P. Enhanced recovery after surgery in emergency resection for obstructive colorectal cancer: A systematic review and meta-analysis. *Int. J. Color. Dis.* **2020**, *35*, 1453–1461. [CrossRef] [PubMed]
27. Millan, M.; Renau-Escrig, A.I. Minimizing the impact of colorectal surgery in the older patient: The role of enhanced recovery programs in older patients. *Eur. J. Surg. Oncol.* **2020**, *46*, 338–343. [CrossRef]
28. Morita, T.; Tsunoda, J.; Inoue, S.C.S. The Palliative Prognostic Index: A scoring system for survival prediction of terminally ill cancer patients. *Support Care Cancer* **1999**, *7*, 128–133. [CrossRef]
29. Dignass, A.U.; Gasche, C.; Bettenworth, D.; Birgegård, G.; Danese, S.; Gisbert, J.P.; Gomollon, F.; Iqbal, T.; Katsanos, K.; Koutroubakis, I.; et al. European Consensus on the Diagnosis and Management of Iron Deficiency and Anaemia in Inflammatory Bowel Diseases. *J. Crohn's Colitis* **2015**, *9*, 211–222. [CrossRef]
30. Huang, Y.; Huang, Y.; Lu, M.; Sun, W.; Sun, X.; Chen, X.; Li, L.; Chandoo, A.; Li, L. Controlling Nutritional Status (CONUT) Score is a Predictor of Post-Operative Outcomes in Elderly Gastric Cancer Patients Undergoing Curative Gastrectomy: A Prospective Study. *Cancer Manag. Res.* **2019**, *11*, 9793–9800. [CrossRef]

31. Calvo-Vecino, J.M.; Hernández, E.V.; Rodríguez, J.M.R.; Segurola, C.L.; Trapero, C.M.; Quintas, C.N.; Antolin, A.R.; Cuellar, E.R.; Aguado, J.J.H.; López, P.R.; et al. *Vía Clínica de Recuperación Intensificada en Cirugía Abdominal (RICA)*; Instituto Aragonés de Ciencias de la Salud: Zaragoza, Spain, 2014.
32. Balducci, L.E.M. Management of Cancer in the Older Person: A Practical Approach. *Oncology* **2000**, *5*, 224–237. [CrossRef]
33. Charlson, M.E.; Pompei, P.; Ales, K.L.; Mackenzie, C.R. A new method of classifying prognostic in longitudinal studies: Development and validation. *J. Chron. Dis.* **1987**, *40*, 373–383. [CrossRef]
34. Beutler, E.; Waalen, J. The definition of anemia: What is the lower limit of normal of the blood hemoglobin concentration? *Blood* **2006**, *107*, 1747–1750. [CrossRef] [PubMed]
35. Bonillo, A.; Domenech, J.; Granero, R. *Macro ¡MANS for SPSS Statistics: SPSS*; IBM: Armonk, NY, USA, 2000.
36. Ripollés-Melchor, J.; Abad-Motos, A.; Díez-Remesal, Y.; Aseguinolaza-Pagola, M.; Padin-Barreiro, L.; Sánchez-Martín, R.; Logroño-Egea, M.; Catalá-Bauset, J.C.; García-Orallo, S.; Bisbe, E.; et al. Association between Use of Enhanced Recovery after Surgery Protocol and Postoperative Complications in Total Hip and Knee Arthroplasty in the Postoperative Outcomes Within Enhanced Recovery after Surgery Protocol in Elective Total Hip and Knee Arthroplast. *JAMA Surg.* **2020**, *155*, e196024. [CrossRef] [PubMed]
37. Mohile, S.G.; Dale, W.; Somerfield, M.R.; Schonberg, M.A.; Boyd, C.M.; Burhenn, P.S.; Canin, B.; Cohen, H.J.; Holmes, H.M.; Hopkins, J.O.; et al. Practical Assessment and Management of Vulnerabilities in Older Patients Receiving Chemotherapy: ASCO Guideline for Geriatric Oncology. *J. Clin. Oncol.* **2018**, *36*, 2326–2347. [CrossRef] [PubMed]
38. Quinn, E.M.; Meland, E.; McGinn, S.; Anderson, J.H. Correction of iron-deficiency anaemia in colorectal surgery reduces perioperative transfusion rates: A before and after study. *Int. J. Surg.* **2017**, *38*, 1–8. [CrossRef]
39. Calvet, X.; Gené, E.; Àngelruíz, M.; Figuerola, A.; Villoria, A.; Cucala, M.; Mearin, F.; Delgado, S.; Calleja, J.L. Cost-minimization analysis favours intravenous ferric carboxymaltose over ferric sucrose or oral iron as preoperative treatment in patients with colon cancer and iron deficiency anaemia. *Technol. Health Care* **2016**, *24*, 111–120. [CrossRef] [PubMed]
40. Gustafsson, U.O.; Scott, M.J.; Hubner, M.; Nygren, J.; Demartines, N.; Francis, N.; Rockall, T.A.; Young-Fadok, T.M.; Hill, A.G.; Soop, M.; et al. Guidelines for Perioperative Care in Elective Colorectal Surgery: Enhanced Recovery After Surgery (ERAS®) Society Recommendations: 2018. *World J. Surg.* **2019**, *43*, 659–695. [CrossRef]
41. Vila, M.P.; Hernández, J.Á.; De Lorenzo, A.G.; Pérez, S.C.; Sanz, M.L.; García-Lorda, P.; Brosa, M. The burden of hospital malnutrition in Spain: Methods and development of the PREDyCES® study. *Nutr. Hosp.* **2010**, *25*, 1020–1024.
42. Meybohm, P.; Herrmann, E.; Steinbicker, A.U.; Wittmann, M.; Gruenewald, M.; Fischer, D.; Baumgarten, G.; Renner, J.; van Aken, H.K.; Weber, C.F.; et al. Patient Blood Management is Associated with a Substantial Reduction of Red Blood Cell Utilization and Safe for Patient's Outcome: A Prospective, Multicenter Cohort Study with a Noninferiority Design. *Ann. Surg.* **2016**, *264*, 203–211. [CrossRef]
43. Mörner, M.E.M.; Edgren, G.; Martling, A.; Gunnarsson, U.; Egenvall, M. Preoperative anaemia and perioperative red blood cell transfusion as prognostic factors for recurrence and mortality in colorectal cancer—A Swedish cohort study. *Int. J. Colorectal. Dis.* **2017**, *32*, 223–232. [CrossRef]
44. Bruns, E.R.J.; Argillander, T.E.; Van Den Heuvel, B.; Buskens, C.J.; Van Duijvendijk, P.; Winkels, R.M.; Kalf, A.; Van Der Zaag, E.S.; Wassenaar, E.B.; Bemelman, W.A.; et al. Oral Nutrition as a Form of Pre-Operative Enhancement in Patients Undergoing Surgery for Colorectal Cancer: A Systematic Review. *Surg. Infect.* **2018**, *19*, 1–10. [CrossRef] [PubMed]
45. Romero-Ortuno, R.; Forsyth, D.R.; Wilson, K.J.; Cameron, E.; Wallis, S.; Biram, R.; Keevil, V. The association of geriatric syndromes with hospital outcomes. *J. Hosp. Med.* **2017**, *12*, 83–89. [CrossRef] [PubMed]
46. Panayi, A.C.; Orkaby, A.R.; Sakthivel, D.; Endo, Y.; Varon, D.; Roh, D.; Orgill, D.P.; Neppl, R.L.; Javedan, H.; Bhasin, S.; et al. The American Journal of Surgery Impact of frailty on outcomes in surgical patients: A systematic review and meta-analysis. *Am. J. Surg.* **2019**, *218*, 393–400. [CrossRef] [PubMed]
47. Alhashemi, M.; Fiore, J.F.; Safa, N.; Al Mahroos, M.; Mata, J.; Pecorelli, N.; Baldini, G.; Dendukuri, N.; Stein, B.L.; Liberman, A.S.; et al. Incidence and predictors of prolonged postoperative ileus after colorectal surgery in the context of an enhanced recovery pathway. *Surg. Endosc.* **2019**, *33*, 2313–2322. [CrossRef] [PubMed]
48. Harris, A.D.; Mcgregor, J.C.; Perencevich, E.N.; Furuno, J.P.; Zhu, J.; Peterson, D.E.; Finkelstein, J. The Use and Interpretation of Quasi-Experimental Studies in Medical Informatics. *J. Am. Med. Inform. Assoc.* **2006**, *13*, 16–23. Available online: https://academic.oup.com/jamia/article/13/1/16/780729 (accessed on 15 September 2022). [CrossRef] [PubMed]

International Journal of
Environmental Research and Public Health

MDPI

Article

Identification of Early Knee Osteoarthritis Based on Knee Joint Trajectory during Stair Climbing

Ami Ogawa [1,*], Hirotaka Iijima [2,3] and Masaki Takahashi [1]

[1] Department of System Design Engineering, Faculty of Science and Technology, Keio University, Yokohama 223-8522, Japan
[2] Institute for Advanced Research, Nagoya University, Nagoya 464-8601, Japan
[3] Biomedical and Health Informatics Unit, Graduate School of Medicine, Nagoya University, Nagoya 461-8673, Japan
* Correspondence: ogawa@sd.keio.ac.jp; Tel.: +81-45-566-1844

Abstract: Patients with knee osteoarthritis show low stair climbing ability, but a diagnosis of stair performance time is not enough to identify the early stages of knee osteoarthritis. Therefore, we developed an indicator named range of the knee joint trajectory (RKJT) as a kinematic parameter to express more detailed characteristics than stair performance time. To achieve this, we used our developed "IR-Locomotion", a markerless measurement system that can track the knee joint trajectory when climbing stairs. This study aimed to test whether the RKJT effectively identifies patients with early knee osteoarthritis even after controlling stair performance time. Forty-seven adults with moderate to severe knee pain (mean age 59.2 years; 68.1% women) underwent the radiographic examination (Kellgren and Lawrence grade) of both knees and a stair climbing test on 11 stairs. The RKJT during the stair climbing test was calculated by "IR-Locomotion". A generalized linear mixed model was used to evaluate the discriminative capability of RKJT on early knee osteoarthritis (i.e., Kellgren and Lawrence grade of 1). As expected, patients with early knee osteoarthritis showed larger RKJT than non-radiographic controls (95% confidence interval: 1.007, 1.076). Notably, this finding was consistent even after adjusting stair performance time.

Keywords: IR-Locomotion; non-contact markerless measurement system; knee joint trajectory; early knee osteoarthritis; stair climbing

Citation: Ogawa, A.; Iijima, H.; Takahashi, M. Identification of Early Knee Osteoarthritis Based on Knee Joint Trajectory during Stair Climbing. *Int. J. Environ. Res. Public Health* **2022**, *19*, 15023. https://doi.org/10.3390/ijerph192215023

Academic Editors: Francisco José Tarazona Santabalbina, Sebastià Josep Santaeugènia Gonzàlez, José Augusto García Navarro and José Viña

Received: 18 October 2022
Accepted: 8 November 2022
Published: 15 November 2022

1. Introduction

Knee osteoarthritis (KOA) is one of the most common arthropathies in the elderly, causing pain and limited range of motion in the knee joint [1], eventually bringing total knee arthroplasty. There has been increasing interest in identifying the early stages of KOA, as no effective disease modifying treatments are available to postpone or prevent KOA. Early diagnosis improves treatment outcomes for patients with KOA [2,3].

Previous researchers have investigated the diagnostic ability of knee joint moments for early KOA during level walking, but these variables were unable to effectively identify early KOA [4–6]. Poor diagnostic ability may be due, at least in part, to level walking that is biomechanically easy for patients with early KOA. Biomechanically and physiologically, more challenging tasks may be suitable for identifying early KOA. Stair walking is known as one of the most demanding tasks in daily activities [7]. For example, the knee joint load during stair walking is greater than that during level walking [8–10]. Our previous study showed that patients with early KOA take longer to climb stairs [11]. This evidence suggests that stair climbing performance time may identify early KOA, but the prediction accuracy is inadequate [11]. In addition to stair climbing performance time, other parameters are required to effectively identify early KOA. Biomechanical outcomes such as joint movement are effective not only in diagnosing but also in understanding the mechanism of symptom progression. However, there is little research which is related to the early KOA [12].

An early KOA detection system is assumed to be installed in houses to daily diagnose the activities of people because few people will bother to seek medical attention in the unconscious early stages of the disease. As measuring activities at the house is delicate, an infrared-based depth sensor, which is non-contact and markerless, to measure kinematic parameters is reasonable at the point of considering privacy protection and body constraint-free. Our previously proposed measurement system, "IR-Locomotion" [13] acquires body joint trajectory based on a depth sensor. In this study, we used a modified version of "IR-Locomotion" [14] to reveal the kinematic characteristics of early KOA patients during stair climbing using for feasibility.

Knee flexion angle in the sagittal plane is known as a significant parameter that reflects the differences between KOA patients and controls [15]. However, the calculation of knee flexion angle using hip, knee, and ankle joint position has a risk of significant error because the depth data around ankle joint position has often noises as feet are always near stairs and are difficult to extract by the depth sensor. Thus, we hypothesized that the knee joint trajectory in the sagittal plane, which is more simply acquired and less affected by body shapes than knee flexion angle, reflects the characteristics of knee movements of early KOA patients. Since patients with KOA are likely to have less knee flexion [15] and lower external knee flexion moment when climbing stairs [16], the knee joint trajectory of early KOA patients is likely different from the controls. When the knee flexion angle, which mainly occurs in the sagittal plane, is slight, the displacement of the knee joint trajectory should be significant because the hip joint must be flexed significantly to keep the leg at a sufficient distance from the nosing line.

Our main aim in this paper was to explore the differences between early KOA patients and control groups based on knee joint trajectory, which can be acquired by "IR-Locomotion," for the development of a system that can be introduced into daily life to screen for early KOA progression as a further aim. In terms of the Kellgren and Lawrence (K&L) grade, which indicates the level of KOA symptoms, the people suspected as early KOA patients with osteophytes or subchondral osteosclerosis are classified into K&L grade of 1, and people without such suspicion are classified into K&L grade of 0. A longitudinal study has shown that patients who are diagnosed with a K&L grade of 1 are more likely to worsen to a K&L grade of 2 or higher, which means that patients with a K&L grade of 1 are in the advanced stages of KOA. The importance of diagnosing KOA at an earlier stage is evident. Throughout the observation of the knee joint trajectories during stair climbing using "IR-Locomotion", we hypothesized that early KOA patients perform a more significant knee joint trajectory than non-radiographic controls in the sagittal plane even after controlling the stair climbing time. This study aimed to examine the hypothesis of identifying early KOA patients using a knee joint trajectories-based indicator during stair climbing.

2. Materials and Methods

2.1. Participants

This study was a secondary cross-sectional baseline analysis of data from a randomized controlled trial [17]. The data of the community-based adults over the age of 50 interested in the measurement were collected via the internet. The inclusion criteria were (1) adults over the age of 50 years; (2) K&L values below grades 0 and 1 were assessed using weight-bearing anteroposterior radiographs for one or both knees; (3) Average pain experienced on a numerical rating scale was greater than three and less than ten over the previous month. The details were based on the previous report [17]. The targets of this study were distinguished into the K&L grade 0 group (control: OA identified without radiography) and the K&L grade 1 group (early KOA), focusing on early KOA detection. In this study, we did not check the type of KOA, so both primary and secondary KOA were mixed.

2.2. Radiographic Severity of Osteoarthritis

A trained examiner (HI) used the original version of the K&L grading system [18] to assess the radiographic severity of the tibiofemoral joints in both knees. Examiner had excellent credibility scores as both intra-examiner (κ: 0.876; 95% CI: 0.829, 0.924) and inter-examiner (κ: 0.845; 95% CI: 0.793, 0.897) [19].

2.3. Instrumentation of Stair Climbing Test

Kinect v2 (Microsoft, Redmond, WA, USA), an RGB-D sensor, was used to measure knee joint positions when climbing stairs. Kinect v2 has a skeleton tracking function that can automatically detect the position of joints based on the machine learning concept. However, it is often impossible to detect the knee joint positions when climbing stairs. The reason is that Kinect v2 does not contain the data acquired by the tilted sensor and needs to be tilted to capture the stair climbing [20]. Therefore, in this study, we used our previously suggested method to obtain the knee joint trajectories with Kinect v2 [13,14]. According to the previous method, we only used depth data captured from behind when climbing stairs. All data were sampled at 20 Hz. The experimental setup is shown in Figure 1. Kinect v2 was set at a distance of 1440 mm from the stairs, a height of 860 mm, and a tilt of 20 degrees as close as possible without interfering with walking so that it would fit within the angle of view. Since the start and end of walking included acceleration and deceleration (i.e., non-steady), the analysis range was set from an arbitrary step to the step before and was extracted from seven steps except the first and the last two steps.

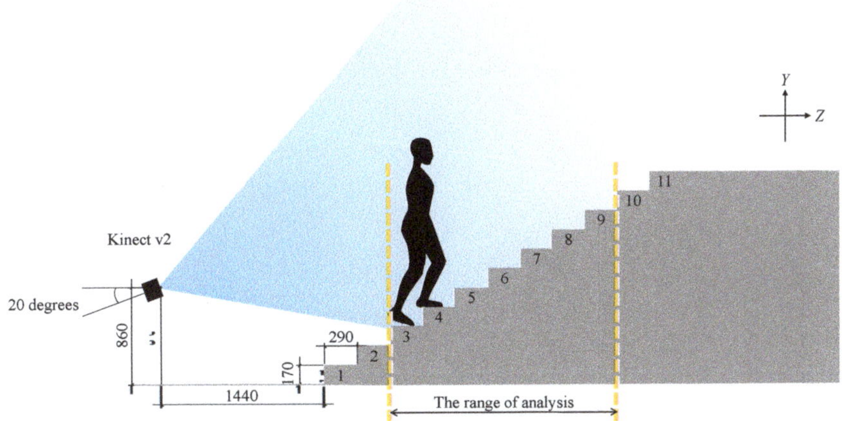

Figure 1. Experimental setup for stair climbing measurements using "IR-Locomotion" as a non-contact markerless system. We measured stairs climbing on an 11 steps staircase. The run length, riser height, and width were 290 mm, 170 mm, and 1350 mm, respectively. Kinect v2, a markerless RGB-D sensor, was set 1440 mm from the stairs at a height of 860 mm and 20 degrees tilted. Since the start and end of walking included acceleration and deceleration, namely a non-steady state, the knee joint trajectory of one step was extracted and analyzed from seven steps, excluding the first and the last two steps.

2.4. Procedure for Eleven-Step Stair Climb Test (11-SCT)

Participants climbed the stairs wearing their clothes with specified typical shoes (LD AROUND M, Mizuno, Tokyo, Japan) as fast as possible and following the method recommended by OARSI [21]. In our 11-SCT, all participants started moving down the stairs due to environmental constraints, while participants normally moved up the stairs in the standard SCT. The 11-SCT contained two trials. The stairs consisted of 11 steps. The run length, riser height, and width of the stairway were 290 mm, 170 mm, and 1350 mm, respectively. Detailed information is provided in the previous report [17].

2.5. Range of Knee Joint Trajectory (RKJT)

Since there was no significant difference in the time of the 11-SCT between ascending and descending in both KOA and the control group [11], we focused on stair ascent in this paper. As an example, as shown in Figure 2, an arbitrary step knee joint trajectory in the sagittal plane was extracted for analysis. During stair walking, spatial position information of the body must be processed based on different reference planes between each step. The range of target steps on the depth (Z) axis was defined as the observed foot contact step, that is, the stair nosing of the step and the next step. One cycle of the knee joint trajectory was extracted for analysis in the range of two steps: the target step, including the stance phase, and the step before it. The stair climbing motion was divided into five sections: two for the swing phase and three for the stance phase, according to the previous method [22]. The swing phase consisted of foot clearance (FCL) and foot placement (FP), while the stance phase started from the initial contact (IC) and consisted of weight acceptance (WA), pull-up (PU), and forward continuation (FCN).

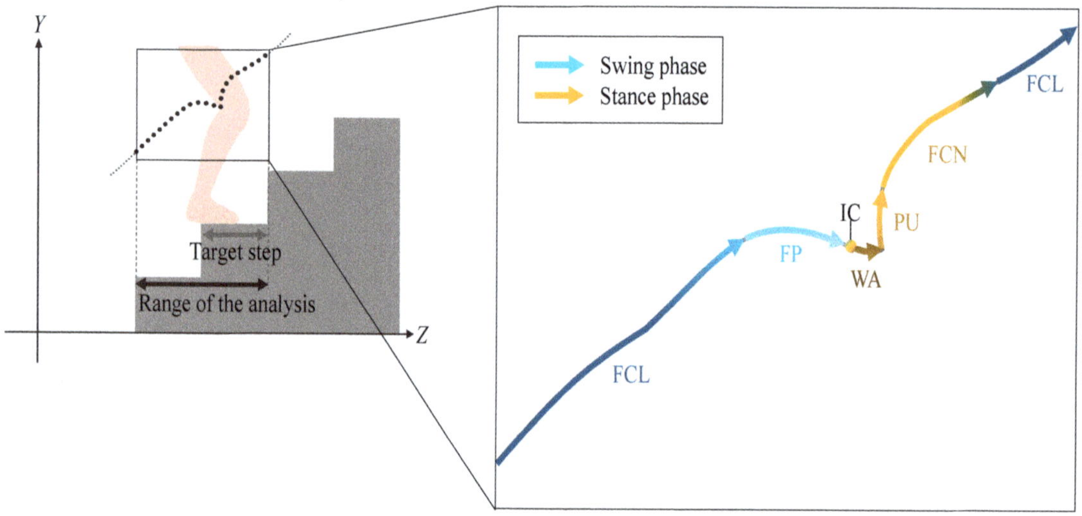

Figure 2. Representative image of knee joint trajectory in one cycle of stair climbing. One cycle of the knee joint trajectory was extracted in the range of two steps. The target step includes the stance phase and the step before it. The swing phase consisted of foot clearance (FCL) and foot placement (FP), while the stance phase started from the initial contact (IC) and consisted of weight acceptance (WA), pull-up (PU), and forward continuation (FCN).

Since RKJT may change by the resolution of the analytical data depending on the distance from the sensor, the characteristics of the subject's gait, and the step positions, it was confirmed in advance that there was no significant difference in the selection of analysis steps between groups. The arbitrarily extracted step positions for analysis did not follow a normal distribution pattern (Shapiro–Wilk test; p-value < 0.05), and the medians were the same at 7 (Mann–Whitney U-test; p-value $= 0.601$).

We proposed a method to analyze the knee joint trajectories using our previous method. Figure 3 shows the procedure of the proposed method. The flow is shown in Figure 3A. First, we calculated the distance from each knee joint position to the nosing line that was lined between stair nosing and defined as d (Figure 3B). Next, the d value was extracted in the analysis range on the Z axis (Figure 3(C1)). The minimum value was only detected when the foot stepped on (Figure 3(C2)), while two maximum values were detected before and after the minimum value. Since this study focused on the ending of FCL to the beginning of PU, where the knee joint moves greater in the d direction, the maximum value occurred during the swing phase before the minimum value was adopted (Figure 3(C3,4)). Finally, the range of knee joint trajectory (RKJT), as the proposed index, was calculated as the difference between the maximum and minimum values of *d* (Figure 3(C5)). When people climbed by stepping on the edge of the steps, the maximum value deviated from the target step, so the range of analysis should include the entire cycle. On the other hand, the wrong maximum value could be adopted in another case because the range of analysis was too wide. Therefore, when the maximum value was obtained near the boundary of the range, the data of ±50 mm from the maximum value was deleted, and the maximum value was obtained again. The threshold was determined in a try-and-error manner. All processes were automated.

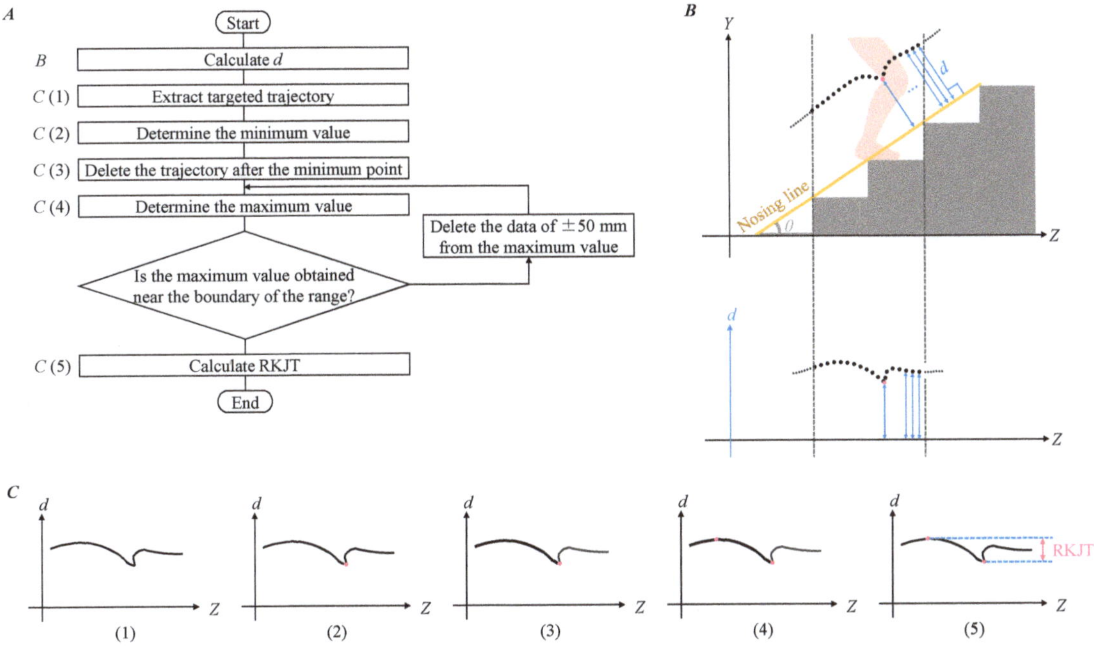

Figure 3. Definition and calculation procedure of range of knee joint trajectory (RKJT). (**A**) The flow of the procedure. (**B**) The distance (*d*) was calculated from the nosing line of the staircase to the knee joint trajectory. The black dotted line represents the knee joint trajectory. The nosing line of the staircase connects the stair nosing of all steps shown by the yellow line. The stair angle (θ) is the angle formed by the nosing line of the staircase and the horizontal lines (Z). (**C**) The description of the procedure is as follows: (1) The value *d* was calculated and extracted in the analytical range; (2) The minimum value was determined (pink dots); (3) The knee joint trajectory before the minimum value was extracted (bold black lines); (4) The maximum value was determined (pink dots); and (5) RKJT was calculated. When the maximum value was obtained near the boundary of the analysis range, the data of ±50 mm was deleted from the maximum value, and phase (4) was repeated.

Of the two trial data, this study used only the second trial data. Case 1 (ICC (1, 2)) of the intraclass correlation coefficient (ICC) was applied to the RKJT data of both trials and assessed in duplicate. The result was 0.759 with good duplicability [23].

2.6. Patients Characteristics and Covariates

Patients self-reported their age, gender, and height. Weight was digitally measured with clothes on and shoes off. Body mass index (BMI) was calculated by dividing weight (kg) by height (m) squared. Knee pain severity and disability levels were evaluated by the Japanese Knee Osteoarthritis Measure (JKOM) subcategories of "pain and stiffness"

(8 questions, 0–32 points) and "activities of daily living" (10 questions, 0–40 points) [24]. JKOM reflects the Japanese social and cultural background to globally standardized indexes. Comparisons with the Western Ontario and McMaster Universities Arthritis Index (WOMAC) [25] and the Medical Outcomes Study 36-Item Short-Form Health Survey (SF-36) [26] have shown adequate reliability and validity [24]. Knee pain during the 11-SCT trial was reported using a visual analog scale (VAS).

2.7. Statistical Analyses

Participants diagnosed with K&L grades of 1 and 0 were selected as the early KOA group and the control group (without radiographically identified KOA), respectively. RKJT were calculated based on their knee joint trajectories. Both left and right limb data from the second trial data were analyzed in this study. Differences in the RKJT mean values of each group were calculated by an appropriate test after performing the normality test and the equivariant test. The results of the Shapiro–Wilk's test rejected the normality hypothesis (p-value = 0.019), and the results of Levene's equal variance test adopted the equal variance hypothesis (p-value = 0.106). Therefore, the Mann–Whitney U-test was used to estimate differences between the mean values.

A generalized linear mixed model (GLMM) was used, and a binomial logistic regression analysis was performed to calculate the odds ratios and their 95% confidence intervals (CI) for the evaluation of the proposed index (continuous). Proposed index, limb side information (0: right; 1: left), age (continuous), gender, BMI (continuous), and VAS pain score during 11-SCT (continuous) were entered as independent variables. The K&L grade (0: K&L grade 0, 1: K&L grade 1) was entered as a dependent variable. Limb side information explained random effects on participant-specific parameters [27,28]. The remaining independent variables explained fixed effects. Age, gender, BMI, and VAS pain scores were entered as covariates. All statistical analyses were performed using IBM SPSS Statistics, version 25 (IBM, Armonk, NY, USA). The p-values less than 0.05 were considered significant.

3. Results

Fifty-nine participants were included in the test. Eight participants were excluded because of wearing oversized pants, which could reduce the accuracy of the data. Both knees were individually considered. This means that 102 samples were calculated from the data of 51 participants. Among them, 90 samples with K&L grade 0 or 1 were analyzed. In addition, due to systematic measurement issues and clothing issues, the data of the first trial were alternatively used in nine samples. According to Table 1, only K&L grade 1 participants were analyzed as early KOA patients (n = 20 participants, n = 35 knees), and K&L grade 0 participants were taken as controls into account (n = 27 participants, n = 55 knees). In four participants, only one knee was analyzed because the K&L grade of the other knee exceeded 1. There were three controls and one early KOA.

Figure 4 shows a visual comparison of the law knee joint trajectory between controls (Figure 4A) and early KOA (Figure 4B) patients with similar height and BMI.

Table 2 compares the RKJT differences between people with and without early KOA. Early KOA people showed that RKJT was 10 mm larger than control group people (95% CI: −17.6, −2.39 mm; p-value = 0.038). Binomial logistic regression shows that a 1 mm increase in the proposed index was significantly associated with a 1.04-fold increase in the odds ratio of early KOA after adjusting for covariates. Early KOA people seem to perform 11-SCT [11] for longer. Therefore, when we considered the stopwatch-based 11-SCT time as an additional covariate, we confirmed whether the identified relationship between RKJT and early KOA was similar. In fact, after adding 11-SCT in covariates (odds ratio: 1.037; 95% CI: 1.001, 1.073; p-value = 0.044), we found that the early KOA displayed a significantly larger RKJT.

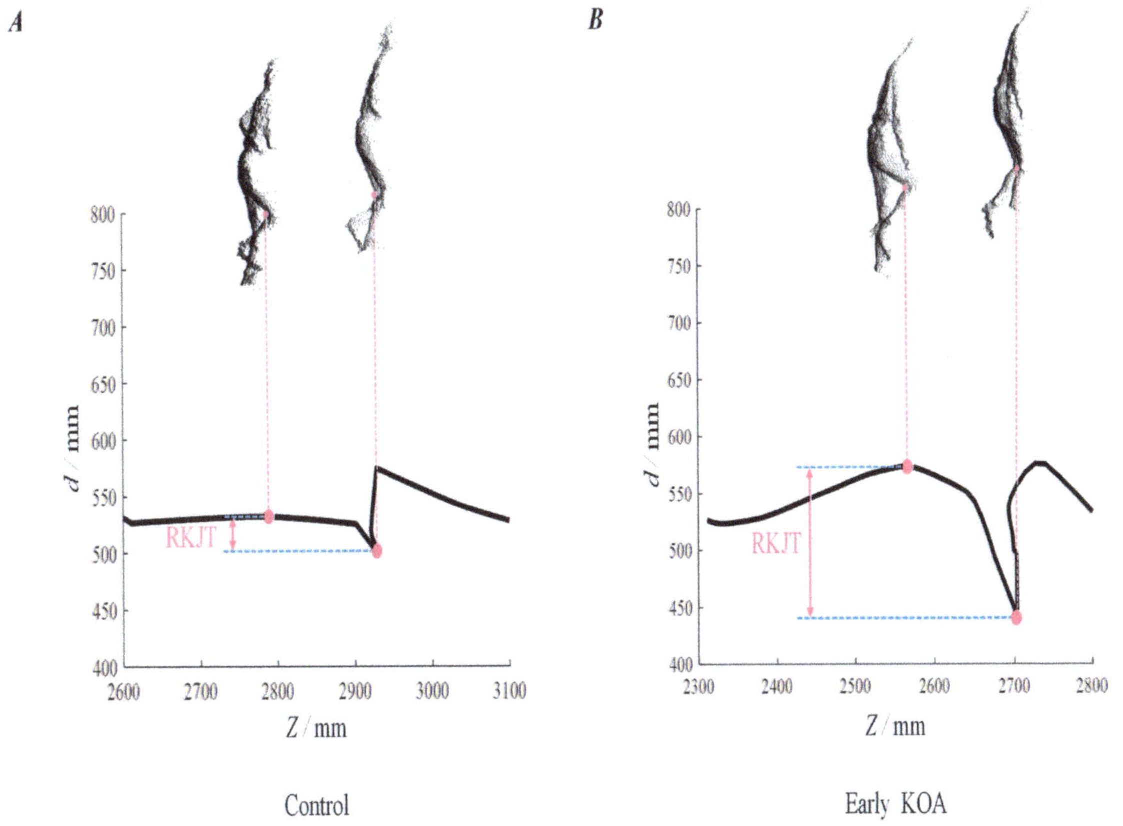

Figure 4. Comparison of RKJT values for K&L grades 0 and 1. (**A**) shows data for participants with K&L grade 0, and (**B**) shows data for K&L grade 1. The black lines show the analysis range, and the pink dots show the maximum and minimum points extracted. The four silhouettes correspond to the maximum and minimum values.

Table 1. Characteristics of participants and their knees.

Person-Level Characteristics	All (*n* = 47 Participants)	Control (*n* = 27 Participants)	Early KOA (*n* = 20 Participants)
Age, years	59.2 ± 5.98	58.7 ± 6.18	60.0 ± 5.79
Female, no. (%)	32 (68.1)	17 (63.0)	15 (75.0)
Height, m	1.61 ± 0.0817	1.62 ± 0.0853	1.61 ± 0.0782
Mass, kg	59.2 ± 10.1	57.3 ± 8.75	61.8 ± 11.5
BMI, kg/m^2	22.7 ± 3.04	21.8 ± 2.61	23.9 ± 3.26
Bilateral disease, no. (%) †	16 (34.0)	0 (0.0)	16 (80.0)
VAS pain score during 11-SCT, mm	14.3 ± 16.5; 8 [0, 63] *	10.7 ± 11.3; 4 [0, 34] *	19.2 ± 21.0; 10 [0, 63] *
JKOM, points			
Pain and stiffness	6.70 ± 4.14; 6 [0, 22] *	5.59 ± 3.48; 5 [1, 16] *	8.20 ± 4.56; 7 [0, 22] *
Activities of daily living	2.89 ± 3.30; 2 [0, 14] *	2.00 ± 2.47; 1 [0, 8] *	4.10 ± 3.92; 3 [0, 14] *
Participation in social activities	2.66 ± 2.05; 2 [0, 9] *	2.15 ± 1.68; 2 [0, 7] *	3.35 ± 2.32; 3 [0, 9] *
General health conditions	1.96 ± 1.02; 2 [0, 4] *	1.78 ± 1.05; 2 [0, 3] *	2.20 ± 0.95; 2 [0, 4] *
Total score	14.2 ± 8.22; 12 [3, 49] *	11.5 ± 5.79; 10 [3, 25] *	17.9 ± 9.68; 16 [6, 49] *

Table 1. *Cont.*

Person-Level Characteristics	All (n = 47 Participants)	Control (n = 27 Participants)	Early KOA (n = 20 Participants)
Knee-level characteristics	All (n = 90 knees)	Control (n = 55 knees)	Early KOA (n = 35 knees)
K&L grade, no. (%)			
Grade 0	55 (61.1)	55 (100.0)	0 (0.0)
Grade 1	35 (38.9)	0 (0.0)	35 (100.0)

BMI, body mass index; JKOM, Japanese Knee Osteoarthritis Measure; K&L grade, Kellgren and Lawrence grade; KOA, knee osteoarthritis; VAS, visual analog scale; 11-SCT, 11-step stair climb test. Unless otherwise stated, the values are mean ± SD. * Median [lower range–upper range] was provided because of the scattered distribution of the answered items. † Bilateral disease was defined as K&L grade ≥1 in both knees. Participants diagnosed with at least one K&L grade 1 knee were considered early KOA patients.

Table 2. Results of a binary logistic regression analysis investigating the relationship between RKJT and the presence of early KOA.

Independent Variable	Control (n = 55 Knees)	Early KOA (n = 35 Knees)	Difference between Control and Early KOA	Model 1 † Control vs. Early KOA		Model 2 †† Control vs. Early KOA	
	Mean ± SD	Mean ± SD	Mean (95% CI)	OR (95% CI)	p-Value	OR (95% CI)	p-Value
RKJT, mm	61.7 ± 18.4	71.7 ± 16.5	**10.0 (−17.6, −2.39)**	**1.04 (1.01, 1.08)**	**0.018**	**1.04 (1.00, 1.07)**	**0.044**

RKJT, range of knee joint trajectory; KOA, knee osteoarthritis; OR, odds ratio; 95% CI, confidence interval. † For the presence of early KOA, Model 1 was calculated to indicate the predictive ability of independent variables and simultaneously included (GLMM) age (continuous), female gender, body mass index (continuous), and VAS pain score during 11-SCT (continuous) in the binary logistic regression model. †† For the presence of early KOA, Model 2 was calculated to indicate the predictive ability of independent variables and simultaneously included (GLMM) and the same variables as model 1 and the stopwatch-based 11-SCT time in the binary logistic regression model. Bold fonts represent statistically significant results.

In addition, to infer the function of RKJT in early KOA, we calculated the RKJT correlation between the maximum and minimum values. The correlation between RKJT and the minimum value (Pearson's correlation coefficients: −0.405; 95% CI: −0.624, −0.126; p-value = 0.006) was shown to be higher than the correlation between RKJT and the maximum value (Pearson's correlation coefficients: 0.296; 95% CI: 0.002, 0.542; p-value = 0.049).

The results without the knees which another knee was diagnosed as having more than a K&L grade of 1, showed the same tendency. The same tendency was also shown in the results, including a K&L grade of 2 in the early KOA group.

4. Discussion

This study hypothesized that RKJT in early KOA was greater than the controls during steady stair climbing. Ninety samples of knee joint trajectories with K&L grades of 0 and 1 were analyzed. The average RKJT in early KOA patients was significantly greater than that in controls, supporting the hypothesis. The results showed that RKJT could identify early KOA patients.

4.1. Interpretation of Larger RKJT in People at Early KOA

Patients at early KOA showed larger RKJT than the control group, which means larger knee joint movement in the sagittal plane. This indicates two theories: (1) Early KOA patients bend their lower thighs forward and step deeply in the stance phase. (2) They raise their lower limbs high during the swing phase. The first theory leads to a smaller minimum value of knee joint trajectory, and the second theory leads to a larger maximum value. The correlation coefficient results indicate that the correlation between RKJT and the minimum value was greater than the correlation between RKJT and the maximum value, supporting the first theory. It can be considered that early KOA patients unintentionally adopted a strategy that puts a high knee load on WA.

The larger RKJT in early KOA was independent of these factors, as GLMM included age, gender, height, BMI, and VAS pain scores. Notably, this trend was similar after the addition of 11-SCT time, indicating that poor stair walking performance cannot explain the mechanism of larger RKJT in early KOA.

4.2. Significance and Clinical Impacts of Study

The development of a screening index in early KOA has not been achieved in previous studies [4–6]. In this study, a new finding was that the range of the knee trajectories during stair climbing, perpendicular to the nosing line, was significantly larger in patients with early KOA. The results showed the importance of observing the knee joint trajectories in early KOA screening. In addition, we proposed a simple diagnostic system that enables frequent diagnosis. Kinetic parameters such as knee contact forces and external moments have been considered in previous studies [6], but it was challenging to set up a device to record kinetic parameters, such as force plates that must be embedded at each step. The findings are clinically valuable because a simple measurement system can identify patients at high risk of KOA, either at home or in rehab centers. Identifying patients with early KOA could be a good screening tool for inclusion in clinical trials aiming to postpone its radiological progression.

4.3. Study Limitations and Strengths

It is not easy to interpret the difference between RKJT with these results biomechanically. However, the results show the ability of RKJT to distinguish people with K&L grades of 0 and 1.

As the control group without radiographic KOA markers also complained of knee pain, the results may not be applicable to asymptomatic healthy subjects. However, pain cannot explain the difference between K&L grades 0 and 1 because the VAS pain score was added to the covariates. Additionally, this study could not consider the effect of patellar-femoral joint arthrosis as a confounding factor as we did not have skyline merchant views. Still, at least a part of the effect of patellar-femoral joint arthrosis is considered by adding the VAS pain score related to patellar-femoral joint arthrosis to the covariates, and the results were robust.

The participants may be unable to walk normally because of wearing specified shoes during the test. Instead, the effect of shoes was eliminated by unifying the shoe conditions to reveal the feasibility of our system. On the other hand, we cannot conclude that the effects of the users' clothes or the physical conditions, which we did not unify the requirements, are negligible only from this study.

"IR-Locomotion" has no installation constraints because there is no need to use large equipment or attach devices to the subject. RKJT has strength in its simplicity for acquisition compared with other kinematic and kinetic parameters. Therefore, early diagnosis of KOA using RKJT may be possible in daily life. This study only validated the use of a diagnostic system on one type of staircase. However, walking also depends on the slope of the stairs. Future work will validate staircase designs that match those of the house.

5. Conclusions

To distinguish between early KOA (K&L grade 1) and control (K&L grade 0) patients, we proposed a new index RKJT that uses the knee joint trajectory when climbing stairs. The results showed a significant relationship between RKJT and K&L grades. Furthermore, RKJT was significantly correlated with K&L grade classification independent of 11-SCT time. These findings provide mechanistic insights into early KOA-related changes in biomechanics.

Author Contributions: All authors contributed to the conceptualization, methodology, and review and editing of the final manuscript. The specific contributions of the authors are (1) Conceptualization and methodology: A.O., H.I., M.T. (2) Software development for analysis: A.O. (3) Interpretation of the data: A.O., H.I., M.T. (4) Writing of original draft: A.O. (5) Review and Editing of the article: A.O., H.I., M.T. (6) Obtaining funding: A.O., H.I., M.T. (7) Data collection and assembly: H.I., MT. All authors have read and agreed to the published version of the manuscript.

Funding: This study was supported in part by (1) funding from Omron Healthcare Co., Ltd.; (2) a Grant-in-Aid from the Japan Society for the Promotion of Science for Research Fellows to HI, and (3) JSPS KAKENHI Grant Number JP20K13807. These funding sources had no role in study design, data collection and analysis, decision to publish, or preparation of the manuscript.

Institutional Review Board Statement: The study was conducted in accordance with the Declaration of Helsinki, and approved by the Institutional Review Board of Kyoto University.

Informed Consent Statement: Written informed consent was obtained from all subjects involved in the study.

Data Availability Statement: Not applicable.

Acknowledgments: The authors thank the members of the Aoyama laboratory (Kyoto University, Kyoto) for their assistance with the data collection.

Conflicts of Interest: The authors did not receive financial support or other benefits from commercial sources for the work reported in this manuscript, or any other financial support that could create a potential conflict of interest or the appearance of a conflict of interest with regard to the work. The authors declare that the research was conducted in the absence of any commercial or financial relationships that could be construed as a potential conflict of interest. The authors declare that this study received funding from Omron healthcare Co., Ltd., a Grant-in-Aid from the Japan Society for the Promotion of Science, and. The funder was not involved in the study design, collection, analysis, interpretation of data, the writing of this article or the decision to submit it for publication.

References

1. Guccione, A.A.; Felson, D.T.; Anderson, J.J.; Anthony, J.M.; Zhang, Y.; Wilson, P.W.; Kelly-Hayes, M.; A Wolf, P.; E Kreger, B.; Kannel, W.B. The effects of specific medical conditions on the functional limitations of elders in the Framingham Study. *Am. J. Public Health* **1994**, *84*, 351–358. [CrossRef]
2. Guermazi, A.; Niu, J.; Hayashi, D.; Roemer, F.W.; Englund, M.; Neogi, T.; Aliabadi, P.; McLennan, C.E.; Felson, D. Prevalence of abnormalities in knees detected by MRI in adults without knee osteoarthritis: Population based observational study (Framingham Osteoarthritis Study). *BMJ* **2012**, *345*, e5339. [CrossRef] [PubMed]
3. Bruyere, O.; Cooper, C.; Arden, N.; Branco, J.; Brandi, M.L.; Herrero-Beaumont, G.; Berenbaum, F.; Dennison, E.; Devogelaer, J.-P.; Hochberg, M.; et al. Can we identify patients with high risk of osteoarthritis progression who will respond to treatment? A focus on epidemiology and phenotype of osteoarthritis. *Drugs Aging* **2015**, *32*, 179–187. [CrossRef] [PubMed]
4. Duffell, L.D.; Southgate, D.F.; Gulati, V.; McGregor, A.H. Balance and gait adaptations in patients with early knee osteoarthritis. *Gait Posture* **2014**, *39*, 1057–1061. [CrossRef]
5. Mahmoudian, A.; van Dieën, J.H.; Baert, I.A.; Bruijn, S.M.; Faber, G.S.; Luyten, F.P.; Verschueren, S.M. Changes in gait characteristics of women with early and established medial knee osteoarthritis: Results from a 2-years longitudinal study. *Clin. Biomech.* **2017**, *50*, 32–39. [CrossRef] [PubMed]
6. Meireles, S.; de Groote, F.; Reeves, N.; Verschueren, S.; Maganaris, C.; Luyten, F.; Jonkers, I. Knee contact forces are not altered in early knee osteoarthritis. *Gait Posture* **2016**, *45*, 115–120. [CrossRef] [PubMed]
7. Larsen, A.H.; Puggaard, L.; Hämäläinen, U.; Aagaard, P. Comparison of ground reaction forces and antagonist muscle coactivation during stair walking with ageing. *J. Electromyogr. Kinesiol.* **2008**, *18*, 568–580. [CrossRef] [PubMed]
8. Bergmann, G.; Deuretzbacher, G.; Heller, M.; Graichen, F.; Rohlmann, A.; Strauss, J.; Duda, G. Hip contact forces and gait patterns from routine activities. *J. Biomech.* **2001**, *34*, 859–871. [CrossRef]
9. Kutzner, I.; Heinlein, B.; Graichen, F.; Bender, A.; Rohlmann, A.; Halder, A.; Beier, A.; Bergmann, G. Loading of the knee joint during activities of daily living measured in vivo in five subjects. *J. Biomech.* **2010**, *43*, 2164–2173. [CrossRef] [PubMed]
10. Liikavainio, T.; Isolehto, J.; Helminen, H.J.; Perttunen, J.; Lepola, V.; Kiviranta, I.; Arokoski, J.P.; Komi, P.V. Loading and gait symmetry during level and stair walking in asymptomatic subjects with knee osteoarthritis: Importance of quadriceps femoris in reducing impact force during heel strike? *Knee* **2007**, *14*, 231–238. [CrossRef]
11. Iijima, H.; Eguchi, R.; Shimoura, K.; Aoyama, T.; Takahashi, M. Stair climbing ability in patients with early knee osteoarthritis: Defining the clinical hallmarks of early disease. *Gait Posture* **2019**, *72*, 148–153. [CrossRef]

12. Emery, C.A.; Whittaker, J.L.; Mahmoudian, A.; Lohmander, L.S.; Roos, E.M.; Bennell, K.L.; Toomey, C.M.; Reimer, R.A.; Thompson, D.; Ronsky, J.L.; et al. Establishing outcome measures in early knee osteoarthritis. *Nat. Rev. Rheumatol.* **2019**, *15*, 438–448. [CrossRef] [PubMed]
13. Ogawa, A.; Mita, A.; Yorozu, A.; Takahashi, M. Markerless knee joint position measurement using depth data during stair walking. *Sensors* **2017**, *17*, 2698. [CrossRef]
14. Ogawa, A.; Iijima, H.; Takahashi, M. Staircase design for health monitoring in elderly people. *J. Build. Eng.* **2021**, *37*, 102152. [CrossRef]
15. Hicks-Little, C.A.; Peindl, R.D.; Hubbard, T.J.; Scannell, B.P.; Springer, B.D.; Odum, S.M.; Fehring, T.K.; Cordova, M.L. Lower extremity joint kinematics during stair climbing in knee osteoarthritis. *Med. Sci. Sport. Exerc.* **2011**, *43*, 516–524. [CrossRef]
16. Iijima, H.; Shimoura, K.; Aoyama, T.; Takahashi, M. Biomechanical characteristics of stair ambulation in patients with knee OA: A systematic review with meta-analysis toward a better definition of clinical hallmarks. *Gait Posture* **2018**, *62*, 191–201. [CrossRef]
17. Shimoura, K.; Iijima, H.; Suzuki, Y.; Aoyama, T. Immediate effects of transcutaneous electrical nerve stimulation on pain and physical performance in individuals with preradiographic knee osteoarthritis: A randomized controlled trial. *Arch. Phys. Med. Rehabil.* **2019**, *100*, 300–306.e1. [CrossRef] [PubMed]
18. Kellgren, J.; Lawrence, J. Radiological assessment of osteo-arthrosis. *Ann. Rheum. Dis.* **1957**, *16*, 494. [CrossRef]
19. Iijima, H.; Suzuki, Y.; Aoyama, T.; Takahashi, M. Interaction between low back pain and knee pain contributes to disability level in individuals with knee osteoarthritis: A cross-sectional study. *Osteoarthr. Cartil.* **2018**, *26*, 1319–1325. [CrossRef] [PubMed]
20. Ogawa, A.; Mita, A.; Georgoulas, C.; Bock, T. A Face Recognition System for Automated Door Opening with parallel Health Status Validation Using the Kinect v2, ISARC. In Proceedings of the International Symposium on Automation and Robotics in Construction, Auburn, AL, USA, 18–21 July 2016; pp. 132–140.
21. Dobson, F.; Hinman, R.S.; Hall, M.; Terwee, C.; Roos, E.M.; Bennell, K. Measurement properties of performance-based measures to assess physical function in hip and knee osteoarthritis: A systematic review. *Osteoarthr. Cartil.* **2012**, *20*, 1548–1562. [CrossRef]
22. McFadyen, B.J.; Winter, D.A. An integrated biomechanical analysis of normal stair ascent and descent. *J. Biomech.* **1988**, *21*, 733–744. [CrossRef]
23. Koo, T.K.; Li, M.Y. A guideline of selecting and reporting intraclass correlation coefficients for reliability research. *J. Chiropr. Med.* **2016**, *15*, 155–163. [CrossRef]
24. Akai, M.; Doi, T.; Fujino, K.; Iwaya, T.; Kurosawa, H.; Nasu, T. An outcome measure for Japanese people with knee osteoarthritis. *J. Rheumatol.* **2005**, *32*, 1524–1532.
25. Bellamy, N.; Buchanan, W.W.; Goldsmith, C.H.; Campbell, J.; Stitt, L.W. Validation study of WOMAC: A health status instrument for measuring clinically important patient relevant outcomes to antirheumatic drug therapy in patients with osteoarthritis of the hip or knee. *J. Rheumatol.* **1988**, *15*, 1833–1840.
26. McConnell, S.; Kolopack, P.; Davis, A.M. The Western Ontario and McMaster Universities Osteoarthritis Index (WOMAC): A review of its utility and measurement properties. *Arthritis Care Res. Off. J. Am. Coll. Rheumatol.* **2001**, *45*, 453–461. [CrossRef]
27. Stewart, S.; Pearson, J.; Rome, K.; Dalbeth, N.; Vandal, A.C. Analysis of data collected from right and left limbs: Accounting for dependence and improving statistical efficiency in musculoskeletal research. *Gait Posture* **2018**, *59*, 182–187. [CrossRef] [PubMed]
28. Hannigan, J.; Chou, L.-S. Sex differences in lower extremity coordinative variability during running. *Gait Posture* **2019**, *70*, 317–322. [CrossRef]

International Journal of
Environmental Research and Public Health

MDPI

Article

Fear of Falling in Older Adults Treated at a Geriatric Day Hospital: Results from a Cross-Sectional Study

Eva M. Melendo-Azuela [1,2,*], Julia González-Vaca [3] and Eva Cirera [4]

1 The Doctoral Programme in Comprehensive Care and Health Services, University of Vic-Central University of Catalonia, 08500 Vic, Spain
2 Faculty of Medicine and Health Sciences, School of Nursing, University of Barcelona, 08007 Barcelona, Spain
3 Nursing Research Group (GRIN) from the IDIBELL Translational Medicine Area, University of Barcelona, 08007 Barcelona, Spain; juliagonzalezvaca@ub.edu
4 Sport and Physical Activity Research Group, Center for Health and Social Research, Department of Experimental Sciences and Methodology, Faculty of Health Sciences and Welfare, University of Vic-Central University of Catalonia, 08500 Vic, Spain; eva.cirera@uvic.cat
* Correspondence: eva.melendo@uvic.cat

Abstract: (1) Background: The fear of falling (FOF) is a geriatric syndrome that causes a decrease in daily activities and personal autonomy. Its prevalence is highly variable as are the methodologies used to assess it. This study aimed at estimating the prevalence and describing the main determinants of FOF in older adults attending a geriatric day hospital. (2) Methods: Descriptive, cross-sectional study of individuals aged ≥70 years, who attended an ambulatory functional rehabilitation group in the metropolitan area of Barcelona. FOF was assessed using the Activities-Specific Balance Confidence (ABC) scale. Other recorded outcomes were: sex, age, marital status, living alone, level of education, degree of autonomy, pain, previous falls, visual acuity, and signs of depression. Prevalence was estimated overall and according to the possible determinants. (3) Results: The study included 62 individuals (66.1% women), with a prevalence of fear of falling of 38.7% (95% CI 26.2–51.2%). The identified determinants were pain (OR = 7.4, 95% CI 1.4–39.7), a history of falls (OR = 25.3, 95% CI 2.1–303.4), poor visual acuity (OR = 5.6, 95% CI 1.0–29.8), and signs of depression (OR = 19.3, 95% CI 1.4–264.3). (4) Conclusions: The prevalence and determinants of fear of falling in older adults attending geriatric day hospitals were similar to those described in those dwelling in the community.

Keywords: fear of falling; prevalence; determinants; geriatric day hospital; urban area; Activities-Specific Balance Confidence (ABC)

Citation: Melendo-Azuela, E.M.; González-Vaca, J.; Cirera, E. Fear of Falling in Older Adults Treated at a Geriatric Day Hospital: Results from a Cross-Sectional Study. *Int. J. Environ. Res. Public Health* **2022**, *19*, 8504. https://doi.org/10.3390/ijerph19148504

Academic Editors: Francisco José Tarazona Santabalbina, Sebastià Josep Santaeugènia Gonzàlez, José Augusto García Navarro and José Viña

Received: 18 May 2022
Accepted: 9 July 2022
Published: 12 July 2022

1. Introduction

The fear of falling syndrome is defined as a decrease in usual daily activities caused by said fear both in people with and without a history of falls [1]. The main consequences of this syndrome are: a decrease in functional autonomy and the ability to perform basic daily-life activities and usual physical activity; an increased risk of falling, depression and progressive loss of the quality of life of older adults [2]. These consequences make individuals frail and vulnerable [3].

The fear of falling syndrome is a significant health problem in older adults living in the community. In this population, the prevalence of fear of falling is highly variable, ranging from 20.8 to 85.0%, based on the study methodology and measurement systems used [2]. In Spain, a prevalence between 31.2 and 71.6% has been described [4–6]. On the other hand, the main determinants of fear of falling in older adults living in the community include the female sex, alteration of physical function, use of technical help to walk, a history of falls, frailty, a perception of low quality of health, depression, chronic diseases, limitations of instrumental activities of daily life, pain, level of education, and visual acuity [7–10].

Some of the older adults living in the community go to a day hospital. For example, in Catalonia, there are around 1.1 million people aged 70 years or more [11], of which around 8000 are treated at the 73 geriatric day hospitals that exist in Catalonia [12,13]. These centers provide ambulatory day care to chronic patients with exacerbations and/or difficulty in managing their pathology through a portfolio of services, including comprehensive and interdisciplinary geriatric care, but also the administration of specific treatments such as cognitive stimulation, functional rehabilitation, or prevention of falls. Patients attending geriatric day hospitals can be derived from an intermediate care center to reduce the time spent in hospital, or from primary care to solve a health need that requires specialized geriatric care.

Several recent studies have described the prevalence of fear of falling and its possible determinants in the community population. However, no specific data for day-hospital patients are available—individuals who, although still living in the community and not requiring hospitalization, need specialized health care. Knowing the prevalence and determinants of fear of falling in this population would allow us to adopt specific guidelines and strategies to prevent fear of falling and improve their care. Therefore, the aim of this study was to estimate prevalence and describe the main determinants of fear of falling in a sample of older adults attending an ambulatory geriatric care center in the urban area of Barcelona.

2. Materials and Methods

2.1. Study Design and Population

This was a descriptive, cross-sectional study in community-dwelling individuals aged ≥70 years, who attended a functional rehabilitation group at the geriatric day hospital of L'Hospitalet de Llobregat Healthcare Center (Barcelona) between September 2017 and December 2018. Study participants had to be able to walk 10 m without stopping (with or without technical help) and not present with cognitive deterioration (<2 errors in the Pfeiffer scale [14]). Illiterate individuals were excluded due to their inability to complete self-administered questionnaires.

All participants signed an informed consent document. Data were collected by assessing participants' medical histories and the Activities-Specific Balance Confidence (ABC) questionnaire. If an individual could not complete this questionnaire on their own, he/she could get help from relatives or the nurse on duty. The study protocol was approved by the Clinical Research Ethics Committee of the University Hospital of Bellvitge (reference PR269/17].

2.2. Study Outcomes

The primary outcome of the study was fear of falling, measured using the Activities-Specific Balance Confidence (ABC) Scale (reliability 0.96, validity 0.84, sensitivity 0.65), a 16-item questionnaire scored from 0 to 100 measuring degree of confidence of balance in not falling while performing activities of daily living [15]. An individual is considered to have a fear of falling when his/her score is lower than 67 points [16].

Other variables analyzed were gender, age, marital status, living alone and level of education. Degree of autonomy was analyzed based on the Barthel Index (<21 points, total dependency; 21–60, severe dependency; 61–90, moderate dependency; 91–99, slight dependency; and 100, independence) [17]. Similarly, participants' pain was assessed using the numerical rating scale (NRS): 0 points, no pain; 1–3, mild pain; 4–7, moderate pain; ≥8, severe pain [18]. To dichotomize this outcome, the values 0–3 and 4–10 were used to define the absence and presence of pain, respectively. The history of falls in the last six months was assessed by asking the direct question, "Have you fallen in the last six months?" or by checking participants' medical histories; visual acuity was assessed using the Jaeger chart (<7 points, correct; ≥7 points, incorrect) [19] and depression was assessed using a short version of the Yesavage scale (0–1 points, no signs of depression; ≥2 points, presence of signs of depression) [20]. All these variables were collected from participants' medical histories or by asking them.

2.3. Statistical Analysis

Quantitative variables were described using medians and interquartile ranges (IQRs), and categorical variables were described by absolute and relative frequencies. The prevalence of fear of falling was estimated using a 95% confidence interval (95% CI).

To identify the determinants of fear of falling, a bivariate analysis with logistic regression was performed with the following variables: sex, age, marital status, pain, previous falls in the last six months, signs of depression, degree of autonomy, and visual acuity. Then, a multivariate logistic regression model was carried out, with fear of falling as the dependent variable and, as independent variables, those with a p-value < 0.25 of Wald test in the bivariate analysis, or those that, when excluded from the model, the estimated coefficients for the remaining variables changed markedly in magnitude (>10%) [21]. Given the low number of cases, it was decided not to assess the interactions between possible determinants. Results were presented as the estimated prevalence of fear of falling and the corresponding OR for each category, together with their respective 95% CI. The level of statistical significance was set at a bilateral alpha value of 0.05. All statistical analyses were performed using the SPSS statistics software for Windows, version 22.

3. Results

3.1. Study Participants' Characteristic

The study included a total of 62 individuals, whose socio-demographic and clinical characteristics are shown in Table 1. Approximately two out of three participants were women, and more than half were aged between 75 and 84 years. Most participants had a primary education, lived with someone, and were either married or widowed. Over 60% were independent and did not present with pain, and about one in four had poor visual acuity.

Table 1. Socio-demographic and clinical characteristics of study participants, n (%). n = 62.

Gender	
Female Sex	41 (66.1)
Male Sex	21 (33.9)
Age	
70–74 years	15 (24.2)
75–84 years	35 (56.5)
>84 years	12 (19.4)
Level of education	
Primary (not completed)	4 (6.5)
Primary	57 (91.9)
Secondary	1 (1.6)
Marital status	
Widowed	28 (45.2)
Married	32 (51.6)
Divorced	2 (3.2)
Lives alone	23 (37.1)
Degree of autonomy [a]	
Independence	38 (61.3)
Slight dependency	10 (16.1)
Moderate dependency	14 (22.6)
Severe dependency	0 (0.0)
Total dependency	0 (0.0)

Table 1. *Cont.*

Pain [b]	
No pain	43 (69.4)
Mild pain	2 (3.2)
Moderate pain	12 (19.4)
Severe pain	5 (8.1)
Falls in the last six months	9 (14.5)
Poor visual acuity [c]	16 (25.8)
Signs of depression [d]	10 (16.1)

[a] Based on the Barthel Index score (100, independence; 91–99, slight dependency; 61–90, moderate dependency; 21–60, severe dependency; <21, total dependency). [b] Based on the numerical rating scale (NRS) (0, no pain; 1–3, mild pain; 4–7, moderate pain; ≥8, severe pain). [c] Score ≥ 7 on the Jaeger chart (≥7). [d] Score ≥ 2 on the short version of the Yesavage scale.

3.2. Prevalence of Fear of Falling

The median (IQR) of the ABC scale was 76.57 (37.97). The prevalence of fear of falling in the population sample under study was 38.7% (95% CI 26.2–51.2%). Based on sex, the prevalence in women was 36.6% (95% CI 21.0−52.0%) and, in men, 42.9% (95% CI 20.0−66.0%).

Table 2 shows participants' socio-demographic and clinical characteristics based on whether they had a fear of falling or not. Regarding individuals with fear of falling, 25% were over 84 years old, 50% were independent and 54.2% reported no pain. In addition, 25% had fallen in the last six months, and 33.3% had poor visual acuity and showed signs of depression 37.5%. Regarding individuals without fear of falling, about one in six was over 84 years old, more than 60% were independent, and over 75% reported no pain. Approximately 8% had fallen in the last six months, and 21.1% had poor visual acuity and 3% showed signs of depression.

Table 2. Socio-demographic and clinical characteristics of study participants based on the presence of fear of falling, *n* (%).

	Fear of Falling (*n* = 24)	No Fear of Falling (*n* = 38)
Gender		
Female sex	15 (62.5)	26 (68.4)
Male sex	9 (37.5)	12 (31.6)
Age		
70–74 years	6 (25.0)	9 (23.7)
75–84 years	12 (50.0)	23 (60.5)
>84 years	6 (25.0)	6 (15.8)
Level of education		
Primary (not completed)	1 (4.2)	3 (7.9)
Primary	22 (91.7)	35 (92.1)
Secondary	1 (4.2)	0 (0.0)
Marital status		
Widowed	12 (50.0)	16 (42.1)
Married	10 (41.7)	22 (57.9)
Divorced	2 (8.3)	0 (0.0)
Lives alone	9 (37.5)	14 (36.8)
Degree of autonomy [a]		
Independence	12 (50.0)	26 (68.4)
Slight dependency	5 (20.8)	5 (13.2)
Moderate dependency	7 (29.2)	7 (18.4)
Severe dependency	0 (0.0)	0 (0.0)
Total dependency	0 (0.0)	0 (0.0)

Table 2. *Cont.*

	Fear of Falling (n = 24)	No Fear of Falling (n = 38)
Pain [b]		
No pain	13 (54.2)	30 (78.9)
Mild pain	0 (0.0)	2 (5.3)
Moderate pain	7 (29.2)	5 (13.2)
Severe pain	4 (16.7)	1 (2.6)
Falls in the last six months	6 (25.0)	3 (7.9)
Poor visual acuity [c]	8 (33.3)	8 (21.1)
Signs of depression [d]	9 (37.5)	1 (2.6)

[a] Based on the Barthel Index score (100, independence; 91–99, slight dependency; 61–90, moderate dependency; 21–60, severe dependency; <21, total dependency). [b] Based on the numerical rating scale (NRS) (0, no pain; 1–3, mild pain; 4–7, moderate pain; ≥8, severe pain). [c] Score ≥ 7 on the Jaeger chart (≥7). [d] Score ≥ 2 on the short version of the Yesavage scale.

3.3. Determinants of Fear of Falling

Tables 3 and 4 show the influence of the different outcomes on the percentage of participants with fear of falling. Results from the bivariate analysis showed that the outcomes associated with a higher prevalence of fear of falling were: presence of pain, falls in the last six months, and presence of signs of depression (Table 3). These outcomes remained significant in the multivariate model (Table 4), in a way that individuals that were more likely to have a fear of falling were those that reported pain (OR = 7.42), had fallen in the last six months (OR = 25.33), showed signs of depression (OR = 19.33), or had problems of visual acuity (OR = 5.56).

Table 3. The influence of possible determinants on the prevalence of fear of falling.

	Prevalence (%)	95% CI	OR	95% CI	p
Gender					
Male	42.86	(19.77–65.94)	1.00		
Female	36.59	(21.19–51.98)	0.83	(0.28–2.45)	0.740
Age					
<75	40.00	(11.92–68.08)	1.00		
75–84	34.29	(17.74–50.83)	0.70	(0.20–2.47)	0.575
>84	50.00	(16.82–83.13)	1.6	(0.33–7.85)	0.562
Marital status [a]					
Married	31.25	(14.27–48.23)	1.00		
Widowed	42.86	(23.32–62.40)	1.89	(0.64–5.52)	0.247
Lives alone					
No	38.46	(22.48–54.44)	1.00		
Yes	39.13	(17.55–60.71)	1.22	(0.41–3.64)	0.715
Degree of autonomy [b]					
Independence	31.58	(16.1–47.06)	1.00		
Slight dependency	50.00	(12.3–87.70)	2.60	(0.59–11.49)	0.206
Moderate dependency	50.00	(20.04–79.96)	2.08	(0.59–7.30)	0.251
Pain [c]					
No	28.89	(15.12–42.66)	1.00		
Yes	64.71	(39.38–90.03)	4.60	(1.38–15.20)	**0.013** *
Falls in the last six months					
No	38.71	(20.55–56.87)	1.00		
Yes	50.00	(16.82–83.13)	5.67	(1.04–31.00)	**0.045** *
Correct visual acuity [d]					
Yes	34.78	(20.48–49.08)	1.00		
No	50.00	(22.48–77.52)	2.50	(0.74–8.50)	0.141

Table 3. *Cont.*

	Prevalence (%)	95% CI	OR	95% CI	*p*
Signs of depression [e]					
No	28.85	(16.11–41.58)	1.00		
Yes	90.00	(67.38–112.62)	21.00	(2.44–180.77)	**0.006** *

[a] Divorced individuals were excluded from the analysis due to the low number of cases. [b] Based on the Barthel Index score (100, independence; 91–99, slight dependency; 61–90, moderate dependency; 21–60, severe dependency; <21, total dependency). [c] Based on the numerical rating scale (NRS) (0–3, absence of pain; 4–10, presence of pain). [d] Score ≥ 7 on the Jaeger chart (≥7). [e] Score ≥ 2 on the short version of the Yesavage scale. * $p < 0.05$.

Table 4. Adjusted model of the prevalence of fear of falling.

	OR	95% CI	*p*
Marital status [a]			
Married	1		
Widowed	4.62	(0.88–24.27)	0.071
Pain [b]			
No	1		
Yes	7.42	(1.39–39.69)	**0.019** *
Falls in the last six months			
No	1		
Yes	25.33	(2.12–303.41)	**0.011** *
Correct visual acuity [c]			
Yes	1		
No	5.56	(1.04–29.77)	**0.045** *
Signs of depression [d]			
No	1		
Yes	19.33	(1.41–264.33)	**0.026** *

[a] Divorced individuals were excluded from the analysis due to the low number of cases. [b] Based on the numerical rating scale (NRS) (0–3, absence of pain; 4–10, presence of pain).[c] Score ≥ 7 on the Jaeger chart (≥7). [d] Score ≥ 2 on the short version of the Yesavage scale. * $p < 0.05$.

4. Discussion

In this study, we observed that the prevalence of fear of falling in individuals aged 70 years or more attending geriatric day hospitals was 38.7%. Those outcomes associated with a higher probability of the fear of falling syndrome were: previous falls, presence of signs of depression, presence of pain, and poor visual acuity.

Although studies of the prevalence of fear of falling in older adults yield highly heterogeneous results, our study showed a similar prevalence to that described by other authors in community-dwelling populations [4,5,8,22]. From the start, patients attending day hospitals might be expected to have a higher prevalence of fear of falling than community-dwelling older adults, and a similar prevalence to that observed in pre-frail and frail populations [23,24]; instead, we found a similar prevalence to that of the general population. However, although prevalence scores are similar, we must understand the use of day hospitals as an advantage for these individuals. Having them in the facility for some time gives us the possibility to detect the fear of falling and perform interventions on the identified determinants to reduce this fear.

The variability in the prevalence of fear of falling might be explained through the measurement method. Most prevalence studies are based on a direct question with different levels of responses, consisting of 3–5 points on a Likert scale [4,5,8,22,25,26]; however, Thiamwong and Suwanno considered the use of one simple question as a limitation, since it does not discuss the multifactorial nature of the fear of falling [22]. Other authors use a validated measurement scale, namely the Falls Efficacy Scale (FES) [6,27], which only assesses indoor activities and, therefore, usually applies to individuals with limitations or low mobility. There is a modified version of FES (mFES) expanding four items referring to activities in the open air. Another scale used is the ABC that measures balance confidence in not falling while performing these activities; it is extended to add a further six items related to the instrumental activities of daily living included in the FES, so it can be applied to individuals with more functionality. Due to the type of patients attending day hospitals, we considered it more appropriate to use the ABC scale in our study. According to the systematic review of Alarcón et al., the prevalence of fear of falling is higher in studies using the FES scale than in those using the ABC scale [28].

Regarding the determinants of fear of falling, our results showed a higher probability of fear of falling in those individuals who had fallen in the last six months and in those who presented with signs of depression, which is consistent with results from previous studies performed in the general population [8,25–27]. Likewise, the probability of fear of falling was higher in individuals with sight problems and in those that reported pain, as described by Liu [27] and Stubbs et al. [9], respectively. However, unlike other authors [4–6,8,22,26,27], we did not observe any association between fear of falling and socio-demographic characteristics such as age, level of education, marital status, or gender. In scientific literature, sex is one of the most common determinants of fear of falling, and several studies have shown a higher prevalence of fear of falling in women than in men, being twice as much, in some cases [4–6,8,22,26,27]. Some authors have attributed this difference to women's higher concern with their health [29] or a higher tendency to develop osteoporosis or a weaker musculoskeletal system [30]. However, in our population, the prevalence of fear of falling was similar in both sexes since, compared with other studies, the prevalence in women was lower and that in men was higher [4–6,8,22,26,27]. In this sense, using patients treated in a specialized geriatric center as a starting point, it is possible that men in our study were as concerned about their health and the consequences of a fall as women.

To our knowledge, it is the first time that the prevalence and determinants of fear of falling are described in older adults, who, although still part of the community, already require assessment and follow-up in a specialized healthcare setting. Initially, we expected that a person who uses a geriatric health service may have a pre-fragility situation and probably have higher prevalence of fear of falling, but we observed that they have a similar situation to older people who live in the community, so it can be concluded that the profile of the person treated in a geriatric day hospital is similar to other older people who do not attend such a resource. In addition, when using a validated scale to measure fear of falling, not only do we assess its presence, but also the limitations entailed in the performance of normal activities, since scales measure an individual's confidence to avoid falling during the performance of daily-life activities. However, our study has some limitations. Being a cross-sectional study, we were not able to identify possible determinants of fear of falling; we were only able to analyze the influence of those determinants that were already described in the literature. In addition, the sample of participants was too small to estimate the significance of the effect of the determinants with enough accuracy, obtaining very wide confidence intervals. Another limitation not measured was chronic potential correlates of fear of falling.

Adults attending geriatric day hospitals are followed up between two and three months, which represents an opportunity to screen fear of falling and perform the necessary intervention. Therefore, it would be convenient to include a validated scale of fear of falling within the comprehensive geriatric assessment that is performed on patients of day hospitals, to detect it as early as possible and try to improve their efficacy in their daily lives. Likewise, programs intended to prevent and treat the determinants of fear of falling in geriatric day hospitals with the final objective of preventing dependency and a decrease in quality of life should be applied.

Fear of falling can be prevented with physical activity programs that improve walking and increase the level of confidence to avoid falls during daily-life activities [31]. In fact, a proper physical activity program may be efficient, even in frail, older adults [32]. In addition, due to the multifactorial nature of fear of falling, the effectiveness of these programs improves when combined with health education programs [33], so they should be applied together to improve individuals' confidence in the performance of daily activities and prevent falls. With respect to pain, it might be necessary to perform interventions to reduce it through pharmacological and non-pharmacological treatments [34]. It would also be advisable to use screening methods for depression and sight defects in order to refer the individual to the necessary professional or service. On the other hand, although it is not an outcome analyzed in this study, polypharmacy should be assessed due to its potential effects in older adults [35].

5. Conclusions

In conclusion, in our study, more than a third of older adults living in the community and attending a geriatric day hospital had a fear of falling, which may negatively impact their quality of life. The prevalence and determinants of fear of falling in patients attending day hospitals were similar to those in community-dwelling older adults. Future studies should analyze the possible association of fear of falling with frailty and, particularly, if fear of falling may be a predictor of frailty.

Author Contributions: Conceptualization, E.M.M.-A.; methodology, E.M.M.-A. and E.C.; software, E.M.M.-A.; validation, E.M.M.-A., E.C. and J.G.-V.; formal analysis, E.M.M.-A. and E.C.; investigation, E.M.M.-A.; resources, E.M.M.-A.; data curation, E.M.M.-A.; writing—original draft preparation, E.M.M.-A.; writing—review and editing, E.C. and J.G.-V.; visualization, E.C.; supervision, E.C. All authors have read and agreed to the published version of the manuscript.

Funding: This research received no external funding.

Institutional Review Board Statement: The study was conducted in accordance with the Declaration of Helsinki and approved by the Clinical Research Ethics Committee of the University Hospital of Bellvitge (reference PR269/17).

Informed Consent Statement: Informed consent was obtained from all subjects involved in the study.

Data Availability Statement: Not applicable.

Acknowledgments: We thank the healthcare team of the geriatric day hospital of L'Hospitalet de Llobregat Healthcare Center for their collaboration in collecting data for this study and the i2e3 Biomedical Research Institute for providing medical writing support.

Conflicts of Interest: The authors declare no conflict of interest.

References

1. Tinetti, M.E.; Powell, L. 4 Fear of Falling and Low Self-efficacy: A Cause of Dependence in Elderly Persons. *J. Gerontol.* **1993**, *48*, 35–38. [CrossRef] [PubMed]
2. Scheffer, A.C.; Schuurmans, M.J.; van Dijk, N.; van der Hooft, T.; de Rooij, S.E. Fear of falling: Measurement strategy, prevalence, risk factors and consequences among older persons. *Age Ageing* **2008**, *37*, 19–24. [CrossRef] [PubMed]
3. Sanjoaquín Romero, A.C.; Fernández Arín, E.; Mesa Lampré, M.P.; García-Arilla Calvo, E. *Tratado de Geriatría Para Resi-Dentes*; Capitulo 4; Sociedad Española de Geriatría y Gerontología (SEGG): Madrid, Spain, 2006; ISBN 84-689-8949-5.
4. Martínez, I.P.; Bravo, B.N.; Pretel, F.A.; Muñoz, J.N.D.; Molina, R.P.E.; Hidalgo, J.L.-T. Fear of falling in elderly community-dwelling individuals. *Gac. Sanit.* **2010**, *24*, 453–459. [CrossRef] [PubMed]
5. Lavedán, A.; Viladrosa, M.; Jürschik, P.; Botigué, T.; Nuín, C.; Masot, O.; Lavedán, R. Fear of falling in community-dwelling older adults: A cause of falls, a consequence, or both? *PLoS ONE* **2018**, *13*, e0194967. [CrossRef]
6. Perez-Jara, J.; Olmos, P.; Abad, M.A.; Heslop, P.; Walker, D.; Reyes-Ortiz, C.A. Differences in fear of falling in the elderly with or without dizzines. *Maturitas* **2012**, *73*, 261–264. [CrossRef]
7. Denkinger, M.D.; Lukas, A.; Nikolaus, T.; Hauer, K. Factors Associated with Fear of Falling and Associated Activity Restriction in Community-Dwelling Older Adults: A Systematic Review. *Am. J. Geriatr. Psychiatry* **2015**, *23*, 72–86. [CrossRef]
8. Dierking, L.; Markides, K.; Al Snih, S.; Peek, M.K. Fear of Falling in Older Mexican Americans: A Longitudinal Study of Incidence and Predictive Factors. *J. Am. Geriatr. Soc.* **2016**, *64*, 2560–2565. [CrossRef]
9. Stubbs, B.; Eggermont, L.H.; Patchay, S.; Schofield, P.A. Pain Interference Is Associated With Psychological Concerns Related to Falls in Community-Dwelling Older Adults: Multisite Observational Study. *Phys. Ther.* **2014**, *94*, 1410–1420. [CrossRef]
10. Oh, E.; Hong, G.-R.S.; Lee, S.; Han, S. Fear of falling and its predictors among community-living older adults in Korea. *Aging Ment. Health* **2015**, *21*, 369–378. [CrossRef]

11. Idescat. Anuari Estadístic de Catalunya. Població a 1 de Gener. Per Sexe i Grups D'edat. Available online: https://www.idescat.cat/pub/?id=aec&n=253 (accessed on 26 October 2021).
12. Observatori del Sistema de Salut de Catalunya. *Central de Resultats. Àmbit Sociosanitari. Dades 2016*; Agència de Qualitat i Avaluació Sanitàries de Catalunya, Departament de Salut, Generalitat de Catalunya: Barcelona, Spain, 2017.
13. Memòria Anual 2020—Servei Català de la Salut. Available online: https://memoria.catsalut.gencat.cat/#atencio-sociosanitaria (accessed on 8 December 2021).
14. Pfeiffer, E. A Short Portable Mental Status Questionnaire for the Assessment of Organic Brain Deficit in Elderly Patients. *J. Am. Geriatr. Soc.* **1975**, *23*, 433–441. [CrossRef]
15. Powell, L.E.; Myers, A.M. The Activities-specific Balance Confidence (ABC) Scale. *J. Gerontol. Ser. A* **1995**, *50A*, M28–M34. [CrossRef]
16. Lajoie, Y.; Gallagher, S. Predicting falls within the elderly community: Comparison of postural sway, reaction time, the Berg balance scale and the Activities-specific Balance Confidence (ABC) scale for comparing fallers and non-fallers. *Arch. Gerontol. Geriatr.* **2003**, *38*, 11–26. [CrossRef]
17. Shah, S.; Vanclay, F.; Cooper, B. Modified Barthel Index Sources. Improving the sensitivity of the Barthel Index for stroke rehabilitation. *J. Clin. Epidemiol.* **1989**, *42*, 703–709. [CrossRef]
18. Scott, J.; Huskisson, E.C. Graphic representation of pain. *Pain* **1976**, *2*, 185–195. [CrossRef]
19. Legge, G.E.; Ross, J.A.; Luebker, A.; LaMay, J.M. Psychophysics of Reading. VIII. The Minnesota Low- Vision Reading Test. *Optom. Vis. Sci.* **1989**, *66*, 843–853. [CrossRef] [PubMed]
20. Hoyl, M.T.; Alessi, C.A.; Harker, J.O.; Josephson, K.R.; Pietruszka, F.M.; Koelfgen, M.; Mervis, J.R.; Fitten, L.J.; Rubenstein, L. Development and Testing of a Five-Item Version of the Geriatric Depression Scale. *J. Am. Geriatr. Soc.* **1999**, *47*, 873–878. [CrossRef]
21. Hosmer, D.W.; Lemeshow, S.; Sturdivant, R.X. *Applied Logistic Regression*; John Wiley & Sons Inc.: Hoboken, NJ, USA, 2013; ISBN 978-0-470-58247-3.
22. Thiamwong, L.; Suwanno, J. Fear of Falling and Related Factors in a Community-based Study of People 60 Years and Older in Thailand. *Int. J. Gerontol.* **2017**, *11*, 80–84. [CrossRef]
23. Merchant, R.A.; Chen, M.Z.; Wong, B.L.L.; Ng, S.E.; Shirooka, H.; Lim, J.Y.; Sandrasageran, S.; Morley, J.E. Relationship Between Fear of Falling, Fear-Related Activity Restriction, Frailty, and Sarcopenia. *J. Am. Geriatr. Soc.* **2020**, *68*, 2602–2608. [CrossRef]
24. Makino, K.; Lee, S.; Bae, S.; Chiba, I.; Harada, K.; Katayama, O.; Shinkai, Y.; Makizako, H.; Shimada, H. Prospective Associations of Physical Frailty With Future Falls and Fear of Falling: A 48-Month Cohort Study. *Phys. Ther.* **2021**, *101*. [CrossRef]
25. Chu, C.-L.; Liang, C.-K.; Chow, P.C.; Lin, Y.-T.; Tang, K.-Y.; Chou, M.-Y.; Chen, L.-K.; Lu, T.; Pan, C.-C. Fear of falling (FF): Psychosocial and physical factors among institutionalized older Chinese men in Taiwan. *Arch. Gerontol. Geriatr.* **2011**, *53*, e232–e236. [CrossRef]
26. Kempen, G.I.; Van Haastregt, J.C.; McKee, K.J.; Delbaere, K.; Zijlstra, G.R. Socio-demographic, health-related and psychosocial correlates of fear of falling and avoidance of activity in community-living older persons who avoid activity due to fear of falling. *BMC Public Health* **2009**, *9*, 170. [CrossRef]
27. Liu, J.Y.W. Fear of falling in robust community-dwelling older people: Results of a cross-sectional study. *J. Clin. Nurs.* **2014**, *24*, 393–405. [CrossRef] [PubMed]
28. Alarcón, T.; González-Montalvo, J.I.; Puime, O. Evaluación de los pacientes con miedo a caídas. ¿El método empleado modifica los resultados? Una revisión sistemática. *Aten. Primaria* **2009**, *41*, 262–268. [CrossRef] [PubMed]
29. Chamroonkiadtikun, P.; Ananchaisarp, T.; Wajancomkul, P. The Prevalence and Associated Factors of the Fear of Falling in Elderly Patients at the Primary Care Clinic of Songklanagarind Hospital. *Top. Geriatr. Rehabil.* **2021**, *37*, 44–49. [CrossRef]
30. Tomita, Y.; Arima, K.; Tsujimoto, R.; Kawashiri, S.-Y.; Nishimura, T.; Mizukami, S.; Okabe, T.; Tanaka, N.; Honda, Y.; Izutsu, K.; et al. Prevalence of fear of falling and associated factors among Japanese community-dwelling older adults. *Medicine* **2018**, *97*, e9721. [CrossRef]
31. Kendrick, D.; Carpenter, H.; Morris, R.W.; Skelton, D.A.; Gage, H.; Bowling, A.; Masud, T.; Stevens, Z.; Pearl, M.; Gawler, S.J.K.A. Exercise for reducing fear of falling in older people living in the community (Protocol). *Cochrane Database Syst. Rev.* **2012**. [CrossRef]
32. Kapan, A.; Luger, E.; Haider, S.; Titze, S.; Schindler, K.; Lackinger, C.; Dorner, T. Fear of falling reduced by a lay led home-based program in frail community-dwelling older adults: A randomised controlled trial. *Arch. Gerontol. Geriatr.* **2016**, *68*, 25–32. [CrossRef]
33. Jung, D.; Lee, J.; Lee, S.-M. A Meta-Analysis of Fear of Falling Treatment Programs for the Elderly. *West. J. Nurs. Res.* **2008**, *31*, 6–16. [CrossRef]
34. Schwan, J.; Sclafani, J.; Tawfik, V.L. Chronic Pain Management in the Elderly. *Anesthesiol. Clin.* **2019**, *37*, 547–560. [CrossRef]
35. Wang, R.; Chen, L.; Fan, L.; Gao, D.; Liang, Z.; He, J.; Gong, W.; Gao, L. Incidence and Effects of Polypharmacy on Clinical Outcome among Patients Aged 80+: A Five-Year Follow-Up Study. *PLoS ONE* **2015**, *10*, e0142123. [CrossRef]

International Journal of
Environmental Research and Public Health

Article

Study of a Quasi-Experimental Trial to Compare Two Models of Home Care for the Elderly in an Urban Primary Care Setting in Spain: Results of Intermediate Analysis

Carolina Burgos Díez [1], Rosa Maria Sequera Requero [2], Jose Ferrer Costa [3], Francisco José Tarazona-Santabalbina [4,5,6], Marià Monzó Planella [7], Cristina Cunha-Pérez [8], Sebastià Josep Santaeugènia González [9,10,*] and Grupo ATDOM †

1 Primary Health Care Centre Passeig Maragall, Institut Català de la Salut, 08041 Barcelona, Spain; carolinaburgosdiez@gmail.com
2 Primary Health Care Centre Gran Sol., Institut Català de la Salut, 08914 Badalona, Spain; rmsequera.mn.ics@gencat.cat
3 Primary Health Care Centre Apenins, Badalona Serveis Assistencials, 08917 Badalona, Spain; jfcosta@bsa.cat
4 Geriatric Medicine Department, Hospital Universitario de la Ribera, 46600 Alzira, Spain; tarazona_frasan@gva.es
5 Medical School, Catholic University of Valencia, Sant Vicent Màrtir, 46001 Valencia, Spain
6 Centro de Investigación Biomédica en Red Fragilidad y Envejecimiento Saludable (CIBERFES), 28029 Madrid, Spain
7 Department of Surgery and Medical-Surgical Specialties, Faculty of Medicine, University of Barcelona, 08036 Barcelona, Spain; mmonzo@ub.edu
8 Faculty of Nursing, Catholic University of Valencia San Vicente Mártir, 46001 Valencia, Spain; cristina.cunha@ucv.es
9 Central Catalonia Chronicity Research Group (C3RG), University of Vic, Central University of Catalonia, 08500 Vic, Spain
10 Chronic Care Program, Ministry of Health, Generalitat de Catalunya, 08028 Barcelona, Spain
* Correspondence: sebastia.santaeugenia@gencat.cat
† Membership of the Group Name is provided in the Acknowledgments.

Citation: Burgos Díez, C.; Sequera Requero, R.M.; Ferrer Costa, J.; Tarazona-Santabalbina, F.J.; Monzó Planella, M.; Cunha-Pérez, C.; Santaeugènia González, S.J.; Grupo ATDOM. Study of a Quasi-Experimental Trial to Compare Two Models of Home Care for the Elderly in an Urban Primary Care Setting in Spain: Results of Intermediate Analysis. *Int. J. Environ. Res. Public Health* **2022**, *19*, 2329. https://doi.org/10.3390/ijerph19042329

Academic Editor: Andreas E. Stuck

Received: 28 December 2021
Accepted: 15 February 2022
Published: 17 February 2022

Abstract: Functional dependence is associated with an increase in need for resources, mortality, and institutionalization. Different models of home care have been developed to improve these results, but very few studies contain relevant information. This quasi-experimental study was conducted to evaluate two models of home care (HC) in a Primary Care setting: an Integrated Model (IM) (control model) and a Functional Model (FM) (study model). Material and Methods: Two years follow-up of patients 65 years old and older from two Primary Health Care Centres (58 IM, 68 FM) was carried out, recruited between June-October 2018 in Badalona (Barcelona, Spain). Results of the mid-term evaluation are presented in this article. Health status, quality of care, and resource utilization have been evaluated through comprehensive geriatric assessment, quality of life and perception of health care scales, consumption of resources and complementary tests. Results: A significant difference was detected in the number of hospital admissions (FM/IM 0.71 (1.24)/1.35 (1.90), p: 0.031) in the Accident and Emergency department (FM/IM 2.01 (2.12)/3.53 (3.59), p: 0.006) and cumulative days of admission per year (FM/IM 5.43 (10.92)/14.69 (20.90), p: 0.003). Conclusions: FM offers greater continuity of care at home for the patient and reduces hospital admissions, as well as admission time, thereby saving on costs.

Keywords: home care models; preventive home visits; primary care; geriatric assessment

1. Introduction

For health services, it is a challenge to manage better the care of those with complex needs, the majority of whom are the elderly, as a consequence of increased life expectancy, which leads to more comorbidity, disability and dependency in the population [1–3].

Currently the percentage of the Spanish population aged 65 and over represents 19.6% of overall and will reach a peak of 31.4% in around 2050 [4].

In Spain, patients with multimorbidity represent 1.38% of the total population seen in Primary Care and comprise 5% of those seen in Primary Care over 65 years of age [5]. This leads to a considerable use of healthcare resources, including medical appointments, Accident and Emergency department visits, hospitalization and medication [1–6].

Multiple studies have analyzed different proposals to improve preventive home care for the elderly. There is evidence that Comprehensive Geriatric Assessment (CGA) based interventions for older patients are beneficial [7], and that multidisciplinary teams offer better quality of care and decrease acute care utilization among high-risk older people [8], decrease the number of cumulative days of admission [9,10] and facilitate continued living at home, largely by preventing the need for nursing home admission and reducing falls [11]. However, Mayo-Wilson et al. published in 2014 a systematic review and meta-analysis detecting many discrepancies in the studies reviewed on the impact that preventive home visits may have on patients with good baseline health or frailty, which could be attributed to the design of these studies, sample sizes or different definitions of the variables measured [12]. In any case, information on home care models of primary care in Spain is very scarce in these studies [13–17].

Currently in Spain, patients who cannot attend the Primary Health Care Centre (PHCC) are included in the Home Care (HC) Programme (ATDOM). Patients included in this programme must have been assessed by a doctor, nurse or social worker and meet at least the following criteria: not being able to move to the centre to be cared for, for reasons of health or physical condition or due to their social or environmental situation, temporarily or permanently [18].

The healthcare offered by this programme is carried out by primary care teams and involves health promotion and preventive activities, control of chronic and acute pathologies, treatment, and rehabilitation, with the aim of ensuring that patients achieve a good quality of life, along with their families, while maintaining the greatest possible autonomy [19].

Traditionally a patient's home care in Spain is managed from the Basic Care Unit (BCU) composed by a general practitioner and a nurse who have cared for the patient since he/she first came into contact with the primary care team from the age of 15 years onwards; thus a BCU is responsible for the care of a group of people (around 1350 to 1550 patients), being the same team as that of the primary health care centre that takes care of the assigned patients, both in the centre and in home visits (integrated or traditional home care model, IM). There are other models of home care based on nursing care with the occasional support of the family doctor, or based on hospital health teams that travel to the community [10], or by interdisciplinary teams based on a reorganization of the Primary Care team that involves the creation of a home care team (family doctor and nurse) dedicated exclusively to the Home Care Programme and which are referred to as the Functional Model (FM), also called the Dispensaries Model. To date, no Spanish studies have been published that evaluate and compare the different aspects of health, resource consumption and perception of care received during 2 years of follow-up of this functional model (FM), compared to the traditional or integrated model (IM). The aim of this study is to compare both models in an urban area in caring for patients at home.

2. Materials and Methods

2.1. Study Design and Thical Aspects

This quasi-experimental study [20] compared the outcome of two Home Care (HC) models implemented in two primary health care centres in Badalona (Barcelona, Spain). The control group consisted of patients following the integrated HC model (IM) provided by the PHCC *Gran Sol*, and the study group consisted of patients following the new functional HC model (FM), linked to the PHCC *Apenins*. Both HC teams consist of a general practitioner and a nurse. In the integrated model, HC is given by the same healthcare team providing medical care at the primary care unit, with an average of 1500 inhabitants assigned to each team. By contrast, in the functional model, HC is given by a healthcare team specifically trained in the management of older, frail and multimorbid patients, providing only full-time preventive home visits and connecting to other, further required special provision services. Further details regarding the characteristics of each model are shown in Table 1. The rationale for choosing these two centres relates to the balanced demographic characteristics of the reference population (Table 2).

Table 1. Main characteristics of the two investigated models.

Characteristics of the Healthcare Team	Integrated HC	Functional HC
Team composition	Nurse and family physician	Nurse and family physician.
Team function	The same healthcare team provides HC and manages patients in the primary health care centre independently of their care needs (prevention, health promotion, patients with complex needs, patients in HC program or patients at end of life).	The healthcare team is dedicated exclusively to HC.
Interprofessional communication	Healthcare professionals are part of the healthcare team regularly managing patients in the primary health care centre.	Although not managing patients in the primary health care centre, the HC team is part of the health care staff of the centre and their members participate in the centre meetings as specialists
Training	Regular training of family doctors, including regular stays at mental health and geriatric units. Regular training of nurses.	Regular training of family doctors, including regular stays at mental health and geriatric units. Nursing staff and doctor receives additional training regarding the management of chronic patients, fragility, and palliative care. Continuous updates.
Type of professional in each visit	Nurse, family doctor or both.	Nurse, family doctor or both.
Preventive visits	Visits of nursing staff scheduled based on the monitoring requirements of each disease as established by local guidelines. Visits of physician scheduled at physician's discretion based on the disease progression and clinical status of patients.	Visits of nursing staff scheduled based on the monitoring requirements of each disease as established by local guidelines. Visits of physician scheduled at physician's discretion based on the disease progression and clinical status of patients.
Dedication to the type of care activity	90% Care at the health centre, 10% at home (depending on the organisation of the centres).	100% Home care
Non-urgent acute visits	The patient calls the centre and the physician schedules the visits at home in a deferred way, according to agenda.	During working hours, the patient directly contacts the physician of the HC team. Outside working hours: the patient calls the centre and the physician available at that moment (not always the one regularly visiting the patient at the primary health care centre) visits the patient at home.
Urgent visits	The patient calls the PHCC and a doctor from the centre, who is on call, sees him/her (this may not be the patient's usual doctor).	The patient calls the HC team until 15:00. From 15:00 to 20:00, the patient calls the PHCC and a doctor from the centre, who is on call, sees him/her (this may not be the patient's usual doctor).
Financial approach	All visits are fully covered by the public health system.	All visits are fully covered by the public health system.

HC: home care PHCC: Primary Health Care Centre.

Table 2. Characteristics of the participating centres [a].

	Integrated HC (PHCC Gran Sol)	Functional HC (PHCC Apenins)	*p*
Location	Badalona, Catalonia, Spain	Badalona, Catalonia, Spain	
Professional profile	MDs and nurses specialized in family medicine	MDs and nurses specialized in family medicine	
Reference population [b], No.	19.442	19.043	
Over-Aging index [c],%	11%	9.2%	<0.001
Foreign population [d], n (%)	3499 (17.9%)	3046 (15.9%)	<0.001
≥65 years old, n (%)	3480 (17.9%)	2970 (15.6%)	<0.001
AMG, adjusted indicator (IC 95%)	1.189 (1.173–1.206)	1.178 (1.161–1.195)	–
Mortality, annual (%)	7	5.7	0.143
IT application	eCAP	eCAP	

HC: Home Care, PC Primary Care, MD Medical Doctor, AMG Adjusted Morbidity Groups [21]; IT Information Technology. [a] Differences between PHCC *Gran Sol* and PHCC *Apenins*. [b] Data from Msiq (Generalitat de Catalunya©), period between January and December 2015. [c] The number of persons aged 74 or over per total of persons over 64 years old. [d] The number of subjects with a foreign nationality.

All the HC interventions performed in both programmes are based on current protocols designed following recommendations of the SEMFYC (Spanish Society of Family and Community medicine) and EUROPREV (European Network for Prevention and Health Promotion in Family Medicine and General Practice) inside the Program of Preventive Activities for Health Promotion (PAPPS) [22]. The study protocol was approved by the IDIAP Ethics Committee of the Jordi Gol Foundation (Approval code: P17/121). Patients (or their caregivers) voluntarily signed an informed consent, and all the information gathered was anonymized before conducting any analysis. All data was handled according to the Spanish Data Protection Law (LOPD) 15/1999 and the EU General Data Protection Regulation 2016/679. Considering the routine interventions defined in the study conventional risks were not expected to increase. Registered in ClinicalTrials.gov (Identifier: NCT03461315; 12 March 2018).

2.2. Selection Criteria

All patients aged over 65 years old and enrolled in the long-term HC programme at any of the two participating primary health care centres for at least 3 months were considered for eligibility. Patients were included irrespective of their cognitive status. Exclusion criteria included patients with a life expectancy of less than a month and patients with a score of 5 or more in the Pfeiffer's cognitive impairment test (3–4 mild, 5–7 moderate, 8 or >severe) [23], who did not have a full-time caregiver or who had a part-time one, because a severe cognitive impairment is likely to interfere with the study procedure. Patients that were not registered as Badalona citizens were also excluded because it was assumed that they had temporary status, as well as patients included in a HC program due to their reduced mobility, in order to reduce bias when measuring patient-requested interventions, because the latter could not easily reach the primary health care centre facility.

2.3. Patient Recruitment

All subjects included in the HC programme at the two primary health care centres that met the selection criteria were contacted by phone and offered the chance to participate in the study. Patients willing to do so were scheduled a domiciliary visit to receive the study documents (i.e., the Patient Information Sheet and self-administered questionnaires/scales) and signed the informed consent themselves or, in case of cognitive impairment, via their full-time caregivers.

2.4. Study Conduct

The study started in June 2018 and ended in October 2020. On the first preventive home visit, once informed consents had been accepted, patients, or the caregivers in case of cognitive impairment (defined as subjects scoring 5 or more in the Pfeiffer's test), were given self-administered questionnaires, such as EuroQoL (an instrument to complement other quality-of-life measures and to facilitate the collection of a common data set for reference purposes [24]), IEXPAC (Chronic Patient Experience Assessment Instrument [25]), and Zarit (Dependent Patient Caregiver Overload Assessment Instrument [26]). All were analyzed by the investigator, who assessed the patient's frailty in situ. The self-administered scales were completed again by the patient and/or caregiver at the end of the second year of follow-up, during a preventive home visit. Besides these start and end visits, participants were interviewed by phone every 6 months to solve any issues and find out if any private hospitalizations or daycare centres had been used. All visits requested by either the patient or the reference doctor were also reported in a case report form (CRF). The medical professionals performing the preventive home visits were trained in the use of the scales to ensure consistency and reinforce their application.

All data, irrespective of source, were recorded in an anonymized CRF, in which patients were identified with a study code. The study investigator kept a key table with the study codes and their corresponding medical record identification codes.

In the current article the intermediate analysis of data collected in the first year of follow-up (October 2018 to September 2019 inclusive) is presented.

2.5. Endpoints and Variables

The primary endpoint was the difference in mean days of hospital stay per year between patients included in the integrated and functional HC programs. Secondary endpoints included the assessment of the differences between the two HC models i.e., mortality and hospital admissions, based on the IHI Triple Aim (Better Care, Better Health, Lower Costs) [27]. To this end, variables regarding subjects' health status, quality of care, and resource utilization of patients included in the two models were compared (Table 3). The demographic characteristics of the study participants were also recorded.

Table 3. Socio-demographic variables and baseline CGA [1] outcomes.

	Apenins	**Gran Sol**	*p*
	(Functional Model)	**(Integrated Model)**	
	$n = 68$	$n = 58$	
Average age	86.66 (7.6)	87.2 (6.7)	0.39
Age %:			
Group 1 (between 65 and 74)	5.9	3.4	0.457
Group 2 (between 75 and 84)	33.8	25.9	
Group 3 (>=85 years)	60.3	70.7	
Sex: (%)			
Male	23.5	27.6	0.602
Female	76.5	72.4	
Typologies of patients in the programme ATDOM [2] (%):			
Patients with non-complex medical problems	5.9	10.3	
Chronically complex patient (CCP [3])	80.9	82.8	0.365
Chronically ill patients with advanced disease (MACA [4])	13.2	6.9	
ICIP [5] realizado n (%)	45 (66.2)	36 (62.1)	0.632
ICIP with PDA [6] n (%)	39 (57.4)	32 (55.2)	0.806
Adjusted Morbidity Groups (AMG [7]) n (%)			
Group 1 (1,2,3)	6 (8.8)	15 (25.9)	0.011
Group 2 (4,5)	62 (91.2)	43 (74.1)	
Degree of dependency (average)			
0–1	41 (60.3)	47 (81)	0.011
2–3	27 (39.7)	11 (18.9)	

Table 3. *Cont.*

	Apenins (Functional Model) n = 68	Gran Sol (Integrated Model) n = 58	p
TIRS [8] n (%)	6 (11.5)	12 (26.6)	0.056
No falls n (%)	63 (92.6)	53 (91.3)	0.957
No presence of decubitus ulcers n (%)	62 (91)	52 (89)	0.475
Barthel	55.15 (25,8)	60.5 (21,4)	0.262
Pfeiffer	3.94 (3.2)	2.83 (3.0)	0.078
Braden	17.75 (2.6)	17.64 (2.4)	0.824
Private caregiver No. (%)	22 (32.4)	2 (3.4)	0
Euroqol (subjective assessment)	4.75 (2.32)	4.35 (1.87)	0.291
IEXPAC [9]	5.85 (1.69)	5.98 (1.17)	0.004
Caregiver overburden (Zarit)	58.08 (17.1)	29.27 (27.8)	0.001

[1] CGA: Comprehensive Geriatric Assessment. [2] ATDOM: Home Care Programme. [3] CCP: Chronically Complex Patient. [4] MACA: Chronically Patients with Advanced Disease. [5] ICIP: Individual and Shared Intervention Plan. [6] PDA: Advance Healthcare Directive Plan. [7] AMG: Adjusted Morbidity Groups. [8] TIRS: Social Risk Indicator Scale. [9] IEXPAC: Chronic Patient Experience Assessment Instrument.

Particularly, the baseline health status of study subjects included the Gerontôpole frailty screening tool and the Adjusted Morbidity Groups (AMG) risk assessment tool, which considers the type of disease, number of systems affected, and complexity of each [21,28]. Additionally, a complete baseline Comprehensive Geriatric Assessment (CGA) was performed, including the following assessments: the ability to perform normal daily tasks (Barthel scale: <20 total dependence, 20–35 severe dependence, 40–55 moderate dependence, 60 mild dependence, 100 autonomous) [29], mental health status (Pfeiffer test: ≤2 risk of cognitive impairment, 3–4 mild cognitive impairment, 5–7 moderate cognitive impairment, 8–10 severe cognitive impairment) [23], decubitus ulceration risk (Braden test: <12 high risk, 13–14 moderate risk, 15–16 < 75 years low risk, 15–18 > 75 years low risk) [30], social risk (TIRS: 1 positive indicator = social risk. Yes/no answers) [31], and social status (degree of dependency 0-1-2-3) [32]. (Seen in Table 3). The health-related quality of life of study participants and satisfaction of caregiver were assessed at baseline and at the final follow-up visit using the EuroQoL, IEXPAC, and questionnaires, respectively.

2.6. Statistical Analysis

The sample size calculation was based on an incidence of hospital admission of 40% and a reduction of 10% in the study group, a 2-year follow-up, and a 1:1 ratio for control and intervention groups. Under these constrains, fixed alpha and beta errors of 5%, yielded an estimated size of 581 subjects per group. The statistical power of this sample assumed alpha and beta errors of 5% was 85.3%. An intermediate analysis was performed when more than 100 subjects were recruited, with a calculated power for the sample size, comprising 126 subjects of 35.5%.

All collected variables are described for the overall study sample and for both study groups. Quantitative variables were described as the mean and standard deviation (SD), and as the median and interquartile range (IR) for normally and non-normally distributed variables, as confirmed by the Kolmogorov-Smirnov test. Categorical variables were described as frequencies and percentages. Measures of central tendency were compared using the *t*-test for independent samples or ANOVA, or their non-parametric counterparts, the Mann-Whitney U and Kruskal-Wallis tests. Categorical variables were compared using the Chi-square test or Fisher's exact test. Post hoc analyses were performed using the Bonferroni or the Games-Howell adjustments. Variables with differences in the bivariate analysis at baseline ($p < 0.1$) and those considered clinically relevant for the authors were included in a linear multiple logistic regression to build a multivariate model in order to predict the difference in mean days of hospital stay and costs of patients in the HC program. To address Better Care and Better Health endpoints, the authors applied a binary logistic regression for mortality and hospital admission variables; age, gender, and comorbidity were included as adjustment variables. A backward stepwise regression was used to avoid overfitting of the model obtained.

A significance threshold was set at two-sided alpha value <0.05. The analysis was performed with SPSS (IBM Corp. Released 2012. IBM SPSS Statistics for Windows, Version 21.0. Armonk, NY, USA: IBM Corp.).

3. Results

Of the 354 patients admitted to the HC programme (113 PHCC *Apenins* and 241 PHCC *Gran Sol*) at the beginning of the study, 171 (48%) were rejected due to non-compliance with the inclusion criteria [20]. All 183 patients potentially eligible for the study were asked for consent, and 57 of them refused to participate in the study (18 PHCC *Apenins* and 39 from PHCC *Gran Sol*). A total of 126 patients, 58 (76% of those eligible) belonging to the Primary Health Care Centre with integrated home care model (PHCC *Gran Sol*) and 68 (63.5% of those eligible) attending under the functional model (PHCC *Apenins*) finally confirmed their participation in the study (see Figure 1, Follow-up chart).

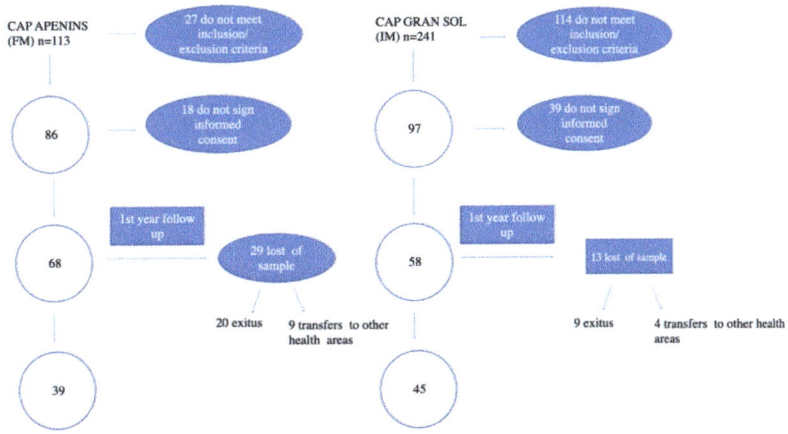

Figure 1. Flow diagram of the studied simple. CAP: PHCC Primary Health Care Centre. LOST OF SAMPLE: loss of patients due to death or transfers to others health areas.

Overall, the sample of 126 patients consisted of 25% men and 75% women, and the average age of the total participants was 86.95 (±7.22) years. Both groups were comparable regarding their basic characteristics as can be seen in Table 3. There were no differences between the two populations in terms of the typology of patients in the ATDOM (home care) programme included in the study: non-complex patients (FM 5.9 vs. IM 10.3%), complex chronic patients (CCP) (FM 80.9%, IM 82.8%), and patients with chronic advanced disease (MACA) (FM: 13.2%, IM: 6.9%), nor in the implementation for these patients of an Individualized and Shared Intervention Plan (ICIP) (FM: 66.2% IM: 62.1%, *p*: 0.632), including advance directives (FM: 57.4%, IM: 55.2%, *p*: 0.806).

Although the two Primary Health Care Centers were chosen due to the similar overall socio-demographic profile of the populations attended (as shown in Table 2), it was detected that the subgroup of the population within the ATDOM (HC) programme presented greater comorbidity in those attending under the FM than in those attending under the IM, to a statistically significant degree: Adjusted Morbidity Groups (AMG) 4.5 N (%):, FM 62 (91.2), IM 43 (74.1), showing a *p*: 0.011. Significant differences were also observed in the degree of functional dependency of the population in both groups (FM/IM% grades 0.1: 60.3/81; grades 2–3: 39.7/18.9, *p*: 0.011).

FM patients were found to have more private (non-family) caregivers than IM patients (FM/IM 22 (32.4)/2 (3.4), *p*: 0.000).

Regarding the self-assessment conducted at the beginning of the study of patients in both groups, no differences were found in terms of the perception of health status (Euroqol) although differences were found in terms of the patients' perceived experience of care (IEXPAC) (Table 3).

At the time of admission to the study, patients received a Comprehensive Geriatric Assessment (CGA) and no differences were observed in the number of patient dimensions assessed (means FM 5.05 and IM 4.36, *p*: 0.131). The results of the Comprehensive Geriatric Assessment in both groups show that both populations are totally comparable in terms of the main variables analyzed, although a more patient-centred assessment of the patient's social needs (TIRS and Zarit) was observed in the FM than in the IM, not allowing for comparability (Table 3).

Concerning mortality at the first year of follow-up, there is no significant difference between the two models, higher in FM 20 (29.4%) compared to IM 9 (15.5%) *p*: 0.089. The multivariate analysis showed no significance differences between models for the crude OR 2.27 (95% CI 0.94–5.48; *p*: 0.069), and adjusting by age, sex, CCP, MACA and GMA there was still no statistical significance: OR 2.18 (95% CI 0.85–5.57; $p = 0.107$) (Data not in table).

There were differences in the referral of these patients to other specialists in the form of virtual consultations (13 (19.1%) FM, 4 (6.9%) IM, *p*: 0.045), but not in person (32 (47.1), 25 (43.1) FM, IM, *p*: 0.657) (Table 4).

Table 4. Health care needs during the first year of follow-up.

	Apenins (Functional Model) $n = 68$	Gran Sol (Integrated Model) $n = 58$	p
Online consultations with a hospital specialist No. (%patients)	13 (19.1)	4 (6.9)	0.045
In-person referrals to hospital specialists No. (%patients)	32 (47.1)	25 (43.1)	0.657
ECG [1] (%patients)	28 (41.2)	0.00 (0.00)	0.000
Conventional XR [2] requested No. (%patients)	44.2%	13.7%	0.053
Ultrasound scans requested No. (%patients)	16.1%	8.6%	0.445
CAT [3] requested No.(%patients)	22%	3.4%	0.021
Blood and urine tests (mean + STD [4])	2.9 (3.9)	1.6 (1.5)	0.020
Prescribed drugs (mean + STD)	10.05 (3.5)	9.81 (5.13)	0.757
Online consultations +G.P. [5] BCU [6] (mean + STD)	9.90 (6.27)	9.07 (6.74)	0.477
In-person consultations G.P. BCU (mean + STD)	6.25 (5.77)	3.98 (3.41)	0.008
In-person consultations G.P. non- BCU (mean + STD)	2.81 (2.55)	0.57 (1.65)	0.000
Online consultations NUR [7] BCU (mean + STD)	1.99 (3.23)	4.90 (5.16)	0.000
In-person consultations NUR BCU (mean + STD)	7.35 (9.50)	4.33 (5.47)	0.028
In-person consultations NUR non- BCU (mean + STD)	3.24 (9.35)	5.33 (10.35)	0.238
MES [8] activation No. (%)	34 (50)	50 (86.2)	0.055
Admissions to private nursing homes No. (%)	3 (4.4)	15 (25.8)	0.003

[1] ECG: Electrocardiogram. [2] XR: X-rays. [3] CAT: Computerized axial Tomography. [4] STD: Standard Deviation. [5] G.P.: General Practitioner. [6] BCU: Basic Care Unit. [7] NUR: nurse. [8] MES: Medical Emergency system.

Patients seen under the FM requested significantly more complementary tests to study their health status during this first year of follow-up: more electrocardiograms (FM/IM 28(41.2)/000 (000), *p*: 0.000), computerized axial tomography (CAT) scans (FM/IM

22%/3.4%, p: 0.021) and blood and urine tests (FM/IM 2.9 (3.9)/1.6 (1.5), p: 0.020). In contrast, no differences were found in the number of drugs prescribed, nor in the number of X-rays or ultrasound scans performed (Table 4).

During the first year of follow-up, significant differences were found in the number of home visits made by both the referring physician (FM/IM 6.25 (5.77)/3.98 (3.41), p: 0.008) and nurse (FM/IM 7.35 (9.50)/4.33 (5.47) p: 0.028). More non-referring physician visits were also detected in the FM than in the IM, FM 2.81 (2.55), IM 0.57 (1.65) significantly (p: 0.000). However, a trend towards less activation of the emergency medical service was observed in the FM population compared to the IM population, although not statistically significant (FM/IM 34 (50)/50 (86), p: 0.055) (Table 4).

As shown in Table 5, during the first year follow-up, patients treated under the FM had a lower rate of institutionalization than IM patients, FM/IM 3 (4.4)/15 (25.8), p: 0.003; however, there was a higher demand for respite care (RESPIR) in the population being cared for under the FM than in the IM (FM/IM 19 (27.9)/7 (12.1), p: 0.028, with no significant differences found in access to teleassistance or home health services between both models of care.

Table 5. Health and social outcomes after one year follow-up.

	Apenins (Functional Model) n = 68	Gran Sol (Integrated Model) n = 58	p
Respite care (up to 30 days) (Respir [1])	19 (27.9)	7 (12.1)	0.028
Teleassistance	60 (88.2)	54 (93.1)	0.353
HHS [2]	53 (77.9)	50 (86.2)	0.231
Admissions to hospital ward (No.)	0.71 (1.24)	1.35 (1.90)	0.031
A&E [3] admissions (No.)	2.01 (2.12)	3.53 (3.59)	0.006
Admission in Intermediate Care Hospital. (No.)	0.21 (0.47)	0.12 (0.32)	0.239
No. of cumulative days of admission (on ward) per year.	5.43 (10.92)	14.69 (20.90)	0.003
No. Admission in Hospital at home	0.01 (0.12)	0.19 (0.68)	0.060
No. Admission in PADES [4]	0.03 (0.17)	0.14 (0.34)	0.033

[1] RESPIR: Limited temporary stays in private residential centres for the elderly or provision of private home care services for the elderly financed by the Barcelona City Council. [2] HHS: Home Help Service. [3] A&E: Accident and Emergency Department. [4] PADES: palliative care support team.

Concerning the consumption of health resources, during the first year of follow-up, the population treated under the FM showed a significantly lower number of hospital ward admissions (FM/IM 0.71 (1.24)/1.35 (1.90), p: 0.031), fewer A&E admissions (FM/IM 2.01 (2.12)/3.53 (3.59), p: 0.006) and fewer cumulative days of ward admission (FM/IM 5.43 (10.92)/14.69 (20.90), p: 0.003) (Table 5), as well as less need for activation of specialized palliative care support teams (PADES) (FM/IM 0.03 (0.17)/0.14 (0.34) p: 0.033). No differences were found in Home Hospitalization activations (FM/IM 0.01 (0.12)/0.19 (0.68) p: 0.060) and Intermediate Care Hospital admissions (FM/IM 0.21 (0.47)/0.12 (0.32) p: 0.239)

A multivariate analysis was performed using ANCOVA adjusting for age, sex and comorbidity categorized by AMG, obtaining a mean difference for cumulative days of hospital admission of 5.57 (SD 10.99) in the functional model compared to 13.88 (SD 16.91), $p < 0.001$. ANCOVA analysis was repeated including, in addition to age, sex and comorbidity, degree of dependency, private caregiver, and overburden as variables. In addition, the new results showed a mean difference for cumulative days of hospital admission of 6.04 (SD 11.38) in the functional model compared to 13.09 (SD 16.44), $p < 0.001$. (Data not in table).

4. Discussion

The present study shows clearly differentiated health outcomes in two populations with similar socio-demographic characteristics (except for a higher comorbidity at the time of inclusion in the FM population), treated under two different models of home care.

Despite being models of healthcare with similar characteristics in terms of patients, based on IGV and individualized care plans, the FM shows a greater intensity of home follow-up, with a greater number of visits and complementary examinations to study the health status of the assigned population. As already evidenced in Stuck's systematic review [33], this fact has clear benefits for the health status of the population under FM. The present study found that the population under FM showed, already in the first year of follow-up (despite having a higher comorbidity than those under IM), a lower risk of institutionalization, admission to acute hospital, emergency care and a higher probability of continuing to live at home after one year despite needing respite care (minimum 30 days per year) [34]. This is possibly related to the initial situation of families cared for under FM, in which the caregiver is more overburdened (see Table 3), although it was not possible to compare this possibility in this first analysis.

The authors consider that the increase in the consumption of intermediate products in the population under FM is associated with the greater comorbidity of these patients (compared to those treated under IM), and probably also with the greater proactivity and follow-up of such patients due to the greater number of follow-up visits made during this first year. This is consistent with different published articles [35,36], to the point of being considered a predictor of healthcare expenditure, since this population, known as high need/high cost [37], is in fact the one that concentrates healthcare expenditure and has the highest risk of mortality. This is confirmed in our study, although there is no statistical significance in terms of mortality.

Recently, a study on the characteristics and resource consumption of PCCs showed the need to find efficient and evaluable models of social and health care [1]. Accordingly, our study shows that the accessibility and intensive follow-up of patients cared for at home under the FM does not lead to an increase in the number of referrals to other specialties or a greater number of pharmacological prescriptions, compared to the IM. Therefore, by being more accessible, the FM could plan the overall care of the person's needs in a more individualized way, adapting this plan to the evolutionary characteristics (comorbidity situation, dependence, family environment, etc.), which results in greater dedication from the team, avoiding unnecessary referral to other specialists and not over-prescribing drugs (exercising a more person-centred vision of prescription, as has been reported in other research). [38].

The FM shows a better resolution of healthcare crises in its population attended at home resulting, as shown in the follow-up at year one, in a lower number of admissions to the A&E and a lower number of admissions to a hospital ward, with a clear trend towards a lower activation of the EMS (p: 0.055), possibly not significant due to the sample size obtained. This result is consistent with the publication of Vila et al. [10] in 2017 in which, with a population of 261 people of similar mean age, it was shown that there was a reduction in the number of cumulative days of admission per year from 3.5 to 1 day ($p < 0.001$) after including patients in a multidisciplinary care programme that included professionals from Primary Care and Hospital Care.

In this mid-term evaluation of this study, it was found that there is a highly significant difference in the number of cumulative days of admission per year between the two models of care, being lower in the FM than in the IM (FM/IM 5.43 (10.92)/14.69 (20.90), p: 0.003) with the repercussion on costs that this represents. In line with what was published this year comparing the institutionalized population with patients who remained at home, revealing important clinical, demographic and mortality differences [39], the present study shows that patients treated in the FM remain at home more often, despite the fact that no greater social risk was detected in the IM than in the FM.

Finally, at the time of recruitment, no differences were found in both models in terms of the perception of quality of life. However, differences were observed in terms of the experience of care received. FM implies a change in the healthcare team from the moment the patient is admitted to the Home Care Programme, leaving aside longitudinally, which is one of the main pillars of primary care and which has been classically associated with better health outcomes than decentralized care. Results will be re-evaluated at the end of the study to find out whether or not the results are consistent with those published by Hogg et al. [40], showing that patients experience an improvement in their subjective quality of life with the introduction of multidisciplinary care models, or with other research, such as that conducted by Marta Gorina et al. [14] in 2013, which concluded that populations cared for under a FM have a higher degree of satisfaction and perceived quality of care, although the tool used to assess this fact is different (IEXPAC in the present study vs. Satisfaction Assessment of Home Care Service (SATISFAD-12).

As mentioned in the calculation of the sample size, due to the characteristics of the patients treated under the primary healthcare domiciliary models, it was expected that there would be a 15% loss of sample over the course of the study. However, in this first year of follow-up, losses amounted to 42.6% in the FM and 22.4% in the IM, of which 31% and 30.7% were due to transfers of the patient to another health area and the rest to exitus, respectively.

Limitations

A limitation of the present study is the impossibility of carrying out a randomized clinical trial due to the type of population and the services provided. Both groups presented significant differences at baseline in comorbidity, dependency, caregiver support and caregiver over-burning that could possibly influence the results obtained, constituting a selection bias. Certainly, both groups are not similar at baseline, but the differences have been minimized via the statistical approach. Although these variables were included as adjustment variables in the multivariate analysis, there is a selection bias.

As this is a comparison study of two healthcare models in two different Primary Healthcare Centres, a quasi-experimental study has been designed with a possible Hawthorne effect (participants in a study may alter their behavior when they are aware of being observed) [41], which is not avoidable.

Likewise, the sample size is reduced to the population included in the Home Care Programme, affecting patient recruitment, as it represents a small percentage of the total population. The *Gran Sol* PHCC is located in an urban area with many architectural barriers that impede patients from reaching the PHCC, and who are therefore attended at home, despite having one or no chronic pathology. To avoid selection bias, patients included in the HC programme were excluded from the study protocol. This justifies the fact that in the Gran sol PHCC there were, initially, 114 patients who did not meet the inclusion/exclusion criteria, as shown in the arrow diagram (Figure 1). Similarly, the study potential is reduced by the fact that some of the patients are MACA, or are in their last days, or in a situation of fragility and refuse to participate in the study.

As this is the frailest and most comorbid segment of the population, many mortality losses were detected.

In this study, two Primary Care Health Centres located in densely populated areas were compared. The authors question whether the results obtained would be expected in other population settings, such as suburban or rural areas. The power of the recruited sample size is low, as expected for an intermediate analysis. Despite this low power, beta error increases, resulting in differences which are more difficult to find, when they do exist. However, the results are conclusive about the advantages of the functional model over the integrated model.

5. Conclusions

In the present comparative study of models of preventive home visits in Primary Care, it is observed that the FM is a more accessible model, of higher effectiveness and greater efficiency in the consumption of acute hospitalization resources, in spite of attending a population with greater morbidity and mortality due to greater comorbidity based on AMG. In addition, it has a favorable impact on permanence at home throughout this first year, avoiding the institutionalization of patients who require a high level of social support at home (as can be seen in the need for respite care or the need to hire a home caregiver) which, when not guaranteed (integrated social and health care at home), leads to greater institutionalization (population under IM). The present study has been carried out in a densely populated city and it would be interesting to test the results also in semi-urban or rural settings. Likewise, due to the high mortality (higher than expected in the initial sample design in both populations), it would be advisable to apply the model to a larger population so that an analysis with greater statistical power could be carried out.

Author Contributions: Conceptualization, C.B.D. and S.J.S.G.; implementation of the design and methodology: C.B.D. and S.J.S.G.; recruitment: R.M.S.R., J.F.C., Grupo ATDOM, software, F.J.T.-S.; validation, C.B.D., F.J.T.-S. and S.J.S.G.; formal analysis, F.J.T.-S.; investigation, C.B.D., R.M.S.R., J.F.C.; resources, C.B.D.; data curation, C.B.D., F.J.T.-S. and S.J.S.G. Writing—original draft preparation, C.B.D.; writing—review and editing, C.B.D., F.J.T.-S. and S.J.S.G.; supervision or mentorship, C.B.D., S.J.S.G., M.M.P.; translation, C.C.-P.; visualization, C.B.D., R.M.S.R., J.F.C., F.J.T.-S., M.M.P. and S.J.S.G. Each author contributed important intellectual content during manuscript drafting or revision and accepts accountability for the overall work by ensuring that questions pertaining to the accuracy or integrity of any portion of the work are appropriately investigated and resolved. All authors have read and agreed to the published version of the manuscript.

Funding: This research received no external funding.

Institutional Review Board Statement: The study was conducted according to the guidelines of the Declaration of Helsinki and approved by the IDIAP Ethics Committee of the Jordi Gol Foundation (Approval code: P17/121) 21 June 2017.

Informed Consent Statement: Informed consent was obtained from all subjects involved in the study. Informed consent was signed by all patients or caregivers before their inclusion. In case of cognitive impairment (i.e., a score of 5 or more in Pfeiffer's test) only caregivers were asked to sign under the statement: "I, (name of caregiver participant), have read the Information Sheet for the study in which my dependent is asked to participate. My questions, if any, have been answered to my satisfaction. Therefore, I consent to participate in the study.

Acknowledgments: We acknowledge the support received by Jose Antonio de la Fuente Cadenas, health technician of the Institut Catala de la Salut and Josep Anton Gonzalez Ares, health technician of Badalona Serveis Assistencials. and the EuroQol Research Foundation for kindly providing the 5D 3L EuroQoL questionnaire as weel as the IEXPAC Working Group University of Alicante, for kindly providing the IEXPAC questionnaire. Grateful thanks to Ruperto Sanz Cantalapiedra for his valuable contributions. Healthcare professionals who care for the patient ATDOM: Ana Cabrera Castro, Costinela Violeta Radoi, Jose Suarez Pelaez, Ana Gil Lopez, Alvaro Lopez Prieto, Xavier Ordoñez, Monica Seisdedos, M Lluïsa Gonzalez Romero, Pilar Marquez Garcia, Jacob Rodriguez Garcia, M Belen Barreiro Bugallo, Aynabel Cosculluela Sánchez, Rosario Ferrero Sanz, Souhel Flayeh Beneyto, Noelia Martinez, Jesus Millan Abella, Ana Novell Guerrero, Anna Pérez Montaña, Rosa M Vicente Montes, Belén Escobar Oliva, Noemí Pérez León.

Conflicts of Interest: The authors declare no conflict of interest.

References

1. Santaeugènia, S.J.; Contel, J.C.; Vela, E.; Cleries, M.; Amil, P.; Melendo-Azuela, E.M.; Gil-Sánchez, E.; Mir, V.; Amblàs-Novellas, J. Characteristics and Service Utilization by Complex Chronic and Advanced Chronic Patients in Catalonia: A Retrospective SevenYear Cohort-Based Study of an Implemented Chronic Care Program. *Int. J. Environ. Res. Public Health* **2021**, *18*, 9473. [CrossRef] [PubMed]

2. Wolff, J.L.; Starfield, B.; Anderson, G. Prevalence, expenditures, and complications of multiple chronic conditions in the elderly. *Arch. Intern. Med.* **2002**, *162*, 2269–2276. [CrossRef] [PubMed]
3. Mitnitski, A.B.; Rutenberg, A.D.; Farrell, S.; Rockwood, K. Aging, frailty, and complex networks. *Biogerontology* **2017**, *18*, 433–446. [CrossRef] [PubMed]
4. Instituto Nacional de Estadística. *Proyecciones de la Población de España a Largo Plazo (2020–2070)*; Ministerio de Asuntos Económicos y Transformación Digital: Madrid, Spain, 2019.
5. Ministerio de Sanidad y Política Social. Unidad de Pacientes Pluripatológicos Estándares y Recomendaciones. *Minist Sanid y Política Soc.* **2009**, 11–12.
6. Gené Badia, J.; Borràs Santos, A.; Contel Segura, J.C.; Ascaso Terén, C.; González Ortega, M.; Gallo De Puelles, P. Factors associated with hospitalization in a cohort of elderly patients included in a home care program. *Med. Clin.* **2012**, *139*, 473–478. [CrossRef] [PubMed]
7. Contel, J.C. Impacto de la atención domiciliaria preventiva en el ingreso en centros sociosanitarios, estado funcional y mortalidad. *FMC* **2002**, *9*, 459.
8. Counsell, S.R.; Callahan, C.M.; Clark, D.O.; Tu, W.; Buttar, A.B.; Stump, T.E.; Ricketts, G.D. Geriatric care management for low-income seniors: A randomized controlled trial. *JAMA* **2007**, *298*, 2623–2633. [CrossRef]
9. Peterson, K.; Helfand, M.; Humphrey, L.; Christensen, V.; Carson, S. *Evidence Brief: Effectiveness of Intensive Primary Care Programs*; 2011-Created: February 2013; Department of Veterans Affairs (US): Washington, DC, USA, 2013.
10. Vila, A.; Villegas, E.; Cruanyes, J.; Delgado, R.; Sabaté, R.A.; Ortega, J.; Araguás, C.; Humet, C. Cost-Effectiveness of a Barcelona Home Care Program for individuals with Multimorbility. *J. Am. Geriatr. Soc.* **2015**, *63*, 1017–1024. [CrossRef]
11. Beswick, A.D.; Rees, K.; Dieppe, P.; Ayis, S.; Gooberman-Hill, R.; Horwood, J.; Ebrahim, S. Complex interventions to improve physical function and maintain independent living in elderly people: A systematic review and meta-analysis. *Lancet* **2008**, *371*, 725–735. [CrossRef]
12. Mayo-Wilson, E.; Grant, S.; Burton, J.; Parsons, A.; Underhill, K.; Montgomery, P. Preventive home visits for mortality, morbidity and Institutionalitation in older Adults: A Systematic Rewiew and meta- Analysis. *PLoS ONE* **2014**, *9*, E89257. [CrossRef]
13. Fusté, J.; Cànovas, J.; Oliete, C.; Fuertes, O. Análisis Comparado de los Modelos de Atención Domiciliaria Sanitarios y Sociales del Sistema Nacional de Salud y del Sistema de Atención a la Dependencia. 29 de Noviembre de 2019 Fundació Unió. Catalana d'hospitals. Available online: https://www.uch.cat/documents/anlisis-comparado-de-los-modelos-de-atencin-domiciliaria-fase-i-i-ii.pdf (accessed on 23 December 2021).
14. Gorina, M.; Limonero, J.T.; Peñart, X.; Jiménez, J.; Gasso, J. Comparación de la satisfacción de los usuarios de atención domiciliaria: Modelo integrado vs. modelo dispensarizado. *Aten. Primaria* **2014**, *46*, 276–282. [CrossRef] [PubMed]
15. Yolanda, A.V.; Antonio, S.R.; Marta, P.A.; Julia, F. Relación de la Calidad de vida con Diferentes Modelos de Atención Domiciliaria en Enfermos Oncológicos Terminales de un Área Sanitaria de Madrid. *Rev. Esp. Salud Publica* **2003**, *77*, 567–579.
16. Morales Asencio, J.M. Efectividad de un Nuevo Modelo de Atención Domiciliaria en Andalucía para Poblaciones Vulnerables en atención Primaria. Estudio Quasi-Experimental, Controlado, Multicéntrico. Doctoral Thesis, University of Malaga, Málaga, Spain, 9 June 2006. Available online: https://www.educacion.gob.es/teseo/mostrarRef.do?ref=39392 (accessed on 2 February 2022).
17. Phyllis Pinnegar, H.; Paredes Millan, M.K.; Plaza Cerrato, L.; Palomino Español, E.; Baena Diez, J.M. Estudio comparativo de dos modelos de atención domiciliaria: Modelo integrado frente a modelo dispesarizado. Comunicaciones y ponencias semFYC: 2022; Comunicaciones: Resultados de investigación. In Proceedings of the XXXVIII Congreso de la semFYC, Barcelona, Spain, 10–12 May 2018. ISSN: 2339-9333.
18. Programa de Millora de L'atenció al Domicili des de L'atenció Primària de Salut. Institut Català de la Salut. Generalitat de Catalunya Departament de Salut. April 2010. Available online: http://ics.gencat.cat/web/.content/documents/assistencia/protocols/atdom.pdf (accessed on 27 December 2021).
19. Cegri Lombardo, F.; Aranzana Martínez, A. Atención domiciliaria: Caminando hacia la excelencia. *Aten. Primaria* **2007**, *39*, 34. [CrossRef]
20. Burgos Diez, C.; Sequera Requero, R.; Tarazona-Santabalbina, F.J.; Contel-Segura, J.C.; Monzó Planella, M.; Santaeugènia-González, S.J. Study protocol of a quasi-experimental trial to compare two models of home care for older people in the primary setting. *BMC Geriatr.* **2020**, *20*, 101. [CrossRef] [PubMed]
21. Monterde, D.; Vela, E.; Clèries, M. Grupo colaborativo GMA. Los grupos de morbilidad ajustados: Nuevo agrupador de morbilidad poblacional de utilidad en el ámbito de la atención primaria. *Aten. Primaria* **2016**, *48*, 674–682. [CrossRef]
22. Spanish Society of Family and Community Medicine (SEMFYC). Program of preventive activities for Health promotion (PAPPS) Actualización 2020. *Aten. Primaria* **2020**, *52*, 1–172.
23. Pfeiffer, E. A short portable mental status questionnaire for the assessment of organic brain deficit in elderly patients. *J. Am. Geriatr. Soc.* **1975**, *23*, 433–441. [CrossRef]
24. The EuroQol Group. EuroQol, a new facility for the measurement of health-related quality of life. *Health Policy* **1990**, *16*, 199–208. [CrossRef]
25. Instrumento de Evaluación de la Experiencia del Paciente Crónico. Available online: www.iexpac.org.www.iemac.org/iexpac (accessed on 23 December 2021).
26. Zarit, S.H.; Reever, K.E.; Bach-Peterson, J. Relatives of the impaired elderly: Correlates of feeling of burden. *Gerontologist* **1980**, *20*, 649–654. [CrossRef]

27. Berwick, D.M.; Nolan, T.W.; Whittington, J. The triple aim: Care, health, and cost. *Health Aff. (Millwood)* **2008**, *27*, 759–769. [CrossRef]
28. Bases para un Modelo Catalán de Atención a las Personas con Necesidades Complejas. Conceptualización e Introducción a los Elementos Operativos. (Versión 6.0 del 27 de marzo de 2017). Programa de Prevenció i Atenció a la Cronicitat. Generalitat de Catalunya. 2017. Available online: https://scientiasalut.gencat.cat/bitstream/handle/11351/3305/bases_modelo_catalan_atencion_personas_necesidades_complejas_2017_cas.pdf?sequence=5&isAllowed=y. (accessed on 23 December 2021).
29. Fi, M.; Dw, B. Functional evaluation: The barthel index. *Md. State Med. J.* **1965**, *14*, 56–61.
30. Braden, B.; Bergstrom, N. A conceptual schema for the study of the etiology of pressure sores. *Rehabil. Nurs.* **1987**, *12*, 8–12. [CrossRef] [PubMed]
31. González, D.C.; Caicoya, A.M.; Sánchez, A.F.; García, V.A.; González, J.G.; Palacios, E.D.; García, A.S. Evaluación de la fiabilidad y validez de una escala de valoración social en el anciano. *Aten. Primaria* **1999**, *23*, 434–440.
32. Institut Català de la Salut. El Treball Social Sanitari en l'atenció al Pacient Crònic. [Internet]. 2013. Available online: https://treballsocialsanitariics.files.wordpress.com/2015/11/el-treball-social-sanitari-en-latencic3b3-al-pacientcrc3b2nic-document-de-metodologia-deprocc3a9s-2013-ics.pdf (accessed on 3 March 2020).
33. Stuck, A.E.; Egger, M.; Hammer, A.; Minder, C.E.; Beck, J.C. Home Visits to Prevent Nursing Home Admission and Functional Decline in Elderly People: Systematic Review and Meta-regression Analysis. *J. Am. Med. Assoc.* **2002**, *287*, 1022–1028. [CrossRef] [PubMed]
34. Available online: https://www.diba.cat/es/web/respir/respir-residencial-per-a-gent-gran (accessed on 23 December 2021).
35. Vela, E.; Clèries, M.; Vella, V.A.; Adroher, C.; García-Altés, A. Análisis poblacional del gasto en servicios sanitarios en Cataluña (España): ¿qué y quién consume más recursos? [Population-based analysis of the Healthcare expenditure in Catalonia (Spain): What and who consumes more resources?]. *Gac. Sanit.* **2019**, *33*, 24–31. [CrossRef]
36. Coderch, J.; Sánchez-Pérez, I.; Ibern, P.; Carreras, M.; Pérez-Berruezo, X.; Inoriza, J.M. Predicción del riesgo individual de alto coste sanitario para la identificación de pacientes crónicos complejos. *Gac. Sanit.* **2014**, *28*, 292–300. [CrossRef] [PubMed]
37. Hochman, M.; Asch, S.M. Disruptive Models in Primary Care: Caring for High-Needs, High-Cost Populations. *J. Gen. Intern. Med.* **2017**, *32*, 392–397. [CrossRef]
38. Espaulella Panicot, J.; Molist Brunet, N.; Sevilla Sánchez, D.; González-Bueno, J.; Amblàs Novellas, J.; Sola Bonada, N.; Codina Jané, C. Modelo de prescripción centrado en la persona para mejorar la adecuación y adherencia terapéutica en los pacientes con multimorbilidad. *Rev. Española Geriatría Gerontol.* **2017**, *52*, 278–281. [CrossRef]
39. Amblàs-Novellas, J.; Santaeugènia, S.J.; Vela, E.; Clèries, M.; Contel, J.C. What lies beneath: A retrospective, population-based cohort study investigating clinical and resource-use characteristics of institutionalized older people in Catalonia. *BMC Geriatr.* **2020**, *20*, 2–8. [CrossRef]
40. Hogg, W.; Lemelin, J.; Dahrouge, S.; Liddy, C.; Armstrong, C.D.; Legault, F.; Dalziel, B.; Zhang, W. Randomized controlled trial of anticipatory and preventive multidisciplinary team care: For complex patients in a community-based primary care setting. *Can. Fam. Physician* **2009**, *55*, e76–e85.
41. Manterola, C.; Otzen, T. Los Sesgos en Investigación Clínica. *Int. J. Morphol.* **2015**, *201*, 1156–1164. Available online: http://www.scielo.cl/scielo.php?script=sci_arttext&pid=S0717-95022015000300056&lng=es (accessed on 9 February 2022). [CrossRef]

International Journal of
Environmental Research and Public Health

MDPI

Article

Health Service Needs from a Household Perspective: An Empirical Study in Rural Empty Nest Families in Sinan and Dangyang, China

Xueyan Cheng [1,2] and Liang Zhang [2,3,*]

1 Shanghai Health Development Research Center (Shanghai Medical Information Center), Shanghai 200032, China; xycheng6972@hust.edu.cn
2 School of Medicine and Health Management, Tongji Medical College, Huazhong University of Science and Technology, Wuhan 430030, China
3 School of Political Science and Public Administration, Wuhan University, Wuhan 430030, China
* Correspondence: zhanglianghust@126.com

Abstract: This study aimed to explore the health service needs of empty nest families from a household perspective. A multistage random sampling strategy was conducted to select 1606 individuals in 803 empty nest households in this study. A questionnaire was used to ask each individual about their health service needs in each household. The consistency rate was calculated based on their consistent answers to the questionnaire. We used a collective household model to analyze individuals' public health service needs on the family level. According to the results, individuals' consistency rates of health service needs in empty nest households, such as diagnosis and treatment service (H1), chronic disease management service (H2), telemedicine care (H3), physical examination service (H4), health education service (H5), mental healthcare (H6), and traditional Chinese medicine service (H7) were 40.30%, 89.13%, 98.85%, 58.93%, 57.95%, 72.84%, and 63.40%, respectively. Therefore, family-level health service needs could be studied from a family level. Health service needs of H1, H3, H4, H5, and H7 for individuals in empty nest households have significant correlations with each other ($r = 0.404, 0.177, 0.286, 0.265, 0.220$, $p < 0.001$). This will be helpful for health management in primary care in rural China; the concordance will alleviate the pressure of primary care and increase the effectiveness of doctor–patient communication. Health service needs in empty nest households who took individuals' public needs as household needs ($n = 746$) included the H4 (43.3%) and H5 (24.9%) and were always with a male householder (94.0%) or at least one had chronic diseases (82.4%). Health service needs in empty nest households that considered one member's needs as household needs ($n = 46$) included the H1 (56.5%), H4 (65.2%), H5 (63.0%), and H7 (45.7%), and the member would be the householder of the family (90.5%) or had a disease within two weeks (100.0%). In conclusion, family members' roles and health status play an important role in health service needs in empty nest households. Additionally, physical examination and health education services are the two health services that are most needed by empty nest households, and are suitable for delivering within a household unit.

Keywords: health services need; household unit; empty nest household; collective model; empirical research

Citation: Cheng, X.; Zhang, L. Health Service Needs from a Household Perspective: An Empirical Study in Rural Empty Nest Families in Sinan and Dangyang, China. *Int. J. Environ. Res. Public Health* **2022**, *19*, 628. https://doi.org/10.3390/ijerph19020628

Academic Editors: Francisco José Tarazona Santabalbina, Sebastià Josep Santaeugènia Gonzàlez, José Augusto García Navarro and José Viña

Received: 22 November 2021
Accepted: 31 December 2021
Published: 6 January 2022

1. Introduction

The family plays an important role in the care and rehabilitation of individuals [1,2]. The family function is complex and focuses on the whole system rather than on the individuals, including roles (e.g., family structure), relationships, well-being, and belonging, and significantly takes place in the context of some public concerns, such as health problems [3,4]. In fact, the role of the family has been widely concerned in primary care in many countries. In the United States, family-centered care has incorporated into the standards and requirements in the performance system [5]. A Canadian study developed a Calgary Family Assessment and Intervention Model to evaluate family functioning in health research and family care [6]. Moreover, family functioning can influence health literacy among members. For example, children have e reported good health literacy within a better family context (better education, higher economic level, etc.) [7]. It is suggested that individuals in the same family could play their respective roles, achieve their practical goals, and maintain the relationships with each other; these are supportive factors to promote individual health [8,9]. In general, understanding the family experience of health and illness within the family's social and cultural context helps health professionals provide targeted health services [10]. In the COVID-19 context, the pandemic poses an acute threat to the well-being of the whole family, which emphasizes the importance of family health [11,12].

Individuals' health could affect other members' health service needs in the same households. For example, for patients with chronic disease, their family caregivers take responsibility for maintaining treatment compliance, supervising medication intake, and providing emotional and economic support. However, family caregivers often show poorer health than those in families without a patient, and they are more likely to suffer from a host of problems, such as anger, fear, and depression [13,14]. People's health is also associated with those with whom they are living. Support from the patients' families, especially their spouse's support, can affect their health-seeking behavior. For example, support from husbands plays an essential role in encouraging women's health [15]. Additionally, because of the critical role of family members, some countries such as Canada, have tried to expand their efforts to actively involve patients' family members in health service improvement and system redesign initiatives [16]. Hughes and Waite [17] found that married couples (age 51–61) living alone or with children are the most advantaged in health, but single women living with children are disadvantaged on all health outcomes. In China, old adults living alone and living with their children showed both advantages and disadvantages in health, while those with a spouse in the household provided the best health protection [18].

In China, the limited doctor–patient interactions are often confined to the brief consultation time and care facilities. However, family members' behavior concordance and daily support could compensate for the situation [19], and family involvement could be critical for health management [20]. Therefore, many health services that could be delivered based on a household unit. In 2016, the Chinese Government proposed a family doctor system, in which general practitioners would establish a long-term service relationship with families that signed a contract with them. The family doctors would offer 95% of the family's primary care and play a vital role as the primary health gatekeeper [21–23]. However, due to a lack of several essential uniform features such as health insurance support, appropriate incentive mechanisms, objective evaluation methods, and an effective way of delivering service, so there is some difficulty implementing the family doctor contracting services in China [24]. The family-centered care model has been promoted as a contemporary model of health service delivery, and evidence has shown that family-centered care for older adults is positive [25,26]. In the Chinese community-based healthcare setting, family-centered care showed a positive effect on seniors with diabetes [27]. Moreover, some health-promotion programs have tried to consider families, but due to a lack of funding and policy, it is difficult to maintain family-centered interests in associating interventions [28]. Therefore, it is crucial to clarify health service needs from a household perspective, which would be indicators for services, such as a family doctor contracting service.

In China, an empty nest household usually refers to households without children or whose children have left their parents' home [29,30]. The accelerated urbanization and inequity of economic development in urban and country areas have resulted in the empty nest becoming the main family pattern in rural China. In 2016, empty nesters accounted for 51.1% of the elderly in China, and this proportion will reach 90% by 2030 according to China's National Committee on Aging [31]. Often associated with low income, poor living conditions, and the lack of social and emotional support, residents in empty nest households are more likely to be vulnerable to different health problems and irreversible decreases in functional capacity [30,32,33]. Moreover, empty nest individuals are usually older adults who are likely to suffer from a high prevalence of chronic conditions and disability [34–36]. When their children move out of their homes, their empty nest parents are more likely to suffer from empty nest syndrome, resulting in loneliness, anxiety, frustration, etc. [37–39]. An investigation conducted in Sichuan, a western province in China, found that 30.11% of elderly empty nesters had anxiety-related symptoms [40,41]. These negative emotions are consistently tied to a subjective feeling of increased pain, disease, and tiredness [42–44]. In general, members in empty nest families are often concerned with poor health statuses, poor mobility, and a high risk of chronic diseases. Therefore, primary care facilities in rural areas are always under significant pressure in health service delivery, such as diagnosis and treatment, chronic disease management, physical examination, and mental health. Otherwise, with the advancement of the internet and social media, older adults have a high demand for remote assessment, such as telemedicine services, to help them become more independent in daily living activities [45,46]. Moreover, some empty nesters use complementary medicine, such as traditional Chinese medicine, to replace other kinds of treatment for economic or preference reasons [47].

Therefore, we assume that health services delivered based on a household unit should be more efficient, especially for empty nest households who have more health service needs but limited support. However, there is little concern about health service needs from a household perspective, the difficulty to collect sufficient family and individual information, and a lack of proper methods to analyze individuals' health service needs from a household perspective. In this study, we conducted empirical research to determine the health service needs in empty nest households in Sinan and Dangyang in China from a household perspective.

2. Methods

2.1. Setting

This study was conducted in Sinan County in Guizhou Province and Dangyang County in Hubei Province, located in western and central China. Both counties are at the first-class economic level (GDP ranked 2/10, 4/13 in their cities in 2020). Sinan County has 7 townships, and Sinan County has 17 townships. Both countries are located in relatively flat areas, and the distribution of households is dense. Households in these counties are equipped with at least one village clinic, one township hospital, and county hospitals, and residents can access different health services from any health facility.

2.2. Study Design and Data Collection

With a 95% confidence level, the calculation is based on the requirement that the absolute sampling error does not exceed 3%. Due to the use of multistage complex sampling, the design effect will generally be between 2 and 2.5 [48,49]. This study considers the design effect at 2.5.

$$n_i = u_a^2 \cdot p \cdot (1 - p) / \delta^2 \tag{1}$$

$$N_i = n_i \cdot DEFF \tag{2}$$

Equation (1) calculates the sample size of simple random sampling. n_i represents the number of samples required for the i stage. The u_a corresponds to the inspection level of u value, and δ is the allowable error. Equation (2) calculates the sample size of multistage stratified random cluster sampling. N_i is the sample size of multistage stratified random cluster sampling. $DEFF$ is sampling efficiency, referring to how many samples in this sampling process can provide the information that one sample could in a simple random sampling. The absolute sampling error is 21.338% of the chronic disease prevalence of the whole population based on the number of patients in China according to the 2013 National Health Service Survey, while the is 1.96 with a 95% confidence level [50]. Sampling efficiency could be affected by the intra-group correlation coefficient (ICC), sample stratification, the average number of respondents in each set, internal heterogeneity. Among them, the design efficiency of the National Census in China is 1.4, which serves as a reference value in this study. Therefore, the sample size in each county was 3584 individuals. According to the Fifth National Health Service Survey, the average amount of individuals per household was 2.9 [34]. Lastly, at least 1200 households in each county should be investigated. The sampling process could be found in Figure 1.

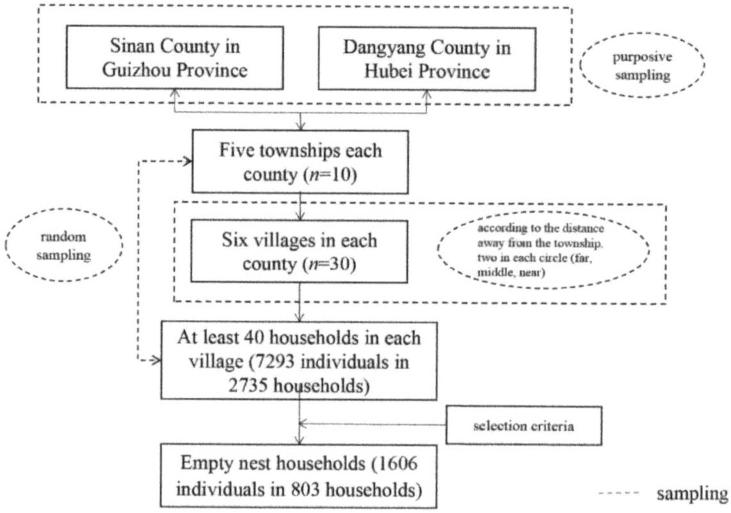

Figure 1. Selection process of empty nest households.

This study was conducted in Sinan County in Guizhou Province and Dangyang County in Hubei Province. The counties were selected with a purposive sampling strategy in central and western China. A multistage stratified random sampling strategy was used to select the households. Five townships in each county were selected randomly, and six villages were considered according to their distance away from each township, with two villages selected randomly far, medium distance, and near the central township. We conducted face-to-face interviews with around 40 households in each village, and each individual was questioned. Finally, we investigated 7293 individuals in 2735 households. The following criteria were used to select the empty nest households: (1) households with only two members in the house first entered the study; (2) households with two members must be spouses; (3) respondents must have lived at the survey site for at least six months; (4) at least one member in an empty nest household could participate in the study via face-to-face interview. Finally, 1606 empty nest people in 803 empty nest households were included ($803 \times 2 = 1606$). Informed consent was obtained from all participants in this study.

2.3. Health Service Needs

According to Nobile [51], the WHO definition of health is a dynamic state of wellbeing characterized by a physical, mental, and social potential, satisfying the demands of life commensurate with age, culture, and personal responsibility. Accordingly, this study divided health service needs into physical health service needs and mental health service needs generally. In China, the health system is always concerned with healthcare and public health [52]. The healthcare system is designed to satisfy people's medical needs, such as diagnosis and treatment and chronic disease management [50]. Public health is the collective action for sustained population-wide health improvement, such as health surveillance and preventive care [53,54]. With the advancement of social media, telemedicine has played an important role in providing health services in China, which should be considered when studying health service needs [55]. Therefore, we finally made the health service needs in this study as diagnosis and treatment services, chronic disease management service, telemedicine, physical examination service, health education service, mental healthcare, and traditional Chinese medicine service.

2.4. Health Service Needs Consistency Rates

According to individuals' answers to the questionnaire, consistency rates of individuals' health service needs could be calculated as:

$$\text{Consistency rate} = \text{COUNT } n\ (x_{1ij} = x_{2ij}, i = 1,2,\ldots\ldots,N; j = 1,2,\ldots\ldots,8)/N \quad (3)$$

x_{1ij} refers to the health service need of one member in an empty nest household, x_{2ij} and refers to the other member's health service need in the same empty nest household. When both selections of two individuals were the same, then this empty nest household would be counted into this study. i is the code of empty nest households, and j refers to different health service needs. n is the number when $x_{1ij} = x_{2ij}$, and N is the total count of empty nest households in this study. If individuals' health service needs in empty nest households do have consistency to some extent, then it is feasible to explore the health service needs from a household perspective.

2.5. Health Service Needs from a Household Perspective

Collective Household Model

Becker first illustrated the collective household model, in which the household is characterized as a collection of individuals. This model assumes that family consumption decisions result from multi-person decision making. An intrinsic feature of the collective model is the sharing rule, which governs the within-household distribution of household capitals. This sharing rule is always an indicator of the bargaining power of individual household members. The ultimate consumption decision on each good or service is always dependent on household characteristics, income levels, etc. However, these factors only affect different weights of individuals' bargaining power in a household's model but not the preferences of individual household members [56,57].

When there are two individuals (1 and 2) in the same household consuming a set of services, the health service needs in a household unit would be [57]:

$$\mathrm{u}(Q,qf,qm) = \max(Q,qf,qm)\{b_1\, uf(Q,qf,qm) + b_2\, um(Q,qf,qm)\} \quad (4)$$

Q refers to the public health service needs and qf and qm are the private health service needs. uf and um are the expected health utilities people would obtain from different health services. In this study, we investigated people's subjective health score, and the gap between their status to full-health status is the utility they would get from receiving health services. b_1 and b_2 refer to individuals' bargaining power in the same empty nest household. It can be affected by individuals' education level, role in their family (householder or not) and their objective health status valued by EQ-5D.

This study took the mean score of different factors as individuals' bargaining power. If the educational level was divided into three levels, the higher level of an individual, the stronger their bargaining power. If individual 1 is a householder, then $b_1 = 1, b_2 = 0$. The value estimated by EQ-5D refers to the objective health status.

3. Results

3.1. Consistency Rates of Different Health Service Needs in Empty Nest Households

In this study, individuals' consistency rates in diagnosis and treatment service, chronic disease management service, telemedicine care, physical examination service, health education service, mental healthcare, and traditional Chinese medicine service were 40.30%,

89.13%, 98.85%, 58.93%, 57.95%, 72.84%, and 63.40%, respectively. Therefore, family level health service needs could be studied based on this concordance.

3.2. Correlations between Individual's Health Service Need in Empty Nest Households

In Table 1, health service needs, diagnosis and treatment service ($r = 0.404$), telemedicine care ($r = 0.177$), physical examination service ($r = 0.286$), health education service ($r = 0.265$), and traditional Chinese medicine service ($r = 0.220$) of individuals in an empty nest household have significant correlations with each other ($p < 0.001$).

Table 1. Correlations between individuals' health service needs (r).

Individual 1 / Individual 2	H1	H2	H3	H4	H5	H6	H7
H1	0.404 **	0.049	0.034	0.063	−0.044	−0.086	−0.045
H2	0.059	0.003	0.004	0.026	−0.036	0.006	0.025
H3	0.059	0.005	0.177 **	0.037	0.002	0.009	0.039
H4	0.003	0.033	−0.004	0.286 **	0.115 **	−0.013	0.015
H5	−0.032	−0.018	0.037	0.104 **	0.265 **	−0.018	0.051
H6	0.005	0.022	0.026	0.015	0.024	0.053	0.036
H7	0.007	0.027	0.031	−0.002	0.085 *	−0.035	0.220 **

Notes: H1: diagnosis and treatment service; H2: chronic disease management service; H3: telemedicine care; H4: physical examination service; H5: health education service; H6: mental healthcare; H7: traditional Chinese medicine service. * $p < 0.05$, ** $p < 0.001$.

3.3. Health Service Needs from a Household Perspective in Empty Nest Households

As shown in Table 2, individuals in empty nest households have similar subjective and objective health mean scores. Individuals with educational levels less than primary school accounted for over 50%.

Table 2. Characteristics for each individual in empty nest households ($n = 803$).

Characteristics	Categories	Individual 1	Individual 2
Subjective health score	Mean score	69.95	67.69
Objective health score	Mean score	0.887	0.865
Educational level	Less than primary school	50.7%	60.0%
	Junior and senior high school	47.8%	36.6%
	More than undergraduate	0.3%	0.8%
Householder	Yes	74.5%	27.9%
	No	25.4%	71.8%

As seen in Table 3, households with public health service needs as their household needs accounted for 93.4% ($n = 746$), and households that take one individual's health service needs as their household needs accounted for 5.7% ($n = 46$). The public needs at the household level mainly include the physical examination service (43.3%) and health education service (24.9%). The individual needs mainly include the diagnosis and treatment service (56.5%), physical examination service (65.2%), health education service (63.0%), and traditional Chinese medical service (45.7%). In general, the health service needs in empty nest households mainly include the diagnosis and treatment service (12.4%), physical examination service (44.2%), health education service (26.9%), and traditional Chinese medicine service (18.9%).

Table 3. Health service needs in empty nest households (*n*, %).

Health Service Needs	Public Needs as Household Needs	Individual Needs as Household Needs			Total
		Individual 1	Individual 2	Total	
H1	73 (9.8)	11 (52.4)	15 (60.0)	26 (56.5)	99 (12.4)
H2	39 (5.2)	0 (0.0)	0 (0.0)	0 (0.0)	39 (4.9)
H3	1 (0.1)	1 (4.8)	0 (0.0)	1 (2.2)	2 (0.3)
H4	323 (43.3)	15 (71.4)	15 (60.0)	30 (65.2)	353 (44.2)
H5	186 (24.9)	16 (76.2)	13 (52.0)	29 (63.0)	215 (26.9)
H6	29 (3.9)	7 (33.3)	1 (4.0)	8 (17.4)	37 (4.6)
H7	130 (18.1)	11 (52.4)	10 (40.0)	21 (45.7)	151 (18.9)

Notes: H1: diagnosis and treatment service; H2: chronic disease management service; H3: telemedicine care; H4: physical examination service; H5: health education service; H6: mental healthcare; H7: traditional Chinese medicine service.

3.4. Characteristics for Health Service Needs in a Household Unit

According to Table 4, the empty nest households who take public health service needs as household needs are mainly male householders, accounting for 94.0% ($n = 803$). Additionally, 69.3% of these households had at least one individual who had an illness within two weeks while we conducted the survey, and 82.4% of households had at least one individual with chronic disease. Therefore, households with at least one individual who had an illness within two weeks or chronic conditions were more likely to take public health needs as household needs.

Table 4. Characteristics for health service needs in a household unit in empty nest households (*n*, %).

			Public Needs as Household Needs	
			n	%
Family characteristics	Householder's gender	Male	701	94.0
		Female	45	6.0
	Two-week prevalence	both	154	20.7
		One individual	335	47.7
		None	236	31.7
	Chronic disease	Both	228	30.7
		One individual	384	51.7
		None	131	17.6

			Individual Needs as Household Needs			
			Individual 1		Individual 2	
			n	%	*n*	%
Individual characteristics	Householder	Yes	19	90.5	0	0.0
		No	2	9.5	25	100.0
	Two-week prevalence	Yes	21	100.0	25	100.0
		No	0	0.0	0	0.0
	Chronic disease	Yes	16	76.2	22	89.0
		No	5	23.8	3	12.0

Households that consider an individual's health service needs as household needs are more concerned with the individual's health. The individual usually had a disease within two weeks (100.0%) or was the householder in the family (90.5%). The results could be found in Figure 2.

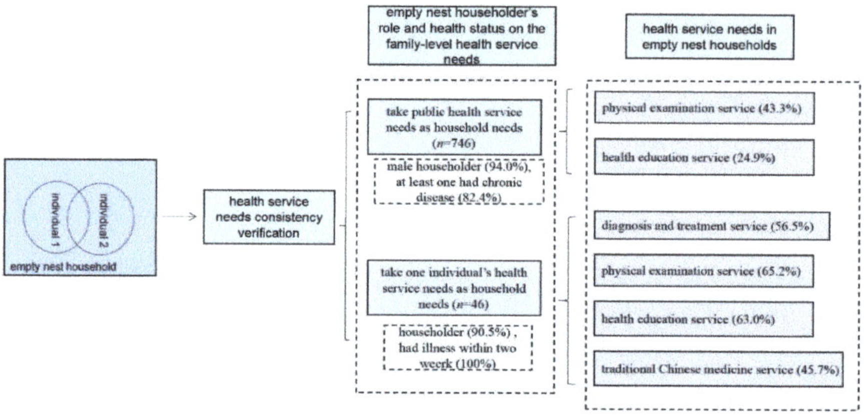

Figure 2. Health service needs in empty nest households.

4. Discussion

In this study, the health service needs of individuals in empty nest households have a high degree of consistency. This may be a credit to the shared family context they have, where their family culture and living habits originate, resulting in individuals in the same household having similar health literacy. For example, Wong et al. [58] found a health literacy information sharing system among family members; the health literacy among family members could be shared and could change individuals' health behaviors. Ishikawa and Kiuchi [59] found that although an individual's ability to achieve health-related literacy is limited, it could be compensated by other family members' abilities. This may lead to the high degree of consistency in their health service needs.

According to the results of this study, the consistency rate was relatively low for diagnosis and treatment service needs. It is believed that people's well-being and severity could be affected by their subjective feeling, so that people may have different health service needs, even though they have the same kind of disease [60]. However, service needs, such as physical examination, health education, and consecutive chronic disease prescription, are commonly considered as essential health service needs, and were added to the family doctor contract services [61], which are suitable to be delivered within a household unit.

In this study, individuals' health service needs in empty nest households have been found to be positively correlated with each other. The environment and lifestyle that individuals share within a household result in similar health service needs [62]. Additionally, when one becomes sick, their spouse usually worries about catching the same illness, leading to a change in the individual's health service needs. Therefore, family-oriented health promotion and disease prevention are promising strategies as family members may support and nurture one another through life stages [28]. In particular, empty nesters may live together with each other for a long time.

There are still difficulties faced in conducting family-centered health services in primary care. In this study, health service needs on a family level in empty nest households mainly include physical examination services, health education services, etc. People in empty nest households are almost always old, suffering from poor health status, and lacking economic support. For these people, only a basic level of health is required, and is hoped for to reduce the children's burden [63]. In China, people over 65 years old could access free physical examination services, which showed positive outcomes in the timely discovery of health problems [64,65]. Health education can popularize common disease prevention knowledge, chronic disease prevention, daily maintenance, and acute disease measures for residents in rural areas with low educational levels. This kind of health education provided by primary care institutions is the main source for empty nest households' health information [66]. However, these health services are not delivered from a family unit, although some are taken into family doctor contracting services [24]. The Chinese Government issued "Guiding Opinions on Regulating the Management of Family Doctor Contracting Services" in 2018, and it proposed that the number of residents signed to each family doctor should not exceed 2000, but there is more demand in reality. For example, the number of contract resident visits for each family doctor from 2013 to 2016 in Pudong, Shanghai, was more than 8000 per year [67].

Family members' roles are essential in producing health service needs. In this study, residents in empty nest households may take householders' health service needs as household needs, especially when the householder is male. This is also proven in other families. For example, Meydanlioglu et al. [68] found father's educational status determined factors associated with their children being overweight and obese. Dongn et al. [69] mentioned that fathers with cognitive empathy could alleviate depression caused by mothers' parental stress. Additionally, parents are critical in guiding children's behavioral changes, such as food choice and physical activity [70,71]. Therefore, some health service needs could be delivered based on this concept. For example, it may be more efficient to provide health services, such as health education to householders in empty nest households, for which other family members would be affected.

Health service needs from a household perspective are dynamic, for the family environment is always changing via family relationships, interactions, beliefs, values, routines, and practices [71]. Any family member's health change will lead to a change in a family's health service needs. When someone has an emergency illness or severe disease, it is more likely that their health service needs should be satisfied first. In contrast, health service needs from a household perspective are also stable. When both individuals in the same households have similar health status, they may consider their common health service needs as their household needs [72], which is common among empty nest households.

According to the United Nations, the first goal of sustainable development is eradicating poverty and the divisive implications of its pathology [73]. Evidence has shown that "poverty due to illness" and "return to poverty due to illness" are the main cause of poverty in remote rural areas in China [74,75]. Empty nest families always have a risk of health-related poverty problems due to their fragility in economy and health. Making health service needs clear from the household perspective would be helpful to locate families with severe health problems that may result in poverty, which will improve sustainable development in rural China.

5. Conclusions

Families are vital to the health of individuals as they promote family members' healthy choices and encourage health behavioral change. Family involvement in health services delivery has been taken into consideration in many countries. However, the relationship of individuals' health services needs is less evident, and few studies have focused on the family-level health service needs. In this study, we took empty nest households in rural China as an example, and explored the health service needs from a household perspective based on a cross-sectional study. A wide range of samples were selected with a multi-level sampling strategy, and a family model was applied to determine family-level health service needs for empty nest households. According to the results, individuals' health service needs in empty nest households are highly consistent and are positively correlated with each other, indicating that health service needs could be studied based on a household unit. It could be served as a reference for policymakers of primary care to improve the effectiveness of health management and reduce the pressure of primary care. It is also a promising strategy to promote doctor–patient communication with limited interaction. This will be helpful for health management in primary care in rural China, for the concordance will alleviate the pressure of primary care and increase the effectiveness of doctor–patient communication. Empty nest households who considered individuals' public needs as household needs were usually with a male householder or were experiencing a chronic condition; the health service needs mainly included physical examination and health education. Health service needs in empty nest households who took one member's needs as household needs included diagnosis and treatment, physical examination, health education, and traditional Chinese medicine services. The member was more likely to be the householder of the family or had a disease within two weeks. Therefore, family members' roles and health status play an important role in health service needs in empty nest households. Additionally, family-level health services needs in empty nest households mainly included physical examination and health education, which could be indicators to deliver health services for a family unit. In addition, figuring out health service needs from a household perspective is helpful to locate health-related poverty families and improve sustainable development in rural China.

Author Contributions: L.Z. and X.C. designed this study. X.C. drafted and reviewed the manuscript. X.C. analyzed the data and interpreted the results. L.Z. modified the manuscript. All authors have read and agreed to the published version of the manuscript.

Funding: This study was funded by the National Natural Science Foundation of China (Grant No: 71734003).

Institutional Review Board Statement: The study was conducted according to the guidelines of the Declaration of Helsinki, and approved by the Ethics Committee of Tongji Medical College, Huazhong University of Science and Technology (IORG No: IORG0003571).

Informed Consent Statement: Informed consent was obtained from all subjects involved in the study.

Data Availability Statement: The data presented in this study are available on request from the corresponding author.

Acknowledgments: We would like to express our great appreciation to the participants in this study. We also thank all of the local government officers and medical staff on the assistance of organization. Moreover, we truly appreciate the help of research assistants on the data entry and management.

Conflicts of Interest: The authors declare no conflict of interest.

References

1. Benzein, E.; Johansson, P.; Arestedt, K.F.; Berg, A.; Saveman, B.I. Families' importance in nursing care: Nurses' attitudes—An instrument development. *J. Fam. Nurs.* **2008**, *14*, 97–117. [CrossRef]
2. Allred, R.H. Nurses and Families: A Guide to Family Assessment and Intervention. *J. Community Health Nurs.* **1984**, *1*, 69–71. [CrossRef]
3. Epstein, N.B.; Bishop, D.S.; Levin, S. The McMaster model of family functioning. *J. Marital Fam. Ther.* **1978**, *4*, 19–31. [CrossRef]
4. McGoldrick, M.; Carter, E.A.; Garcia-Preto, N. *The Expanded Family Life Cycle: Individual, Family, and Social Perspectives*; Pearson Allyn & Bacon: Boston, MA, USA, 2013.
5. The Joint Commission. *Advancing Effective Communication, Cultural Competence, and Patient- and Family-Centered Care: A Roadmap for Hospitals*; The Joint Commission: Oakbrook Terrace, IL, USA, 2010.
6. Leahey, M.; Wright, L.M. Application of the Calgary Family Assessment and Intervention Models: Reflections on the Reciprocity between the Personal and the Professional. *J. Fam. Nurs.* **2016**, *22*, 450–459. [CrossRef] [PubMed]
7. Lopes, R.T.; Neves, É.T.B.; Dutra, L.D.C.; Gomes, M.C.; Paiva, S.M.; Abreu, M.H.N.G.; Ferreira, F.M.; Granville-Garcia, A.F. Socioeconomic status and family functioning influence oral health literacy among adolescents. *Rev. Saude Publica* **2020**, *54*, 30. [CrossRef]
8. Çuhadar, D.; Savaş, H.A.; Ünal, A.; Gökpınar, F. Family Functionality and Coping Attitudes of Patients with Bipolar Disorder. *J. Relig. Health* **2015**, *54*, 1731–1746. [CrossRef]
9. Astedt-Kurki, P.; Tarkka, M.T.; Rikala, M.R.; Lehti, K.; Paavilainen, E. Further testing of a family nursing instrument (FAFHES). *Int. J. Nurs. Stud.* **2009**, *46*, 350–359. [CrossRef]
10. Montoro-Gurich, C.; Garcia-Vivar, C. The Family in Europe: Structure, Intergenerational Solidarity, and New Challenges to Family Health. *J. Fam. Nurs.* **2019**, *25*, 170–189. [CrossRef]
11. Prime, H.; Wade, M.; Browne, D.T. Risk and resilience in family well-being during the COVID-19 pandemic. *Am. Psychol.* **2020**, *75*, 631–643. [CrossRef]
12. Trougakos, J.P.; Chawla, N.; McCarthy, J.M. Working in a pandemic: Exploring the impact of COVID-19 health anxiety on work, family, and health outcomes. *J. Appl. Psychol.* **2020**, *105*, 1234–1245. [CrossRef]
13. Kilic, S.T.; Oz, F. Family Caregivers' Involvement in Caring with Cancer and their Quality of Life. *Asian Pac. J. Cancer Prev.* **2019**, *20*, 1735–1741. [CrossRef]

14. Meng, N.; Chen, J.; Cao, B.; Wang, F.; Xie, X.; Li, X. Focusing on quality of life in the family caregivers of patients with schizophrenia from the perspective of family functioning: A cross-sectional study. *Medicine* **2021**, *100*, e24270. [CrossRef]
15. Ohashi, A.; Higuchi, M.; Labeeb, S.A.; Mohamed, A.G.; Chiang, C.; Aoyama, A. Family support for women's health-seeking behavior: A qualitative study in rural southern Egypt (Upper Egypt). *Nagoya J. Med. Sci.* **2014**, *76*, 17–25.
16. Abelson, J.; Humphrey, A.; Syrowatka, A.; Bidonde, J.; Judd, M. Evaluating Patient, Family and Public Engagement in Health Services Improvement and System Redesign. *Healthc. Q.* **2018**, *21*, 61–67. [CrossRef]
17. Hughes, M.E.; Waite, L.J. Health in household context: Living arrangements and health in late middle age. *J. Health Soc. Behav.* **2002**, *43*, 1–21. [CrossRef] [PubMed]
18. Li, L.W.; Zhang, J.; Liang, J. Health among the oldest-old in China: Which living arrangements make a difference? *Soc. Sci. Med.* **2009**, *68*, 220–227. [CrossRef]
19. Rosland, A.M.; Piette, J.D. Emerging models for mobilizing family support for chronic disease management: A structured review. *Chronic Illn.* **2010**, *6*, 7–21. [CrossRef] [PubMed]
20. Ris, I.; Schnepp, W.; Mahrer Imhof, R. An integrative review on family caregivers' involvement in care of home-dwelling elderly. *Health Soc. Care Community* **2019**, *27*, e95–e111. [CrossRef]
21. Fu, P.; Wang, Y.; Liu, S.; Li, J.; Gao, Q.; Zhou, C.; Meng, Q.; Sylvia, S. Analysing the preferences for family doctor contract services in rural China: A study using a discrete choice experiment. *BMC Fam. Pract.* **2020**, *21*, 148. [CrossRef] [PubMed]
22. Shang, X.; Huang, Y.; Li, B.; Yang, Q.; Zhao, Y.; Wang, W.; Liu, Y.; Lin, J.; Hu, C.; Qiu, Y. Residents' Awareness of Family Doctor Contract Services, Status of Contract with a Family Doctor, and Contract Service Needs in Zhejiang Province, China: A Cross-Sectional Study. *Int. J. Environ. Res. Public Health* **2019**, *16*, 3312. [CrossRef]
23. Gu, L.; Zhu, R.; Li, Z.; Zhang, S.; Li, J.; Tian, D.; Sun, Z. Factors Associated with Rural Residents' Contract Behavior with Village Doctors in Three Counties: A Cross-Sectional Study from China. *Int. J. Environ. Res. Public Health* **2020**, *17*, 8969. [CrossRef]
24. Yuan, S.; Wang, F.; Li, X.; Jia, M.; Tian, M. Facilitators and barriers to implement the family doctor contracting services in China: Findings from a qualitative study. *BMJ Open* **2019**, *9*, e032444. [CrossRef]
25. Bamm, E.L.; Rosenbaum, P. Family-centered theory: Origins, development, barriers, and supports to implementation in rehabilitation medicine. *Arch. Phys. Med. Rehabil.* **2008**, *89*, 1618–1624. [CrossRef]
26. Liu, Y.; Xiao, X.; Peng, C.; Zhao, T.; Wu, Y.; Yu, W.; Ou, L.; Chen, X.; Wu, X.; Xu, D.R.; et al. Development and Implementation of Couple-Based Collaborative Management Model of Type 2 Diabetes Mellitus for Community-Dwelling Chinese Older Adults: A Pilot Randomized Trial. *Front. Public Health* **2021**, *9*, 686282. [CrossRef] [PubMed]
27. Tu, J.; Liao, J. Primary care providers' perceptions and experiences of family-centered care for older adults: A qualitative study of community-based diabetes management in China. *BMC Geriatr.* **2021**, *21*, 438. [CrossRef] [PubMed]
28. Barnes, M.D.; Hanson, C.L.; Novilla, L.B.; Magnusson, B.M.; Crandall, A.C.; Bradford, G. Family-Centered Health Promotion: Perspectives for Engaging Families and Achieving Better Health Outcomes. *Inquiry* **2020**, *57*, 46958020923537. [CrossRef] [PubMed]
29. Gao, M.; Li, Y.; Zhang, S.; Gu, L.; Zhang, J.; Li, Z.; Zhang, W.; Tian, D. Does an Empty Nest Affect Elders' Health? Empirical Evidence from China. *Int. J. Environ. Res. Public Health* **2017**, *14*, 463. [CrossRef]
30. Lv, X.-L.; Jiang, Y.-H.; Sun, Y.-H.; Ren, C.-Z.; Sun, C.-Y.; Sun, L.; Wu, Z.-Q.; Zhao, X. Short form 36-Item Health Survey test result on the empty nest elderly in China: A meta-analysis. *Arch. Gerontol. Geriatr.* **2013**, *56*, 291–297. [CrossRef]
31. Su, D.; Wu, X.-N.; Zhang, Y.-X.; Li, H.-P.; Wang, W.-L.; Zhang, J.-P.; Zhou, L.-S. Depression and social support between China' rural and urban empty-nest elderly. *Arch. Gerontol. Geriatr.* **2012**, *55*, 564–569. [CrossRef]
32. Liang, Y.; Wu, W. Exploratory analysis of health-related quality of life among the empty-nest elderly in rural China: An empirical study in three economically developed cities in eastern China. *Health Qual. Life Outcomes* **2014**, *12*, 59. [CrossRef]
33. Zhang, H. Who will care for our parents? Changing boundaries of family and public roles in providing care for the aged in urban China. *Care Manag. J.* **2007**, *8*, 39–46. [CrossRef]
34. Chatterji, S.; Byles, J.; Cutler, D.; Seeman, T.; Verdes, E. Health, functioning, and disability in older adults—Present status and future implications. *Lancet* **2015**, *385*, 563–575. [CrossRef]
35. Yang, T.; Chu, J.; Zhou, C.; Medina, A.; Li, C.; Jiang, S.; Zheng, W.; Sun, L.; Liu, J. Catastrophic health expenditure: A comparative analysis of empty-nest and non-empty-nest households with seniors in Shandong, China. *BMJ Open* **2016**, *6*, e010992. [CrossRef] [PubMed]
36. Jing, Z.; Li, J.; Fu, P.P.; Wang, Y.; Yuan, Y.; Zhao, D.; Hao, W.; Yu, C.; Zhou, C. Catastrophic health expenditure among single empty-nest elderly with multimorbidity in rural Shandong, China: The effect of co-occurrence of frailty. *Int. J. Equity Health* **2021**, *20*, 23. [CrossRef] [PubMed]
37. Chang, Y.; Guo, X.; Guo, L.; Li, Z.; Yang, H.; Yu, S.; Sun, G.; Sun, Y. Comprehensive Comparison between Empty Nest and Non-Empty Nest Elderly: A Cross-Sectional Study among Rural Populations in Northeast China. *Int. J. Environ. Res. Public Health* **2016**, *13*, 857. [CrossRef]
38. Fahrenberg, B. Coping with the empty nest situation as a developmental task for the aging female—An analysis of the literature. *Z. Gerontol.* **1986**, *19*, 323–335. [PubMed]
39. Pillay, A.L. Midlife depression and the "empty nest" syndrome in Indian women. *Psychol. Rep.* **1988**, *63*, 591–594. [CrossRef] [PubMed]

40. Wang, Z.; Shu, D.; Dong, B.; Luo, L.; Hao, Q. Anxiety disorders and its risk factors among the Sichuan empty-nest older adults: A cross-sectional study. *Arch. Gerontol. Geriatr.* **2013**, *56*, 298–302. [CrossRef]
41. Cheng, P.; Jin, Y.; Sun, H.; Tang, Z.; Zhang, C.; Chen, Y.; Zhang, Q.; Zhang, Q.; Huang, F. Disparities in prevalence and risk indicators of loneliness between rural empty nest and non-empty nest older adults in Chizhou, China. *Geriatr. Gerontol. Int.* **2015**, *15*, 356–364. [CrossRef]
42. Pressman, S.D.; Gallagher, M.W.; Lopez, S.J. Is the emotion-health connection a "first-world problem"? *Psychol. Sci.* **2013**, *24*, 544–549. [CrossRef]
43. Geisser, M.E.; Roth, R.S.; Theisen, M.E.; Robinson, M.E.; Riley, J.R. Negative affect, self-report of depressive symptoms, and clinical depression: Relation to the experience of chronic pain. *Clin. J. Pain* **2000**, *16*, 110–120. [CrossRef] [PubMed]
44. Liu, J.; Tian, J.; Yue, P.; Wang, Y.; Du, X.; Chen, S. Living experience and care needs of Chinese empty-nest elderly people in urban communities in Beijing, China: A qualitative study. *Int. J. Nurs. Sci.* **2015**, *2*, 15–22. [CrossRef]
45. Yuan, Y.; Liu, Y.; Gong, L.; Chen, H.; Zhang, S.; Kitayama, A.; Takashi, E.; Liang, J. Demand Analysis of Telenursing for Community-Dwelling Empty-Nest Elderly Based on the Kano Model. *Telemed. E-Health* **2021**, *27*, 414–421. [CrossRef] [PubMed]
46. Wu, Y.; Liu, Y.; Su, Z.; Sun, S.; Liu, C.; Ding, W.; Gao, Y. Demands for Telenursing-Based Long-Term Care Among Disabled Older Adults in Qingdao, China: A Cross-Sectional Study. *Patient Prefer. Adherence* **2021**, *15*, 1981–1990. [CrossRef] [PubMed]
47. Xu, W.; Li, Z.; Pan, Z.; He, R.; Zhang, L. Prevalence and associated factors of self-treatment behaviour among different elder subgroups in rural China: A cross-sectional study. *Int. J. Equity Health* **2020**, *19*, 32. [CrossRef] [PubMed]
48. Kaiser, R.; Woodruff, B.A.; Bilukha, O.; Spiegel, P.B.; Salama, P. Using design effects from previous cluster surveys to guide sample size calculation in emergency settings. *Disasters* **2006**, *30*, 199–211. [CrossRef] [PubMed]
49. Liu, J.; Zhang, G. Research on the application of design effect in complex sampling-taking China's household survey as an example. *J. Appl. Stat. Manag.* **2015**, *34*, 628–635.
50. Center for Health Statistics and Information, NHFPC. *The Fifth National Health Service Survey and Analysis Report*; Chinese Union Medical University Press: Beijing, China, 2015.
51. Nobile, M. The WHO Definition of health: A critical reading. *Med. Law* **2014**, *33*, 33–40.
52. Meng, Q.; Mills, A.; Wang, L.; Han, Q. What can we learn from China's health system reform? *BMJ* **2019**, *365*, l2349. [CrossRef]
53. Liu, Y. China's public health-care system: Facing the challenges. *Bull. World Health Organ.* **2004**, *82*, 532–538. [PubMed]
54. Beaglehole, R.; Bonita, R.; Horton, R.; Adams, O.; McKee, M. Public health in the new era: Improving health through collective action. *Lancet* **2004**, *363*, 2084–2086. [CrossRef]
55. Cui, F.; He, X.; Zhai, Y.; Lyu, M.; Shi, J.; Sun, D.; Jiang, S.; Li, C.; Zhao, J. Application of Telemedicine Services Based on a Regional Telemedicine Platform in China from 2014 to 2020, Longitudinal Trend Analysis. *J. Med. Internet Res.* **2021**, *23*, e28009. [CrossRef] [PubMed]
56. Becker, G.S. Altruism in the Family and Selfishness in the Market Place. *Economica* **1981**, *48*, 1–15. [CrossRef]
57. Cherchye, L.; De Rock, B.; Lewbel, A.; Vermeulen, F. Sharing Rule Identification for General Collective Consumption Models. *Econometrica* **2015**, *83*, 2001–2041. [CrossRef]
58. Wong, J.Y.H.; Wai, A.K.C.; Zhao, S.; Yip, F.; Lee, J.J.; Wong, C.K.H.; Wang, M.P.; Lam, T.H. Association of Individual Health Literacy with Preventive Behaviours and Family Well-Being during COVID-19 Pandemic: Mediating Role of Family Information Sharing. *Int. J. Environ. Res. Public Health* **2020**, *17*, 8838. [CrossRef] [PubMed]
59. Ishikawa, H.; Kiuchi, T. Association of Health Literacy Levels between Family Members. *Front. Public Health* **2019**, *7*, 169. [CrossRef] [PubMed]
60. Kaprio, H.; Suominen, A.L.; Lahti, S. Association between subjective oral health and regularity of service use. *Eur. J. Oral Sci.* **2012**, *120*, 212–217. [CrossRef]
61. National Medical Reform Office Notice on Promoting Guidance for Family Doctor Contract Services. Available online: http://www.mohrss.gov.cn/SYrlzyhshbzb/shehuibaozhang/zcwj/yiliao/201606/t20160615_241854.html (accessed on 20 December 2021). (In Chinese)
62. Luttik, M.L. Family Nursing: The family as the unit of research and care. *Eur. J. Cardiovasc. Nurs.* **2020**, *19*, 660–662. [CrossRef] [PubMed]
63. Zhai, Y.; Hu, M. Association of empty nest with depressive symptom in a Chinese elderly population: A cross-sectional study. *J. Affect. Disord.* **2015**, *187*, 218–223. [CrossRef]
64. Ge, D.; Chu, J.; Zhou, C.; Qian, Y.; Zhang, L.; Sun, L. Rural-urban difference in the use of annual physical examination among seniors in Shandong, China: A cross-sectional study. *Int. J. Equity Health* **2017**, *16*, 86. [CrossRef]
65. Sun, X.; Chen, Y.; Tong, X.; Feng, Z.; Wei, L.; Zhou, D.; Tian, M.; Lv, B.; Feng, D. The use of annual physical examinations among the elderly in rural China: A cross-sectional study. *BMC Health Serv. Res.* **2014**, *14*, 16. [CrossRef]
66. Xu, L.J.; Meng, Q.; He, S.W.; Yin, X.L.; Tang, Z.L.; Bo, H.Y.; Lan, X.Y. The effects of health education on patients with hypertension in China: A meta-analysis. *Health Educ. J.* **2014**, *73*, 137–149. [CrossRef]
67. Liu, S.; Liu, Y.; Zhang, T.; Wang, L.; Huang, J.; Liang, H.; Chen, G.; Liu, C.; Zhang, Y. The developing family doctor system: Evidence from the progress of the family doctor signing service from a longitudinal survey (2013–2016) in Pudong New Area, Shanghai. *BMC Fam. Pract.* **2021**, *22*, 11. [CrossRef] [PubMed]
68. Meydanlioglu, A.; Akcan, A.; Oncel, S.; Adibelli, D.; Cicek Gumus, E.; Sarvan, S.; Kavla, I. Prevalence of obesity and hypertension in children and determination of associated factors by CHAID analysis. *Arch. Pediatr.* **2021**, in press. [CrossRef] [PubMed]

69. Dongn, S.; Dong, Q.; Chen, H. Mothers' Parenting Stress, Depression, Marital Conflict, and Marital Satisfaction: The Moderating Effect of Fathers' Empathy Tendency. *J. Affect. Disord.* **2022**, *299*, 682–690. [CrossRef]
70. Crandall, A.; Novilla, L.K.B.; Hanson, C.L.; Barnes, M.D.; Novilla, M.L.B. The Public Health Family Impact Checklist: A Tool to Help Practitioners Think Family. *Front. Public Health* **2019**, *7*, 331. [CrossRef]
71. Woodman, J.; Simon, A.; Hauari, H.; Gilbert, R. A scoping review of 'think-family' approaches in healthcare settings. *J. Public Health* **2020**, *42*, 21–37. [CrossRef]
72. Samanta, T.; Chen, F.; Vanneman, R. Living Arrangements and Health of Older Adults in India. *J. Gerontol. Ser. B Psychol. Sci. Soc. Sci.* **2014**, *70*, 937–947. [CrossRef]
73. United Nations. Transforming Our World: The 2030 Agenda for Sustainable Development. 2015. Available online: https://sustainabledevelopment.un.org/post2015/transformingourworld/publication (accessed on 20 December 2021).
74. Hu, C.; Yu, W.; Lv, Y.; Chen, H.; Deng, Q.; Zhang, L. Study on the Health Status and Health Service Utilization of the Elderly of a Remote and Poor Village in a Mountainous Area in Jinzhai, Anhui. *Int. J. Environ. Res. Public Health* **2017**, *14*, 408. [CrossRef]
75. Li, Z.; Zhang, L. Poverty and health-related quality of life: A cross-sectional study in rural China. *Health Qual. Life Outcomes* **2020**, *18*, 153. [CrossRef]

International Journal of
Environmental Research and Public Health

MDPI

Article

The Paradoxical Effect of Living Alone on Cognitive Reserve and Mild Cognitive Impairment among Women Aged 60+ in Mexico City

Antonio R. Villa [1], Elsa Guerrero [2], Ana M. Villa [1], Rosalinda Sánchez-Arenas [3], María Araceli Ortiz-Rodríguez [4], Vania Contreras-Sánchez [5], María Alonso-Catalán [5], Benjamín Guerrero-López [5], Ingrid Vargas-Huicochea [5] Germán E. Fajardo-Dolci [6] and Claudia Díaz-Olavarrieta [5,*]

1 Research Division, Faculty of Medicine, National Autonomous University of Mexico, 3000 Ave. Universidad, Copilco Universidad, Coyoacán, Mexico City 04510, Mexico; avilla@unam.mx (A.R.V.); nut.avilla@gmail.com (A.M.V.)
2 University Program of Health Research, National Autonomous University of Mexico, 3000 Ave. Universidad, Copilco Universidad, Coyoacán, Mexico City 04510, Mexico; elsaguerrero2013@gmail.com
3 Epidemiology and Health Services Research Unit, CMN Siglo XXI, Mexican Institute of Social Security, Mexico City 06720, Mexico; felicitasarenas@gmail.com
4 Faculty of Nutrition, Autonomous University of the State of Morelos, Cuernavaca 62350, Mexico; araceli.ortiz@uaem.mx
5 Department of Psychiatry and Mental Health, Faculty of Medicine, National Autonomous University of Mexico, 3000 Ave. Universidad, Copilco Universidad, Coyoacán, Mexico City 04510, Mexico; vaniacontreras@comunidad.unam.mx (V.C.-S.); mariaalonsocatalan@gmail.com (M.A.-C.); jguerrero@facmed.unam.mx (B.G.-L.); vargashuicochea.ingrid@gmail.com (I.V.-H.)
6 Faculty of Medicine, National Autonomous University of Mexico, 3000 Ave. Universidad, Copilco Universidad, Coyoacán, Mexico City 04510, Mexico; german.fajardo@unam.mx
* Correspondence: dra_claudiadiazolavarrieta@facmed.unam.mx

Citation: Villa, A.R.; Guerrero, E.; Villa, A.M.; Sánchez-Arenas, R.; Ortiz-Rodríguez, M.A.; Contreras-Sánchez, V.; Alonso-Catalán, M.; Guerrero-López, B.; Vargas-Huicochea, I.; Fajardo-Dolci, G.E.; et al. The Paradoxical Effect of Living Alone on Cognitive Reserve and Mild Cognitive Impairment among Women Aged 60+ in Mexico City. *Int. J. Environ. Res. Public Health* **2021**, *18*, 10939. https://doi.org/10.3390/ijerph182010939

Academic Editors: Francisco José Tarazona Santabalbina, Sebastià Josep Santaeugènia Gonzàlez, José Augusto García Navarro and José Viña

Received: 19 September 2021
Accepted: 8 October 2021
Published: 18 October 2021

Abstract: An elderly person who lives alone must often be autonomous and self-sufficient in daily living activities. We explored if living alone and marital status were associated with mild cognitive impairment and low cognitive reserve in a sample of Mexican women aged 60+ attending continuing education courses using a cross-sectional design. Objective cognitive functions were assessed using the MMSE and Blessed Dementia Scale. We administered the Cognitive Reserve Questionnaire. Independence skills were assessed with the Katz index and Lawton index. Multivariate logistic regression analysis was used. We recruited 269 participants ($\bar{x} = 69.0 \pm 5.8$ years). Single, widowed, separated, and divorced women comprised 73% of the participants. A third lived alone and 84% had completed high school. Mild cognitive deficit was observed among 24.5–29.0%; the upper range for cognitive reserve was 61.7%. Living alone versus living with someone was associated with cognitive impairment (OR = 0.51, $p = 0.04$) and with low to medium cognitive reserve (OR = 0.51, $p = 0.02$) after adjusting for confounding variables. Living alone was an independent factor associated with a lower probability of displaying mild cognitive impairment and a higher probability of displaying high cognitive reserve. Women living alone in this study had a more robust cognitive framework and had built their own support networks.

Keywords: cognitive reserve; living alone; cognitive impairment; elderly women; Mexico

1. Introduction

In 2000, Mexico's annual growth rate among the elderly was 5.1%, a figure that if kept stable would place the current older adult population at 7.6%. If this were to double every 19 years, by 2050, this age group will represent 21% of the total population [1]. In addition, for every 100 young adults (under 15 years), Mexico City will have approximately 209.7 inhabitants aged 65 and over [2].

Among the elderly, a supportive social network has a positive effect on their emotional wellbeing and self-esteem [3,4]. Lack of social support has also been documented as a consistent risk factor for poor health and quality of life among those who live alone [5]. However, a person who lives alone may also need to be autonomous and self-sufficient in daily living activities, both basic and instrumented [6,7]; the latter requires the need to develop more robust memory skills, which in turn brings about a higher cognitive reserve [8,9]. Cognitive reserve (CR) is defined as the ability a person has to adapt to age-related brain changes or damage caused by certain brain pathologies [10]. It is a difficult to measure construct; education and verbal intelligence have been widely used as proxies of CR, since direct measurement of an individual' s CR is elusive [11]. CR is construed as an active skill because it is dynamic and can be modified by circumstances and cognitive experiences. It is determined by several modifiable factors, including occupational attainment, engagement in leisure activities, physical activity, social engagement, or brain-challenging tasks. CR has been shown to mitigate the negative impact of aging on cognitive function and it is worth noting that there are modifiable factors that can increase it [12].

Gender, a history of chronic disease, displaying cognitive and functional impairment, personality traits, experiencing stressful life events, family history of depression, experiencing sensory losses, social losses, insufficient physical activity, increased levels of dependency, lack of support, and loneliness are among the predisposing factors associated with depression [13]. Caring for an ailing partner has also been shown to have an impact on the burden of care, as this increases over time. This burden is greater for female than for male caregivers. However, psychosocial stressors also increase the severity of caregiving activities among both genders. [14,15].

Some individuals are more vulnerable to developing feelings of loneliness in response to environmental triggers than others [16]. Elderly men face widowhood under more challenging circumstances compared to elderly women, especially in traditional contexts such as Mexico's, where the burden of household chores falls mainly on women and where a generational effect is also markedly prevalent [17–19]. A widowed woman may have the ability and resources to acquire the adaptative tools that living alone entails [20,21]. It is therefore important to make a distinction between experiencing loneliness and living alone. The first represents a feeling of abandonment and lack of contact with someone, while the second implies autonomy. Loneliness has been associated with an increased predisposition to illness, greater cognitive decline, and even premature mortality. On the other hand, people living alone can seek and receive emotional support outside the home and build their own support networks [22,23].

We explored if living alone in a megacity such as Mexico City (population: 21,782,000) or living with someone was associated with displaying mild cognitive impairment and low cognitive reserve. We surveyed a sample of Mexican women aged 60 and older who had on average completed high school or college.

2. Materials and Methods

We carried out a cross-sectional study among elderly women who were enrolled in the University of the Elderly (*Universidad de la Tercera Edad*), a public educational institution in Mexico City that offers continuing education courses. During 2017–2019, we recruited subjects in a consecutive way as they agreed to participate in the study and signed a written informed consent form. All interviews were conducted face to face. The approximate student enrollment at the time the study was fielded was estimated at 2000. The main reason women declined to participate was that the study instruments took on average 1.5 h to administer and they did not have sufficient time to respond.

The inclusion criteria were being 60 years or older, being female, not being totally deaf or blind, and not presenting with severe mental illness (i.e., schizophrenia) or cognitive decline (i.e., dementia). Participants needed to be able to respond to the survey questions by themselves without the help of a carer (caregiver) or proxy. Those who did not comply, did not complete the survey, or did not agree to participate were excluded.

The sociodemographic variables in the survey included age, marital status (single, married, cohabitating, widowed, separated, or divorced), the person or people that the participant was living with (alone, with a partner, with offspring, with another relative or a non-relative), current occupation (unemployed, homemaker, worker, retired, or receiving public funds (*pensionado*), and other), the financial independence of the participant (does or does not get any financial support), comorbidities (overweight, obesity, diabetes, systemic arterial hypertension, vascular cerebral disease, chronic pulmonary disease, hypothyroidism, history of cancer, current cancer, chronic renal disease, or depression).

Memory Cognitive Reserve was measured using the Cognitive Reserve Questionnaire with the following dimensions [24,25]: (i) academic background (primary, junior, and senior high-school, college/graduate school), (ii) parents' academic background (can read and write, junior and high-school or college/graduate school), (iii) continuing education courses (did not attend, attended 1 to 2, 3 to 5, and >5), (iv) prior occupation (administrative, management, and executive level), (v) musical training (does not play at all, plays a little, received formal musical training), (vi) languages (speaks 1 or more aside from mother tongue, 2, 3, >3),(vii) reading habit measured by number of books read per year (never, occasionally, 2–5 a year, 6–10 a year, >10 a year), and (viii) plays memory-challenge games (never, occasionally, regularly). The final cognitive reserve classification comprised four categories: low range (≤6 points), low-medium range (7–9 points), high-medium range (10–14 points), and upper range (≥15 points). In this study, we added the lowest three categories that were considered as having low cognitive reserve: the lower range in addition to the medium-low range plus the medium-upper range (≤14 points). These values were compared with the upper range category (≥15 points). We decided this because the figures in the first categories were few and we had to create a dichotomous variable.

Objective cognitive functions were assessed using (i) Folstein´s Mini Mental State Evaluation (MMSE), with the following cut-off scores: normal ≥27, suspected cognitive impairment 25–26, and mild cognitive impairment ≤24 [26,27]; (ii) the Blessed Dementia Scale, which measures changes in daily living skills and personal habits (in personality and behavior it is also used to screen for dementia, with the following cut-off scores: normal (<4), mild impairment (4 to 9), moderate (from 10 to 14), and severe (>15) [28]); and (iii) the Clock Test, which assesses visuospatial abilities, visual motor memory programming, and other abilities with the following cut-off scores: unaltered (0 mistakes), slightly altered (1–2 mistakes), moderate (3–4 mistakes), and severe (5 or more mistakes). This test has been shown to correlate with scores in the MMSE [29].

Basic activities of daily living (BADL) were measured using the Katz index as preserved (score = 6 shows total self-feeding, mobility, continence, toileting, dressing, and bathing abilities) or altered in one or more functions [30]. We administered the Lawton instrumental activities of daily living (IADL), which assesses the ability to use the phone, go shopping, prepare meals, perform household chores, use public transportation, and handle prescription drugs and money, and is classified as total independence (=8), slight (6–7), moderate (4–5), mild (2–3), and total dependency (0–1) [31]. In addition to this, gait speed was calculated by measuring the time in seconds it took the patient to walk 8 m in a straight line (4 m back and forth) [32,33]·

The statistical analysis included the description of continuous variables through the mean and standard deviation as well as categorical variables through relative frequencies. The comparison of independent variables (sociodemographic, comorbidities, cognitive functions, functionality, and gait speed) among women with low cognitive reserve compared with participants that scored in the normal range in cognitive reserve was performed through logistic regression analysis deriving odds ratios (OR) by the exponential of regression coefficients. We reported the values of statistical significance with 95% confidence intervals and p-values. Multivariate models were built to test the main effect of living alone versus living with someone on cognitive impairment and cognitive reserve after adjusting by confounding variables. All analyses were carried out with SPSS/PC v25.0. The research protocol was submitted and registered by the Institutional Review Board (Research and

Ethics Committee) at the Faculty of Medicine of the National Autonomous University of Mexico (FMED/CI/GRD/014/2011).

3. Results

We collected information on 269 women aged 60 and older. The mean age was 69.0 ± 5.8 years (minimum 60 to maximum 86). From the total sample, women that were single, widowed, separated, and divorced comprised 73%. A third of the participants (33%) lived alone, and this was our main effect variable to test. A total of 225 women (84%) had a high level of schooling (had completed high school, college, or a graduate degree). They were currently retired (58%) or homemakers (32%) and the majority received non-governmental financial aid (87%). Their prior work history included administrative (33%), management (32%), and executive level jobs (27%). Among the prevalence of chronic diseases, overweight and obesity were the most common (both added up to 71%), followed by hypertension (36%), hypothyroidism (17%), and diabetes (13%). This latter figure is lower than the one expected for their age group in Mexico. The prevalence of mild or suspected cognitive impairment on the MMSE was 29%. The prevalence of cognitive impairment (low, moderate, and severe) with the Blessed Orientation Memory Concentration Test was 24.5%. According to the Cognitive Reserve Questionnaire, 62% of women scored in the highest category (upper range) of cognitive reserve. Two-thirds reported reading as their main hobby. Over 40% of the participants frequently solved memory-challenge games, more than half played a musical instrument, and a third spoke two or more languages. It is worth noting the percentage of women that fulfilled Fried's criteria for prefrailty (76%) and frailty (13%) despite having a reasonably sound health status. The percentage of women with total independence for instrumental activities of daily living was 89%, while 36% reported some degree of dependence when carrying out basic activities of daily living (see Table 1).

Table 1. Characteristics of functionally independent study participants aged 60 years.

Variable	***n* = 269**	
Age (years)	69.0 ± 5.8	
BMI (kg/m^2)	27.1 ± 3.9	
Gait speed test for 8 m (seconds)	4.6 ± 1.1	
	N°	%
Marital status		
Single	109	40.5
Married or common-law	72	26.8
Widowed	64	23.8
Separated	11	4.1
Divorced	13	4.8
Who does she currently live with?		
Alone	88	33
Partner	71	26.6
Children	78	29.2
Other: relative	24	9
Other: non-relative	6	2.2
Did not respond	2	-
Level of schooling		
Elementary and middle school	44	16.4
High school	96	35.7
College	113	42
Graduate school	16	5.9

Table 1. *Cont.*

	N°	%
Current occupation		
Retired	156	58.2
Homemaker	86	32.1
Self employed	17	6.3
Unemployed	3	1.1
Other	7	2.2
Financial dependency		
Does not get aid	34	12.7
Gets aid	234	87.3
Did not respond	1	-
Comorbidity		
Overweight	135	50.8
Obesity	53	19.9
Diabetes	34	12.7
Hypertension	98	36.4
Cerebrovascular disease	9	3.3
Chronic obstructive pulmonary disease	11	4.1
Hypothyroidism	45	16.7
Cancer	28	10.5
Chronic kidney disease	5	1.9
Depression	15	5.6
Exhaustion		
Fatigue (by self-report)	37	13.8
Cognitive Reserve Questionnaire (CRQ)		
Level of schooling		
Elementary-middle school	44	16.4
High-School	96	35.7
College or Graduate school	129	47.9
Parents´ level of schooling		
Can read and write	19	7.1
Junior high-school	137	50.9
High-school or College	113	42
Attended continuing education courses		
None	51	19
1 to 2	7	2.6
3 to 5	23	8.6
>5	188	69.9
Prior occupation (work history)		
Administrative work	111	41.3
Middle management	85	31.6
Executive level	73	27.1
Musical training		
Does not play at all	118	43.9
Plays a little	140	52
Formal musical training	11	4.1
Languages		
1 (mother tongue only)	179	66.5
2	78	29
3	7	2.6
>3	5	1.9

Table 1. *Cont.*

	N°	%
Reading activity		
Never	33	12.3
Occasionally	56	20.8
2–5 books in one year	106	39.4
6–10 books in one year	45	16.7
>10 books in one year	29	10.8
Memory-challenge games		
Never plays	78	29
Occasionally	81	30.1
Regularly	110	40.9
Cognitive Reserve Questionnaire (CRQ)		
		15.2 ± 3.2
Score	1	0.4
	16	5.9
Low range	86	32
Medium-low range	166	61.7
Medium-high range		
Upper range		
Cognitive Function (Mini Mental State Evaluation)		
Score		
Normal		27.3 ± 2.1
Suspected cognitive impairment	191	71
Mild cognitive impairment	53	19.7
	25	9.3
Blessed Dementia Scale		
Score		2.02 ± 3.2
Normal	203	75.5
Mild impairment	56	20.8
Moderate impairment	8	3
Severe impairment	2	0.7
Clock-Test		
Unaltered	184	68.4
Mild alteration	50	18.6
Moderate alteration	30	11.2
Severe alteration	5	1.9
Fried's frailty criteria		
Not frail	28	10.4
Prefrail	205	76.2
Frail	36	13.4
Basic activities of daily living (Katz index)		
Total independence	171	63.6
Some dependence	98	36.4
Instrumental activities of daily living (Lawton index)		
Total independence	240	89.2
Slight dependency	28	10.4
Moderate dependency	1	0.4
Moderate dependency		

Living alone had a statistical trend of association with a lower probability of displaying mild, moderate, or severe impairment as measured by the Blessed Dementia Scale (OR = 0.52, p = 0.05) as well as with a lower probability of having cognitive impairment or suspected cognitive impairment as measured by the MMSE (OR = 0.57, p = 0.07). The above results seem to indicate that a woman in our study who lived alone had 92% (inverse value of 0.52) and 75% (inverse value of 0.57), respectively, lower probability of having some degree of cognitive impairment compared with a woman who lived with someone. Furthermore, living alone was also statistically associated with a lower probability of having a low, medium-low, or medium-high range of cognitive reserve (OR = 0.52, $p = 0.02$), namely, a 92% lower probability of displaying decreased cognitive reserve. The clock test did not yield statistical significance (see Table 2).

Table 2. Association between living alone, displaying cognitive impairment (MMSE and Blessed), and cognitive reserve.

Dependent Variable	Living Alone N = 88		Living with Someone N = 181		OR	95% CI	*p*-Value
	N°	%	N°	%			
MMSE Mild cognitive impairment or suspected cognitive impairment	19	21.6	59	32.6	0.57	0.31–1.03	0.07
Blessed Dementia Scale Mild, moderate, or severe impairment	15	17	51	28.2	0.52	0.28–0.99	0.05
Cognitive Reserve Questionnaire Low, medium-low, or medium-high range	25	28.4	78	43.1	0.52	0.30–0.91	0.02
Clock test Mild, moderate, or severe alteration	24	27.3	61	33.7	0.74	0.42–1.29	0.33

When we compared the variables associated with social support, mild or moderate physical activities including walking outside the home, perception of satisfaction and memory status compared to their peers, and the prevalence of comorbidity, we did not find statistically significant differences among women who lived alone compared to those who lived with someone. There was a trend towards being more physically active (walking outside the home or gardening) among women who lived with someone and less physically active among women who lived alone. However, there was a significant difference in the percentage of women who lived alone and were primary caregivers (8%) compared to women who lived with someone and cared for someone (21%) ($p = 0.008$) (see Table 3).

Table 3. Association between living alone or with someone, social and physical activities, and comorbidity variables.

Variable	Lives Alone *n* = 88		Lives with Someone *n* = 181		*p*-Value
	N°	%	N°	%	
Do you have someone to take care you?	2	2.3	5	2.8	1.00
Do you receive financial aid?	79	89.8	155	86.1	0.44
Do you carry out activities outside the home?	85	97.7	177	98.3	0.66
Do you feel satisfied with life?	77	87.5	160	88.4	0.84
Do you prefer to stay home rather than go out?	14	15.9	34	18.8	0.61
Do you think you have more memory problems than most people?	9	10.2	14	7.7	0.49
Do you think other people have a better sense of well being compared to you?	3	3.4	7	3.9	1.00
In the last 7 days, did you carry out activities outside the home?	83	95.4	179	98.9	0.09
In the last 7 days, did you practice gardening at home?	19	21.8	60	33.1	0.06
In the last 7 days, were you someone else's primary caregiver?	7	8.0	38	21.0	0.008
Do you suffer from hypertension?	29	33.0	69	38.1	0.42
Do you suffer from diabetes?	7	8.0	27	15.0	0.12
Do you suffer from chronic kidney disease?	3	3.4	2	1.1	0.34
Do you suffer from cerebrovascular disease?	3	3.4	6	3.3	1.00
Do you suffer from chronic obstructive pulmonary disease?	5	5.7	6	3.3	0.35
Do you suffer from hypothyroidism?	15	17.0	30	16.6	1.00

For further analysis, we tested the main effect of low cognitive reserve associated with the probability of displaying cognitive impairment with an OR = 1.9, 95% CI: 1.05–3.43, and $p = 0.03$, regardless of the effect of age, living alone, functional dependence (measured by the Lawton scale), schooling (had completed college or more vs. an incomplete college degree or less), and caring for someone.

Table 4 shows that living alone is an independent factor associated with a lower probability of displaying mild, moderate, and severe impairment, and low, medium-low, and medium-high cognitive reserve. The main effect of living alone was adjusted in multivariate models by age (years), the number of Fried's criteria for frailty syndrome, history of stroke, fatigue (by self-report), and gait speed (seconds).

Table 4. Multivariate models to test the main effect of living alone on cognitive impairment (MMSE and Blessed) and cognitive reserve.

Model 1—MMSE *			**Model 2—Blessed ***			**Model 3—CRQ ***		
Dependent variable: mild or suspected cognitive impairment vs. no cognitive impairment			Dependent variable: cognitive impairment (mild, moderate, and severe) vs. no cognitive impairment			Dependent variable: low, medium-low, and medium-high cognitive reserve vs. high cognitive reserve		
Main effect: living alone vs. living with someone			Main effect: living alone vs. living with someone			Main effect: living alone vs. living with someone		
OR	95% CI	*p*-value	OR	95% CI	*p*-value	OR	95% CI	*p*-value
0.57	0.31–1.03	0.06	0.51	0.26–0.97	0.04	0.51	0.29–0.89	0.02

* All models were adjusted for age (years) and number of Fried's criteria for frailty syndrome, history of stroke, fatigue (by self-report), and gait speed (seconds).

4. Discussion

Our results show that in our sample of Mexican elderly women, living alone is an independent factor associated with a lower probability of displaying mild cognitive impairment as assessed by two standardized international tests and with a higher probability of showing high cognitive reserve. When we tested the main effect of low cognitive reserve on the probability of displaying cognitive impairment, we found that it was a risk factor independently associated with different covariates, including living alone. Women that were single, widowed, separated, or divorced did not show an association with the same variables of cognitive decline or cognitive reserve. In our study, marital status did not necessarily imply women lived alone; for example, they could report being widowed and still live with a family member. However, women that reported being widowed, divorced, and separated did not have a higher risk of displaying mild cognitive impairment compared with married women. We may speculate that a woman who lives alone comprises one factor representing different dimensions [20]. They are able and used to carrying out household chores by themselves, and they may have overcome widowhood and/or divorce with perhaps more resilience than men [19,20,34,35]. These conditions may favor their sense of autonomy, independence, and functionality, and thus increase their cognitive reserve and decrease their risk of developing cognitive impairment [8,36].

Different social theories have been put forth to explain the ageing process [37]. Successful ageing and healthy ageing are two well-known examples. However, in 2002 the World Health Organization (WHO) made a call to consider another construct, that of active ageing [38]. Under this paradigm, active ageing entails the optimization of resources and opportunities to further a person's health which in turn will translate into an enhanced quality of life as we age. Because ours is an aging world, it is of paramount importance that we find the mechanisms that allow people not to stagnate. Societies and health care systems need to recognize what their aging population's needs are and satisfy them based on the complexities involved in growing old. Furthermore, this process could potentiate different abilities, including physical, social, and mental, both at the individual and collec- [illegible] results seem to highlight the ways in which social participation becomes relevant for women and men to build supportive social networks by attending continuing education courses with a curriculum that also included arts and drama. As Teater and Chonody illustrated [37], active ageing is multidimensional and includes "being active and participating in social, economic, cultural, spiritual, and civic issues". Therefore, our findings seem to indicate that women in our sample adhere to WHO's definition of active, successful, and healthy ageing. There is evidence to suggest that social isolation is also associated with higher mortality (more prevalent among widowed men than widowed women) and that the effect of long-term isolation increases the risk of depression and dementia. As an older person ages, their social networks decrease, reducing their cognitive stimulus and lowering their cognitive reserve. Thus, the lack of social interaction together with loneliness have negative effects on cognitive reserve and impairment. This interaction could be mediated by the capacity for neuronal plasticity in analogous pathways at the level of the hippocampus and the prefrontal region [39].

Our study has several limitations: ours is a sample with characteristics that differ from the "average" Mexican woman aged 60 and older. According to the 2020 National Population and Housing Census, there was a total of 8,276,286 women aged 60+. Of them, only 11% (892,107) had completed higher education (college or graduate school). In our study, participants that comprised this level of educational attainment was 47.9% [40]. It was also a sample of reasonably healthy elderly women whose prevalence of chronic diseases was lower than the that reported at the national level for similarly aged women. According to the 2018–19 National Survey of Nutrition and Health of Mexico [41]: the prevalence of diabetes was 27.1% versus 12.7% in our study, hypertension was 47.8% versus 36.4%, overweight was 38.7% versus 50.8%, and obesity was 33.0% versus 19.9%. We also observed a difference in the functional dependence in instrumental activities of daily living of our participants compared to the one reported in the 2012 National Survey of Nutrition and Health of Mexico [42]: 10.8% versus 28.4%. However, our study sample had a higher prevalence in dependence of basic activities of daily living than reported in the latter survey: 36.4% versus 29.6%. It is also important to note that our subject's mean age was 69 years. These factors may have influenced their memory skills and high cognitive reserve. Nevertheless, we were able to find statistically significant differences between women living alone versus women who reported living with someone as an associated independent factor. It is also possible that women with more independence may have had prior professional development that contributed to this, and while in our study only seven participants reported having an executive level job, most held professional positions (administrative or middle management); thus, we assume the probability of bias is low. However, we are aware that our results cannot be generalized to older women in Mexico City since ours is a selected sample with the characteristics described above. Our choice of screening instruments—the MMSE, the Blessed Dementia Scale, and the Clock Test, in terms of its sensitivity to detect cognitive impairment in a population without severe memory pathology—have been described in the literature and could be considered another study limitation. Our study design (cross-sectional) did not allow us to accurately distinguish whether women who lived with someone displayed cognitive decline or were physically frail and thus required assistance from a caregiver, unlike women who lived alone who were healthier and more independent. However, we think this possibility is not supported by our results, which showed that women who lived alone were not different from those who lived with someone in terms of sociodemographic characteristics, social support, perception of health, and comorbidities. In daily routines such as physical activities (walking outside the home, gardening, etc.) women who lived alone showed a trend towards statistical significance in performing better than women who lived with someone. Being a primary caregiver also stood out when we compared both groups of women; it was higher (21%) among women who lived with someone compared to those who lived alone (8%). This finding suggests that women who live with someone and are primary caregivers must be physically and cognitively fit, so this may not be a group displaying a bias towards a greater deficit in overall functioning.

5. Conclusions

While our sample of elderly women had higher levels of schooling compared to Mexico's national average of 9.2 years, [43] had a low prevalence of some of the most common and epidemic chronic conditions that plague this country (diabetes, overweight, and obesity) [41,42], and belonged to a middle income bracket, living alone was an independent factor associated with a lower probability of displaying mild cognitive impairment assessed with a higher probability of showing high cognitive reserve. Our study participants could be construed as a "privileged" sample in terms of social determinants of health; however, they were also self-motivated and disciplined to further their education by attending continuing education courses at a public university. They played a musical instrument, spoke more than one language, read often, played memory-challenge games, and most probably created a robust social network while attending these intellectual and artistic activities. When we enquired if they received financial aid from the government (in the form of a pension), a very small percentage did (13%), which meant they perhaps led frugal lives. Since we did not find statistical differences between women who lived alone and those who lived with someone in terms of the prevalence of chronic diseases and activity outside the home, except for having cared for someone in the last week, we can assume that the former group of women displayed a more robust cognitive framework and were able to build their own support networks.

Author Contributions: Conceptualization, E.G., A.M.V., R.S.-A., M.A.O.-R., B.G.-L. and C.D.-O.; data curation, A.R.V., M.A.O.-R. and B.G.-L.; formal analysis, A.R.V., A.M.V., R.S.-A. and M.A.O.-R.; funding acquisition, C.D.-O.; research, R.S.-A., M.A.O.-R., B.G.-L. and C.D.-O.; methodology, A.R.V., E.G. and M.A.O.-R.; project administration, A.R.V., V.C.-S., I.V.-H., C.D.-O. and M.A.-C.; resources, B.G.-L., I.V.-H. and G.E.F.-D.; writing—original draft, V.C.-S. and C.D.-O.; writing—review and editing, V.C.-S., I.V.-H., M.A.-C. and C.D.-O. All authors have read and agreed to the published version of the manuscript.

Funding: This work was partially supported by PAPIIT (Programa de Apoyo a Proyectos de Investigación e Innovación Tecnológica, PAPIIT # IA304221, Diaz-Olavarrieta C, Villa AR).

Institutional Review Board Statement: The research protocol was submitted and registered by the Institutional Review Board (Research and Ethics Committee) at the Faculty of Medicine of the National Autonomous University of Mexico (FMED/CI/GRD/014/2011).

Informed Consent Statement: We recruited subjects in a consecutive way as they agreed to parti cipate in the study and signed a written informed consent form.

Acknowledgments: The authors wish to extend their gratitude to the women who kindly participated in this study and to the University of the Elderly (Universidad de la Tercera Edad) for allowing us to carry out data collection.

Conflicts of Interest: The authors declare no conflict of interest.

References

1. Partida-Bush, V. Proyecciones de la Población de México 2005–2050. Consejo Nacional de Población y Vivienda. México. 2006. Available online: http://www.alianzacivica.org.mx/guia_transparencia/Files/pdf/desarrollo/14_PROYECCIONESDELAPOBLACIONDEMEXICO/14_PROYECCIONESDELAPOBLACIONDEMEXICO.pdf (accessed on 29 May 2021).
2. Consejo Nacional de Población. Cuadernillos Estatales de las Proyecciones de la Población de México y de las Entidades Fe-derativas, 2016–2050, República Mexicana. Primera edición. July 2019; ISBN 978-607-427-320-5. Available online: https://www.gob.mx/conapo/documentos/cuadernillos-estatales-de-las-proyecciones-de-la-poblacion-de-mexico-y-de-las-entidades-federativas-2016-2050-208243?idiom=es (accessed on 16 July 2021).
3. Cornwell, B.; Laumann, E.O. The health benefits of network growth: New evidence from a national survey of older adults. *Soc. Sci. Med.* **2015**, *125*, 94–106. [CrossRef] [PubMed]
4. Werner-Seidler, A.; Afzali, M.H.; Chapman, C.; Sunderland, M.; Slade, T. The relationship between social support networks and depression in the 2007 National Survey of Mental Health and Well-being. *Soc. Psychiatry Psychiatr. Epidemiol.* **2017**, *52*, 1463–1473. [CrossRef] [PubMed]

5. Kim, J.; Lee, J.-E. Social Support and Health-Related Quality of Life Among Elderly Individuals Living Alone in South Korea: A Cross-Sectional Study. *J. Nurs. Res.* **2018**, *26*, 316–323. [CrossRef] [PubMed]
6. Sun, W.; Watanabe, M.; Tanimoto, Y.; Shibutani, T.; Kono, R.; Saito, M.; Usuda, K.; Kono, K. Factors associated with good self-rated health of non-disabled elderly living alone in Japan: A cross-sectional study. *BMC Public Health* **2007**, *7*, 297. [CrossRef] [PubMed]
7. Henning-Smith, C.; Gonzales, G. The Relationship Between Living Alone and Self-Rated Health Varies by Age: Evidence From the National Health Interview Survey. *J. Appl. Gerontol.* **2019**, *39*, 971–980. [CrossRef]
8. Cabeza, R.; Albert, M.; Belleville, S.; Craik, F.I.M.; Duarte, A.; Grady, C.L.; Lindenberger, U.; Nyberg, L.; Park, D.C.; Reuter-Lorenz, P.A.; et al. Maintenance, reserve and compensation: The cognitive neuroscience of healthy ageing. *Nat. Rev. Neurosci.* **2018**, *19*, 701–710. [CrossRef]
9. Stern, Y.; Barnes, C.A.; Grady, C.; Jones, R.N.; Raz, N. Brain reserve, cognitive reserve, compensation, and maintenance: Operationalization, validity, and mechanisms of cognitive resilience. *Neurobiol. Aging* **2019**, *83*, 124–129. [CrossRef]
10. Stern, Y. Cognitive reserve in ageing and Alzheimer's disease. *Lancet Neurol.* **2012**, *11*, 1006–1012. [CrossRef]
11. Kwak, S.; Shin, M.; Kim, H.; Cho, B.; Ha, J.H.; Han, G.; Kim, H.; Koo, Y.; Kwon, S.; Lee, C.; et al. Moderating effect of cognitive reserve on the association between grey matter atrophy and memory varies with age in older adults. *Psychogeriatrics* **2020**, *20*, 87–95. [CrossRef]
12. Amanollahi, M.; Amanollahi, S.; Anjomshoa, A.; Dolatshahi, M. Mitigating the negative impacts of aging on cognitive function; modifiable factors associated with increasing cognitive reserve. *Eur. J. Neurosci.* **2021**, *53*, 3109–3124. [CrossRef]
13. Isik, K.; Başoğul, C.; Yildirim, H. The relationship between perceived loneliness and depression in the elderly and influencing factors. *Perspect. Psychiatr. Care* **2021**, *57*, 351–357. [CrossRef]
14. Swinkels, J.C.; Broese van Groenou, M.I.; de Boer, A.; Tilburg, T.G.V. Male and female partner-caregivers' burden: Does it get worse over time? *Gerontologist* **2019**, *16*, 1103–1111. [CrossRef]
15. Swinkels, J.; Tilburg, T.V.; Verbakel, E.; Broese van Groenou, M. Explaining the gender gap in the caregiving burden of partner caregivers. *J. Gerontol. B Psychol. Sci. Soc. Sci.* **2019**, *10*, 309–317.
16. O' Luanaighy, C.; Lawlorz, B.A. Loneliness and the health of older people. *Int. J. Geriatr. Psychiatry* **2008**, *23*, 1213–1221. [CrossRef] [PubMed]
17. Förster, F.; Pabst, A.; Stein, J.; Röhr, S.; Löbner, M.; Heser, K.; Miebach, L.; Stark, A.; Hajek, A.; Wiese, B.; et al. Are older men more vulnerable to depression than women after losing their spouse? Evidence from three German old-age cohorts (AgeDifferent.de platform). *J. Affect. Disord.* **2019**, *1*, 650–657. [CrossRef] [PubMed]
18. Monserud, M.A.; Wong, R. Depressive Symptoms among Older Mexicans: The Role of Widowhood, Gender, and Social Integration. *Res. Aging* **2015**, *37*, 856–886. [CrossRef]
19. Monserud, M.A. Marital Status and Trajectories of Depressive Symptoms among Older Adults of Mexican Descent. *Int. J. Aging Hum. Dev.* **2018**, *88*, 22–45. [CrossRef]
20. Hahn, E.A.; Cichy, K.E.; Almeida, D.M.; Haley, W. Time Use and Well-Being in Older Widows: Adaptation and Resilience. *J. Women Aging* **2011**, *23*, 149–159. [CrossRef]
21. King, B.M.; Carr, D.; Taylor, M.G. Depressive Symptoms and the Buffering Effect of Resilience on Widowhood by Gender. *Gerontol.* **2019**, *59*, 1122–1130. [CrossRef]
22. Shawa, R.J.; Cullena, B.; Grahama, N.; Lyalla, D.M.; Mackaya, D.; Okolieb, C.; Pearsalla, R.; Warda, J.; Johnb, A.; Smitha, D.J. Living alone, loneliness and lack of emotional support as predictors of suicide and self-harm: A nine year follow up of the UK Biobank cohort. *J. Affect. Disord.* **2021**, *279*, 316–323. [CrossRef]
23. O'Súilleabháin, P.S.; Gallagher, S.; Steptoe, A. Loneliness, Living Alone, and All-Cause Mortality: The Role of Emotional and Social Loneliness in the Elderly During 19 Years of Follow-Up. *Psychosom. Med.* **2019**, *81*, 521–526. [CrossRef] [PubMed]
24. Rami, L.; Valls-Pedret, C.; Bartrés-Faz, D.; Caprile, C.; Solé-Padullés, C.; Castellvi, M.; Olives, J.; Bosch, B.; Molinuevo, J.L. Cognitive reserve questionnaire. Scores obtained in a healthy elderly population and in one with Alzheimer's disease. *Rev. Neurol.* **2011**, *16*, 195–201.
25. Vásquez-Amézquita, M. Factores predictores de la reserva cognitiva en un grupo de adultos mayores. *Rev. Chil. Neuropsicol.* **2016**, *11*, 5–11.
26. Folstein, M.F.; Folstein, S.E.; McHugh, P.R. "Mini-mental state": A practical method for grading the cognitive state of patients for the clinician. *J. Psychiatr. Res.* **1975**, *12*, 189–198. [CrossRef]
27. Torres-Castro, S.; Mena-Montes, B.; González-Ambrosio, G.; Zubieta-Zavala, A.; Torres-Carrillo, N.; Acosta-Castillo, G.; Espinel-Bermúdez, M. Escalas de tamizaje cognitivo en habla hispana: Una revisión crítica. *Neurología* **2018**, in press. [CrossRef]
28. Blessed, G.; Tomlinson, B.E.; Roth, M. The Association between Quantitative Measures of Dementia and of Senile Change in the Cerebral Grey Matter of Elderly Subjects. *Br. J. Psychiatry* **1968**, *114*, 797–811. [CrossRef] [PubMed]
29. Manos, P.J.; Wu, R. The Ten Point Clock Test: A Quick Screen and Grading Method for Cognitive Impairment in Medical and Surgical Patients. *Int. J. Psychiatry Med.* **1994**, *24*, 229–244. [CrossRef]
30. Katz, S.; Ford, A.B.; Moskowitz, R.W.; Jackson, B.A.; Jaffe, M.W. Studies of illness in the aged. The index of adl: A standardized measure of biological and psychosocial function. *JAMA* **1963**, *21*, 914–919. [CrossRef]
31. Lawton, M.P.; Brody, E.M. Assessment of older people: Self-maintaining and instrumental activities of daily living. *Gerontologist* **1969**, *9*, 179–186. [CrossRef]

32. Varela-Pinedo, L.; Ortiz-Saavedra, P.J.; Chávez-Jimeno, H. Gait speed as an indicator of fragility in community-dwelling elders in Lima, Peru. *Rev. Esp. Geriatr. Gerontol.* **2010**, *45*, 22–25.
33. Graham, J.E.; Ostir, G.V.; Kuo, Y.-F.; Fisher, S.R.; Ottenbacher, K.J. Relationship Between Test Methodology and Mean Velocity in Timed Walk Tests: A Review. *Arch. Phys. Med. Rehabil.* **2008**, *89*, 865–872. [CrossRef]
34. Zebhauser, A.; Hofmann-Xu, L.; Baumert, J.; Häfner, S.; Lacruz, M.; Emeny, R.T.; Döring, A.; Grill, E.; Huber, D.; Peters, A.; et al. How much does it hurt to be lonely? Mental and physical differences between older men and women in the KORA-Age Study. *Int. J. Geriatr. Psychiatry* **2013**, *29*, 245–252. [CrossRef]
35. Springfield, S.; Qin, F.; Hedlin, H.; Eaton, C.B.; Rosal, M.C.; Taylor, H.; Staudinger, U.M.; Stefanick, M.L. Resilience and CVD-protective Health Behaviors in Older Women: Examining Racial and Ethnic Differences in a Cross-Sectional Analysis of the Women's Health Initiative. *Nutrients* **2020**, *12*, 2107. [CrossRef] [PubMed]
36. Petkus, A.J.; Resnick, S.M.; Rapp, S.R.; Espeland, M.A.; Gatz, M.; Widaman, K.F.; Wang, X.; Younan, D.; Casanova, R.; Chui, H.; et al. General and domain-specific cognitive reserve, mild cognitive impairment, and dementia risk in older women. *Alzheimer's Dement. Transl. Res. Clin. Interv.* **2019**, *5*, 118–128. [CrossRef]
37. Teater, B.; Chonody, J. Promoting Actively Aging: Advancing a Framework for Social Work Practice with Older Adults. *Fam. Soc. J. Contemp. Soc. Serv.* **2017**, *98*, 137–145. [CrossRef]
38. WHO. Active Ageing A Policy Framework [Online]. WHO/NMH/NPH/02.8. A Contribution of the World Health Organization to the Second United Nations World Assembly on Ageing, Madrid, Spain, April 2002. Available online: https://apps.who.int/iris/bitstream/handle/10665/67215/WHO_NMH_NPH_02.8.pdf (accessed on 29 May 2021).
39. Bzdok, D.; Dunbar, R.I. The Neurobiology of Social Distance. *Trends Cogn. Sci.* **2020**, *24*, 717–733. [CrossRef] [PubMed]
40. INEGI. Censo Nacional de Población y Vivienda. 2020. Available online: https://www.inegi.org.mx/programas/ccpv/2020/#Tabulados (accessed on 29 May 2021).
41. Shamah-Levy, T.; Vielma-Orozco, E.; Heredia-Hernandez, O.; Romero-Martinez, M.; Mojica-Cuevas, J.; Cuevas-Nasu, L.; Santaella-Castell, J.A.; Rivera-Dommarco, J. Encuesta Nacional de Salud y Nutricion 2018–19: Resultados Nacionales. Cuernavaca, Mexico: Instituto Nacional de Salud Publica. 2020. Available online: https://ensanut.insp.mx/encuestas/ensanut2018/doctos/informes/ensanut_2018_informe_final.pdf (accessed on 29 May 2021).
42. Gutierrez, J.P.; Rivera-Dommarco, J.; Shamah-Levy, T.; Villalpando-Hernandez, S.; Franco, A.; Cuevas-Nasu, L.; Romero-Martinez, M.; Hernandez-Avila, M. Encuesta Nacional de Salud y Nutricion. 2012. Resultados Nacionales. 2a. ed. Cuernavaca, Mexico: Instituto Nacional de Salud Publica (MX). 2013. Available online: https://ensanut.insp.mx/encuestas/ensanut2012/doctos/informes/ENSANUT2012ResultadosNacionales.pdf (accessed on 29 May 2021).
43. INEGI. México—Encuesta Nacional Sobre Salud y Envejecimiento en México (ENASEM) 2018—Información General. Available online: https://www.inegi.org.mx/rnm/index.php/catalog/619 (accessed on 29 May 2021).

International Journal of
Environmental Research and Public Health

MDPI

Article

Characteristics and Service Utilization by Complex Chronic and Advanced Chronic Patients in Catalonia: A Retrospective Seven-Year Cohort-Based Study of an Implemented Chronic Care Program

Sebastià J. Santaeugènia [1,2], Joan C. Contel [1,2], Emili Vela [3,4], Montserrat Cleries [3,4], Paloma Amil [2], Eva M. Melendo-Azuela [1,2], Esther Gil-Sánchez [2], Victoria Mir [2] and Jordi Amblàs-Novellas [1,2,5,*]

1 Central Catalonia Chronicity Research Group (C3RG), Centre for Health and Social Care Research (CESS), University of Vic—Central University of Catalonia (UVIC-UCC), 08500 Barcelona, Spain; sebastia.santaeugenia@gencat.cat (S.J.S.); jccontel@gencat.cat (J.C.C.); emelendo_ext@gencat.cat (E.M.M.-A.)
2 Chronic Care Program, Department of Health, Generalitat de Catalunya, 08020 Barcelona, Spain; pamil@gencat.cat (P.A.); egils@gencat.cat (E.G.-S.); victoriamir@gencat.cat (V.M.)
3 Unitat d'Informació i Coneixement, Servei Català de la Salut, 08006 Barcelona, Spain; evela@catsalut.cat (E.V.); mcleries@catsalut.cat (M.C.)
4 Digitalization for the Sustainability of the Healthcare System (DS3), Sistema de Salut de Catalunya, 08006 Barcelona, Spain
5 Chair and Department of Palliative Care, University of Vic—Central University of Catalonia (UVIC-UCC), 08500 Barcelona, Spain
* Correspondence: jordiamblas@gmail.com

Citation: Santaeugènia, S.J.; Contel, J.C.; Vela, E.; Cleries, M.; Amil, P.; Melendo-Azuela, E.M.; Gil-Sánchez, E.; Mir, V.; Amblàs-Novellas, J. Characteristics and Service Utilization by Complex Chronic and Advanced Chronic Patients in Catalonia: A Retrospective Seven-Year Cohort-Based Study of an Implemented Chronic Care Program. *Int. J. Environ. Res. Public Health* **2021**, *18*, 9473. https://doi.org/10.3390/ijerph18189473

Academic Editor: Paul B. Tchounwou

Received: 28 June 2021
Accepted: 6 September 2021
Published: 8 September 2021

Abstract: The Chronic Care Program introduced in Catalonia in 2011 focuses on improving the identification and management of complex chronic (CCPs) and advanced chronic patients (ACPs) by implementing an individualized care model. Its first stage is their identification based on chronicity, difficult clinical management (i.e., complexity), and, in ACPs, limited life prognosis. Subsequent stages are individual evaluation and implementation of a shared personalized care plan. This retrospective study, including all CCPs and ACPs identified in Catalonia between 2013 and 2019, was aimed at describing the characteristics and healthcare service utilization among these patients. Data were obtained from an administrative database and included sociodemographic, clinical, and service utilization variables and morbidity-associated risk according to the Adjusted Morbidity Groups (GMA) stratification. During the study period, CCPs' and ACPs' prevalence increased and was higher in lower-income populations; most cases were women. CCPs and ACPs had all comorbidities at higher frequencies, higher utilization of healthcare services, and were more frequently at high risk (63% and 71%, respectively) than age-, sex-, and income level-adjusted non-CCP (23%) and non-ACP populations (30%). These results show effective identification of the program's target population and demonstrate that CCPs and ACPs have a higher burden of multimorbidity and healthcare needs.

Keywords: chronic care; integrated care; geriatric care; palliative care; primary health care; multimorbidity; complexity; healthcare services utilization; complex needs; advanced chronic patients

1. Introduction

Catalonia (Northeast of Spain) is one of the regions with the oldest population in the world due to its ever-increasing life expectancy and its lower fertility rate (83.5 years and 35.6%, respectively, in 2019) [1,2]. In 2016, public health expenditure per capita was 2137 (USD PPP), total health expenditure was 7.6% of the gross domestic product (GDP), and hospital beds supply was 1.7 per 1000, below the 3.7 per 1000 average of EU15. Currently, 19% of the Catalan population is aged >65 years, and this figure is projected to increase to >33%, with 12–15% of the population >80 years old by 2050. It is also one of the most intensively aging populations in the world [3,4]. This demographic evolution is associated with a higher prevalence of people with chronic diseases, who are currently a major healthcare and social concern, and constitute a burden for healthcare systems [5–7].

The national Catalan healthcare system provides free universal coverage, except for pharmaceuticals, which require a user co-payment, to a population of 7.6 million. One of its main distinguishing features is the separation of planning and financing functions, allowing for commissioning of healthcare services from public and private-owned centers, including acute care hospitals (67 centers and 12,776 beds), intermediate care hospitals (98 post-acute and long-term care centers with 8261 beds), mental health (40 acute mental health centers (3805 beds) and 129 community mental health network facilities) and primary care system facilities (377 primary care centers). The health plan is the main strategic planning instrument for all health interventions of the Government of Catalonia and, for the 2011–2015 period, was focused on improving care for patients who used it most: persons with chronic conditions and complex needs (CCPs) and those with advanced chronic conditions (ACPs) [8].

Compared with the general population, CCPs and ACPs have higher utilization of healthcare resources, including emergency admissions, consultations, and accident and emergency (A&E) services, higher drug use, and tend to be dependent [9,10]. However, healthcare systems are organized to treat singular diseases and, despite the increasing number of people with multiple chronic conditions and complex needs, still provide fragmented care [11,12]. As a result, standardized plans developed from a disease management perspective fail to fulfill CCPs' and ACPs' needs, increasing the risk of poor outcomes, such as emergency admissions, readmissions, and a higher number of primary healthcare visits [13]. Considering the concern raised by these unadjusted healthcare plans, new integrated and patient-centered care models have been developed worldwide, and these new interventions are being evaluated in other countries [12,14–18].

1.1. Patient Definition, Identification, and Specific Model of Care of CCPs and ACPs in Catalonia

To improve the main outcomes and care for CCPs and ACPs, the Catalan Chronic Care Program of late 2011 focused on developing a specific toolkit to define and identify CCPs and ACPs [10,16,19,20]. For the definition of CCPs, the Catalan program broadened the multimorbidity paradigm used in other regions of Spain and Europe and adopted people with complex care needs as population targets. Complexity was defined, following this new paradigm, in three dimensions, including clinical, social/contextual, and healthcare system complexities, and was treated as the result of the interaction between variables associated with each of the dimensions [21] (Figure 1).

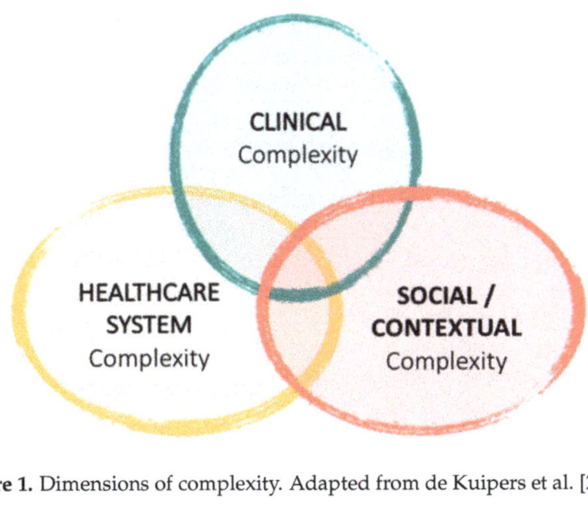

Figure 1. Dimensions of complexity. Adapted from de Kuipers et al. [21].

Patients with chronic conditions and complex care needs (CCPs) were defined as those whose situation reflected the difficulty of their management and care and the need to adopt specific individual plans, owing to concurrent diseases, their utilization of healthcare services, and their context [22]. Complex care needs are the common feature among CCPs, who are estimated to account for 3.5–4% of the population [23]. The identification of CCPs is based on detecting relationships between different clinical, context-related, and health and social care system-related criteria. Clinical criteria include multimorbidity, dynamic and unpredictable outcome, and classification within the 5% at higher risk according to the Adjusted Morbidity Groups (GMA, from Spanish "Grupos de Morbilidad Ajustados") model. Context-related criteria include dysfunctional or risky social situations to meet the person's needs, and health and social care system-related criteria include, among other factors, patient management differences among each healthcare professional settings. Unlike specific screening instruments, a specific, unequivocal guideline to define CCPs has not been established. Despite the lack of a perfect algorithm to identify CCPs, a perceptive decision related to complexity criteria has been established as the basis of the model and is used by healthcare professionals to define a patient as CCP. The definitions of complexity and their associated criteria are summarized in Table S1.

In addition to complex care needs, people with advanced chronic conditions (ACPs) need palliative care and have a limited life prognosis (from a few days up to one year). Considering the previous evidence regarding the benefits of early identification and palliative care in this population (i.e., first end-of-life transition) [24], the Catalan healthcare plan has changed its paradigm of care for people needing palliative care from the classical dichotomic perspective (i.e., curative vs. palliative care) applicable to cancer patients in their last days or weeks in palliative care units, to a dynamic synchronous perspective applicable to any disease and advanced chronic condition, regardless of the person's location [25]. This view of palliative care is different compared to other countries and seeks to identify patients in end-of-life transition needing early palliative care. To facilitate the early identification of ACPs in Catalonia, the Department of Health and the Chair of Palliative Care of Vic University developed a screening tool for the early identification of the need for palliative care among individuals with limited life expectancy: the NECesidades PALiativas tool (NECPAL-CCOMS-ICO) [26]. The NECPAL tool is a validated instrument to screen and identify people with palliative care needs that combines the surprise question ('Would you be surprised if this patient dies in the next year?') with other items, including the request for palliative care by the patient or family and the need for palliative care as identified by professionals, general clinical indicators, psychosocial factors, multimorbidity, use of resources, and specific indicators to evaluate disease severity and progression. Previous studies in the Catalan setting have shown a 1–1.5% prevalence of APCs, with advanced frailty and/or dementia (55%), advanced organ disease (32%), and cancer (13%) [26].

Providing optimal care to CCPs and ACPs requires combining the points of view of healthcare systems, responsible for facilitating a better response to these patients through healthcare planning and resource management, and healthcare professionals responsible for providing care based on the multidimensional needs of people. Furthermore, success relies on incorporating the population view, based on care and organizational models to respond to the needs of this population, and the individual view, based on patient-centered models providing individualized care, considering that the main goal is to obtain good results from the patients' perspective [27]. Along these lines, the Department of Health of Catalonia developed the Chronic and Integrated Healthcare program, an individualized care model that integrated the population perspective, aimed at improving provision of services and integrating care, which revolves around primary care teams, with the individual perspective, aimed at customizing care for each individual patient (Figure 2). This individual care model included four stages, of which patient identification as CCPs and ACPs is the first.

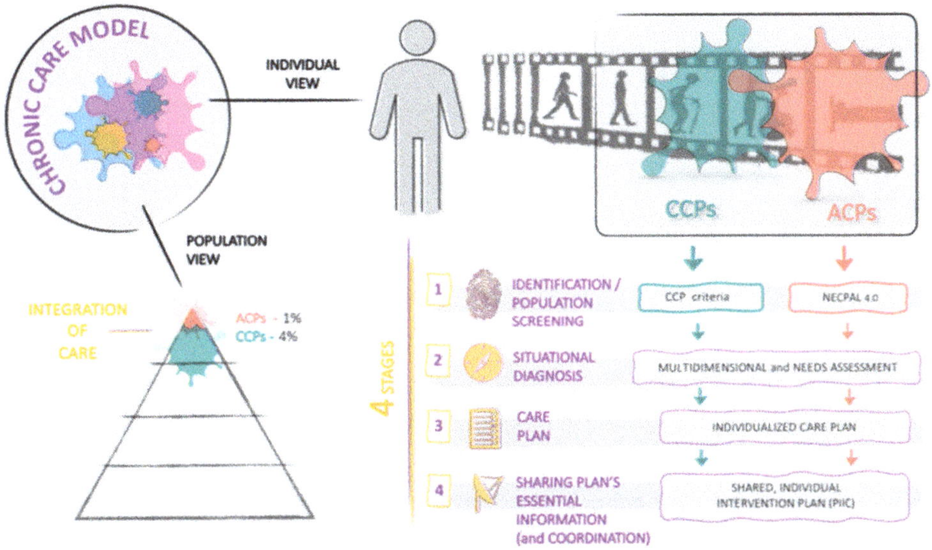

Figure 2. Diagram depicting the model of care for persons with complex chronic conditions and advanced chronic conditions of the Catalan health program 2011–2015 from the individualized and population perspectives. The model is organized in four main stages. ACPs, advanced chronic patients; CCPs, complex chronic patients.

1.2. Justification and Aims of This Study

Even though the Chronic and Integrated Healthcare program has been implemented for several years, the populations identified as CCPs and ACPs remain to be analyzed. Furthermore, few studies have evaluated other initiatives aimed at providing integrated care for people with chronic conditions and complex needs, often with unexpected results [28]. The healthcare system of Catalonia registers patient stratification based on the GMA categories and identification as CCP or ACP on electronic healthcare records, making this information available for its monitorization and evaluation. In this study, we used the Catalan electronic administrative clinical database to describe the sociodemographic and clinical characteristics of patients identified as CCPs and ACPs with the goal of assessing the first stage of the Chronic Care program in the first few years of its implementation. Additionally, we evaluated the clinical characteristics and utilization of healthcare services among these populations.

2. Materials and Methods

2.1. Study Design, Participants, and Database

This was a retrospective analysis of an administrative database that included all individuals identified as CCPs or ACPs in Catalonia (northwest of Spain) between 2013 and 2019. CCPs and ACPs were identified by primary care specialists in 377 primary care centers based on the criteria previously presented (Table S1). Owing to the use of an electronic database as the data source and the irreversible anonymization of the data extracted, patient informed consent was not applicable in this study. This study was conducted in accordance with the Ethical Principles for Medical Research Involving Human Subjects of the Helsinki Declaration and the local Personal Data Protection Law (LOPD 15/1999); it was approved by the Research Ethics Committee of University of Vic/Central University of Catalonia (UVIC-UCC) reference number 63/2018.

2.2. Data Source

Sociodemographic and clinical data were obtained from the Catalan Health Surveillance System (CHSS) that, since 2011, has been collecting detailed information about the utilization of healthcare by/among the entire population of Catalonia (7,600,000 inhabitants). This record, which has been analyzed in previous publications in other areas [29–31], gathers data recorded in multiple settings, including primary care, acute care hospitals, intermediate care hospitals, mental health centers, outpatient clinics, and emergency services. Furthermore, this record collects information regarding prescriptions and pharmacy expenses and invoices, including outpatient clinics, non-urgent medical transportation, outpatient rehabilitation, home oxygen therapy, and dialysis. No data about private healthcare could be collected because these centers use different codes for patient identification.

2.3. Variables

The sociodemographic variables considered in this study were age, sex, and income level, classified as high (annual income > 100,000 €), intermediate (18,000–100,000 €), low (<18,000 €), and very low (receiving welfare support from the government). Clinical variables were diagnoses, as they appear in the CHSS database according to the usual clinical practice, and coded according to the International Classification of Diseases, ninth revision, Clinical Modification (ICD-9-CM). The multimorbidity burden was stratified based on the Adjusted Morbidity Groups (GMA), which considers the type of disease—acute or chronic—, number of systems affected, and complexity of each disease [32,33]. The GMA enable the classification of all the population into four strata based on their morbidity-associated risk. The four strata are (1) Baseline risk (healthy stage), with an GMA score up to the 50th percentile of the total population; (2) Low risk, with a GMA score between the 50th–80th percentiles; (3) Moderate risk, with a GMA score between the 80–95th percentiles; and (4) High risk, with a GMA score above the 95th percentile [32,34]. Variables associated with the utilization of healthcare services during the first year after identification of CCPs and ACPs were number of (1) visits to primary healthcare centers; (2) outpatient visits; (3) emergency service admissions; (4) acute care hospital admissions and length of stay (days); (5) admissions to intermediate care hospitals and length of stay (days); (6) admissions to psychiatric centers and length of stay (days), and (7) prescribed drugs, according to the different chemical and therapeutic classification groups, and units.

In order to describe the evolution of epidemiological and clinical characteristics among CCPs and ACPs, an annual incidence study was conducted. Health expenditure was calculated according to the standard costs of each service provided by the Department of Health (Generalitat de Catalunya) for each year [35].

2.4. Statistical Methods

Categorical variables were described as frequencies and percentages and quantitative variables as the mean and standard deviation (SD) and/or the median and interquartile range (IQR; Q1, Q3). Incidence and prevalence rates were expressed per 1000 inhabitants and mortality rates per 100. Categorical variables were compared using Pearson's Chi-squared test with Yates' continuity correction. Survival curves were calculated using the Kaplan–Meier estimator and compared using the Gehan test. The statistical significance threshold was set at a bilateral alpha value of 0.05. The utilization of healthcare services and associated expenditure of CCPs and ACPs were compared with the population of patients not identified as complex adjusted by age, sex, and income level, hereinafter referred to as "non-CCP" and "non-ACP", respectively. Comorbidities and healthcare services utilization were compared using the rate ratio by median-unbiased estimation (mid-p), and healthcare services expenditure was compared using the Student's *t*-test. To analyze geographic variability, the Poisson regression was used to calculate cumulative incidence rates for the 2017–2019 period, adjusted by age, sex, morbidity (GMA), and income level. Data after the first few years of implementation of the Chronic Care program (2017–2019) were considered to be more stable and were aggregated to increase the robustness of these analyses. All analyses were performed using the R statistical package (version 4.0.3).

3. Results

3.1. Characteristics of the Overall Cohort (2013–2019)

During the study period, 303 357 individuals with a median (IQR) age of 82 (74.0, 86.0) years were identified as CCPs, and 98,587 persons with a median (IQR) age of 84.0 (75.0–90.0) years were identified as ACPs. Table 1 summarizes the sociodemographic and clinical characteristics of the overall study population according to their identification as CCP and ACP. Sociodemographic and clinical characteristics of CCPs and ACPs identified throughout the study period (2013–2019) were significantly different between groups, although differences were small. ACPs were older and, as expected, were more frequently classified in the high-risk GMA stratification category compared to CCPs. Dementia was more frequent in ACPs compared to CCPs.

Table 1. Sociodemographic and clinical characteristics of the overall study cohort (2013–2019) according to their identification as complex chronic patients and advanced chronic patients, *n* (%).

	CCP	ACP	*p*-Value [a]
	N = 303,357	N = 98,587	
Sociodemographic characteristics			
Sex			
Male	133,454 (44.0)	46,007 (46.7)	<0.001
Female	169,903 (56.0)	52,580 (53.3)	
Age, years			
<15	1020 (0.34)	134 (0.14)	<0.001
15–44	5060 (1.67)	1074 (1.09)	
45–64	25,466 (8.39)	8921 (9.05)	
65–74	45,554 (15.0)	12,871 (13.1)	
75–84	113,495 (37.4)	28,701 (29.1)	
>84	112,762 (37.2)	46,886 (47.6)	
Income level			
High	701 (0.23)	350 (0.36)	<0.001
Medium	46,586 (15.4)	16,439 (16.7)	
Low	244,127 (80.5)	78,429 (79.6)	
Very Low	11,930 (3.93)	3358 (3.41)	
Clinical Characteristics			
GMA stratification			
Baseline risk	1704 (0.56)	304 (0.31)	<0.001
Low risk	19,826 (6.54)	3508 (3.56)	
Moderate risk	119,234 (39.3)	26,512 (26.9)	
High risk	162,593 (53.6)	68,263 (69.2)	
Comorbidities			
Arterial hypertension	247,001 (81.4)	76,284 (77.4)	<0.001
Arthrosis	157,006 (51.8)	47,320 (48.0)	<0.001
Diabetes mellitus	124,762 (41.1)	35,848 (36.4)	<0.001
Heart failure	100,330 (33.1)	35,084 (35.6)	<0.001
Chronic kidney disease	98,985 (32.6)	34,539 (35.0)	<0.001
Chronic obstructive pulmonary disease	87,748 (28.9)	29,954 (30.4)	<0.001
Depression	91,961 (30.3)	28,305 (28.7)	<0.001
Ictus	71,847 (23.7)	25,735 (26.1)	<0.001
Ischemic heart disease	70,247 (23.2)	21,187 (21.5)	<0.001
Dementia	43,957 (14.5)	20,492 (20.8)	<0.001
Osteoporosis	53,707 (17.7)	16,572 (16.8)	<0.001
Arthritis	28,422 (9.37)	9031 (9.16)	0.051
Cirrhosis	6853 (2.26)	3173 (3.22)	<0.001
HIV infection	1617 (0.53)	360 (0.37)	<0.001

Abbreviations: ACP, advanced chronic patients; GMA, adjusted morbidity groups (in Spanish "Grupos de morbilidad ajustados"); CCP, complex chronic patients; HIV, human immunodeficiency virus. [a] Pearson's Chi-squared test with Yates' continuity correction.

Survival analysis showed a significantly decreased probability of survival of ACPs compared to CCPs (Figure 3). One-, 3-, and 5-year survival rates were 82.4% and 37.0%, 58.2% and 14.8%, and 39.3% and 5.4% for CCP and ACP, respectively.

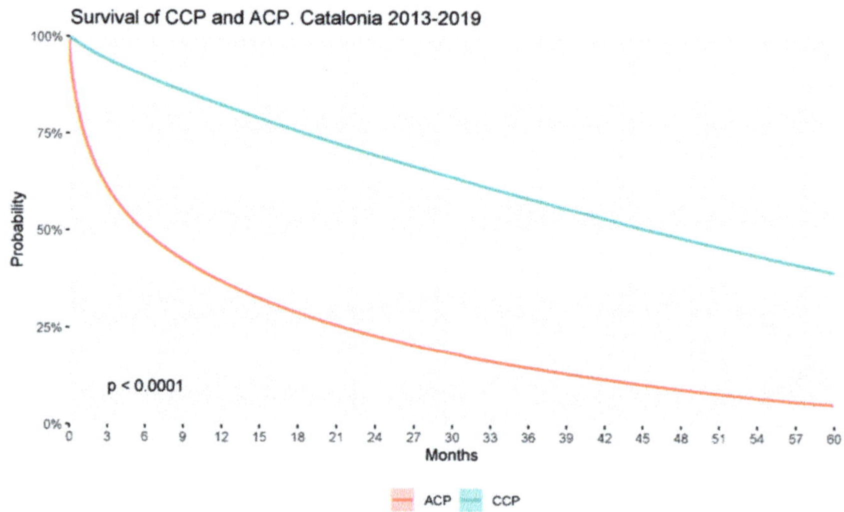

Figure 3. Probability of survival up to 5 years of the overall population of chronic complex patients and advanced chronic patients identified throughout the study period (2013–2019).

3.2. Epidemiological Evolution of the Identification of CCPs and ACPs

During the study period (2013–2019), the prevalence rates of both CCPs and ACPs in the general population increased from 8.8 and 1.2 cases per 1000 people in 2013 to 21.7 and 2.6 cases per 1000 people in 2019, respectively. Conversely, incidence rates decreased from 2013 to 2019 in both populations, from 9.1 to 3.9 new cases per 1000 individuals for CCPs and from 1.8 to 1.6 new cases per 1000 individuals for ACPs, respectively. CCP prevalence and incidence rates were more variable during the study period compared with those of ACP (Figure 4).

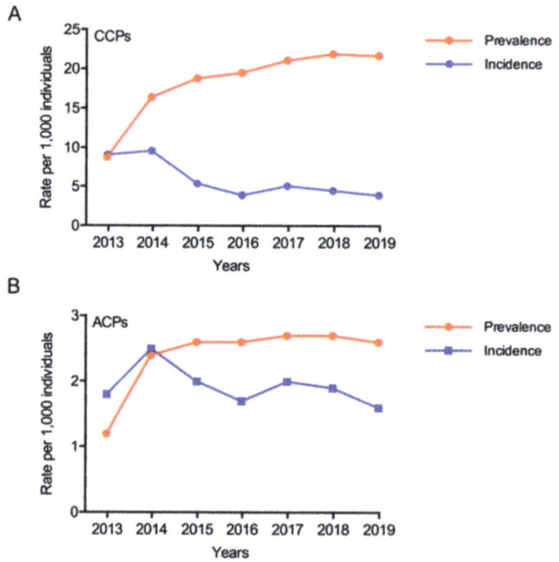

Figure 4. Prevalence and incidence rates of (**A**) CCPs and (**B**) ACPs throughout the study period (2013–2019).

Analysis of survival rates according to the year of identification showed significantly and progressively decreased survival of ACPs and, even though survival curves of CCPs showed a similar significant trend, differences between years were more modest (Figure 5).

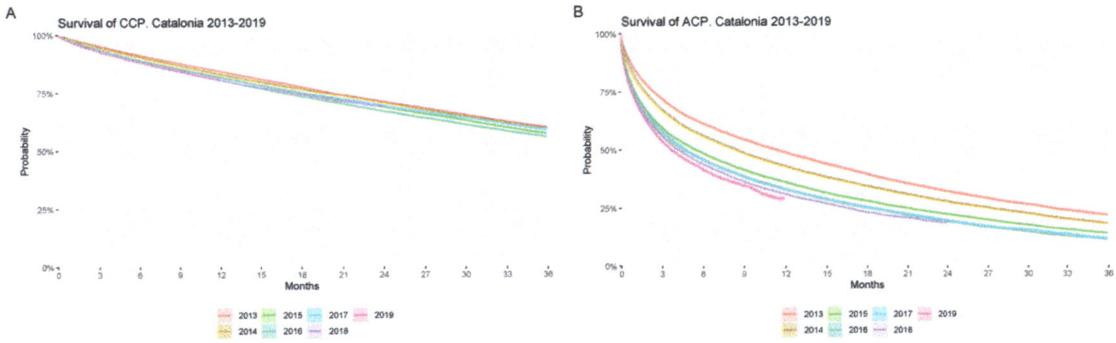

Figure 5. Probability of survival up to 3 years of (**A**) chronic complex patients (CCPs) and (**B**) advanced chronic patients (ACPs) according to year of identification.

From the beginning and until the end of the study period, the mean age of CCP and ACP incident cases increased in men and women. Mean ages of patients identified as CCP and ACP were persistently higher in women throughout the whole study period: Age differences ranged from 3.8 to 4.3 years for CCP and from 4.9 to 5.7 years for ACP. Sociodemographic and clinical characteristics of patients identified as CCP and ACP showed significant changes throughout the study period. The ages of patients identified as CCPs and ACPs increased gradually and slowly (1.5-year and 1.7-year differences for CCPs and ACPs, respectively, in 7 years), and the proportion of patients at a high and very high GMA risk progressively increased, with a concomitant decrease in patients at low and moderate risk. The prevalence of different morbidities throughout the study period significantly changed (Tables 2 and 3).

Table 2. Evolution of the sociodemographic and clinical characteristics of complex chronic patients throughout the study period, *n* (%).

	2013	2014	2015	2016	2017	2018	2019	*p*-Value [a]
	N = 68,382	N = 70,996	N = 39,770	N = 29,098	N = 37,907	N = 33,916	N = 23,288	
Sociodemographic characteristics								
Sex								
Male	30,738 (45.0)	31,081 (43.8)	17,252 (43.4)	12,902 (44.3)	16,664 (44.0)	14,710 (43.4)	10,107 (43.4)	<0.001
Female	37,644 (55.0)	39,915 (56.2)	22,518 (56.6)	16,196 (55.7)	21,243 (56.0)	19,206 (56.6)	13,181 (56.6)	
Age, years								
<15	250 (0.37)	263 (0.37)	126 (0.32)	80 (0.27)	60 (0.16)	104 (0.31)	137 (0.59)	<0.001
15–44	1222 (1.79)	1596 (2.25)	593 (1.49)	453 (1.56)	534 (1.41)	364 (1.07)	298 (1.28)	
45–64	6319 (9.24)	6243 (8.79)	3109 (7.82)	2342 (8.05)	2992 (7.89)	2536 (7.48)	1925 (8.27)	
65–74	10,738 (15.7)	10,766 (15.2)	5787 (14.6)	4231 (14.5)	5714 (15.1)	4950 (14.6)	3368 (14.5)	
75–84	27,234 (39.8)	27,078 (38.1)	14,867 (37.4)	10,634 (36.5)	13,906 (36.7)	11,835 (34.9)	7941 (34.1)	
>84	22,619 (33.1)	25,050 (35.3)	15,288 (38.4)	11,358 (39.0)	14,701 (38.8)	14,127 (41.7)	9619 (41.3)	
Income level								
High	118 (0.17)	129 (0.18)	83 (0.21)	71 (0.24)	110 (0.29)	97 (0.29)	93 (0.40)	0.0000
Medium	9409 (13.8)	9663 (13.6)	5812 (14.6)	4826 (16.6)	6406 (16.9)	5886 (17.4)	4584 (19.7)	
Low	54,858 (80.2)	60,332 (85.0)	31,908 (80.2)	22,858 (78.6)	29,919 (78.9)	26,604 (78.4)	17,648 (75.8)	
Very Low	3991 (5.84)	869 (1.22)	1963 (4.94)	1343 (4.62)	1472 (3.88)	1329 (3.92)	963 (4.14)	
Clinical characteristics								
GMA stratification								
Baseline risk	558 (0.82)	461 (0.65)	248 (0.62)	177 (0.61)	120 (0.32)	71 (0.21)	69 (0.30)	0.0000
Low risk	6296 (9.21)	5285 (7.44)	2584 (6.50)	1628 (5.59)	1841 (4.86)	1285 (3.79)	907 (3.89)	
Moderate risk	29,499 (43.1)	29,241 (41.2)	16,310 (41.0)	11,004 (37.8)	13,834 (36.5)	11,634 (34.3)	7712 (33.1)	
High risk	32,029 (46.8)	36,009 (50.7)	20,628 (51.9)	16,289 (56.0)	22,112 (58.3)	20,926 (61.7)	14,600 (62.7)	

Table 2. *Cont.*

	2013	2014	2015	2016	2017	2018	2019	p-Value [a]
	N = 68,382	N = 70,996	N = 39,770	N = 29,098	N = 37,907	N = 33,916	N = 23,288	
Comorbidities								
Diabetes mellitus	30,682 (44.9)	30,241 (42.6)	15,705 (39.5)	11,326 (38.9)	15,044 (39.7)	12,976 (38.3)	8788 (37.7)	<0.001
Heart failure	24,134 (35.3)	23,634 (33.3)	13,075 (32.9)	9554 (32.8)	11,937 (31.5)	10,672 (31.5)	7324 (31.4)	<0.001
COPD	21,552 (31.5)	21,044 (29.6)	11,076 (27.9)	8083 (27.8)	10,557 (27.8)	9164 (27.0)	6272 (26.9)	<0.001
Arterial hypertension	55,972 (81.9)	57,452 (80.9)	32,304 (81.2)	23,620 (81.2)	31,069 (82.0)	27,682 (81.6)	18,902 (81.2)	<0.001
Depression	18,709 (27.4)	20,893 (29.4)	11,895 (29.9)	9188 (31.6)	12,181 (32.1)	11,217 (33.1)	7878 (33.8)	<0.001
HIV infection	428 (0.63)	526 (0.74)	197 (0.50)	111 (0.38)	140 (0.37)	130 (0.38)	85 (0.36)	<0.001
Ischemic heart disease	17,327 (25.3)	17,116 (24.1)	8937 (22.5)	6499 (22.3)	8322 (22.0)	7189 (21.2)	4857 (20.9)	<0.001
Ictus	15,671 (22.9)	16,442 (23.2)	9478 (23.8)	7004 (24.1)	9107 (24.0)	8383 (24.7)	5762 (24.7)	<0.001
Chronic kidney disease	20,601 (30.1)	21,667 (30.5)	12,603 (31.7)	9745 (33.5)	13,558 (35.8)	12,469 (36.8)	8342 (35.8)	<0.001
Cirrhosis	1559 (2.28)	1640 (2.31)	956 (2.40)	653 (2.24)	845 (2.23)	677 (2.00)	523 (2.25)	0.016
Osteoporosis	10,591 (15.5)	12,082 (17.0)	6900 (17.3)	5260 (18.1)	7491 (19.8)	6752 (19.9)	4631 (19.9)	<0.001
Arthrosis	32,201 (47.1)	35,116 (49.5)	20,336 (51.1)	15,501 (53.3)	21,187 (55.9)	19,258 (56.8)	13,407 (57.6)	0.000
Arthritis	4927 (7.21)	6089 (8.58)	3374 (8.48)	2859 (9.83)	4221 (11.1)	3955 (11.7)	2997 (12.9)	<0.001
Dementia	7846 (11.5)	9288 (13.1)	5841 (14.7)	4424 (15.2)	6051 (16.0)	6174 (18.2)	4333 (18.6)	<0.001

Abbreviations: GMA, adjusted morbidity groups (in Spanish "Grupos de Morbilidad Ajustados"); COPD, chronic obstructive pulmonary disease; HIV, human immunodeficiency virus. [a] Pearson's Chi-squared test with Yates' continuity correction.

Table 3. Evolution of the sociodemographic and clinical characteristics of advanced chronic patients throughout the study period, *n* (%).

	2013	2014	2015	2016	2017	2018	2019	p-Value [a]
	N = 13,206	N = 18,137	N = 14,755	N = 12,918	N = 14,587	N = 14,040	N = 10,944	
Sociodemographic characteristics								
Sex								
Male	6122 (46.4)	8440 (46.5)	6950 (47.1)	6134 (47.5)	6828 (46.8)	6495 (46.3)	5038 (46.0)	0.236
Female	7084 (53.6)	9697 (53.5)	7805 (52.9)	6784 (52.5)	7759 (53.2)	7545 (53.7)	5906 (54.0)	
Age, years								
<15	31 (0.23)	18 (0.10)	15 (0.10)	13 (0.10)	11 (0.08)	27 (0.19)	137 (0.59)	<0.001
15–44	208 (1.58)	213 (1.17)	158 (1.07)	126 (0.98)	152 (1.04)	123 (0.88)	298 (1.28)	
45–64	1310 (9.92)	1623 (8.95)	1292 (8.76)	1167 (9.03)	1333 (9.14)	1286 (9.16)	1925 (8.27)	
65–74	1694 (12.8)	2380 (13.1)	1944 (13.2)	1720 (13.3)	1894 (13.0)	1840 (13.1)	3368 (14.5)	
75–84	4306 (32.6)	5557 (30.6)	4439 (30.1)	3606 (27.9)	4107 (28.2)	3796 (27.0)	7941 (34.1)	
>84	5657 (42.8)	8346 (46.0)	6907 (46.8)	6286 (48.7)	7090 (48.6)	6968 (49.6)	9619 (41.3)	
Income level								
High	40 (0.30)	45 (0.25)	42 (0.28)	45 (0.35)	52 (0.36)	66 (0.47)	19 (0.17)	<0.001
Medium	1829 (13.9)	2563 (14.1)	2310 (15.7)	2294 (17.8)	2717 (18.6)	2600 (18.5)	94 (0.86)	
Low	10,622 (80.5)	15,284 (84.3)	11,791 (79.9)	10,110 (78.3)	1130 (77.5)	10,893 (77.6)	910 (8.32)	
Very Low	708 (5.36)	244 (1.35)	609 (4.13)	469 (3.63)	517 (3.54)	481 (3.43)	1399 (12.8)	
Clinical characteristics								
GMA stratification								
Baseline risk	88 (0.67)	65 (0.36)	60 (0.41)	33 (0.26)	26 (0.18)	19 (0.14)	13 (0.12)	0.0000
Low risk	840 (6.36)	888 (4.90)	615 (4.17)	378 (2.93)	331 (2.27)	276 (1.97)	180 (1.64)	
Moderate risk	4562 (34.5)	5627 (31.0)	4258 (28.9)	3393 (26.3)	3485 (23.9)	3015 (21.5)	2172 (19.8)	
High risk	7716 (58.4)	11,557 (63.7)	9822 (66.6)	9114 (70.6)	10,745 (73.7)	10,730 (76.4)	8579 (78.4)	
Comorbidities								
Diabetes mellitus	4963 (37.6)	6627 (36.5)	5262 (35.7)	4646 (36.0)	5350 (36.7)	4983 (35.5)	4017 (36.7)	0.005
Heart failure	4768 (36.1)	6390 (35.2)	5204 (35.3)	4539 (35.1)	5221 (35.8)	4891 (34.8)	4071 (37.2)	0.002
COPD	3992 (30.2)	5590 (30.8)	4522 (30.6)	3861 (29.9)	4428 (30.4)	4144 (29.5)	3417 (31.2)	0.053
Arterial hypertension	10,060 (76.2)	13,823 (76.2)	11,290 (76.5)	10,048 (77.8)	11,471 (78.6)	10,923 (77.8)	8669 (79.2)	<0.001
Depression	3304 (25.0)	4835 (26.7)	4098 (27.8)	3684 (28.5)	4460 (30.6)	340 (30.9)	3584 (32.7)	<0.001
HIV infection	63 (0.48)	68 (0.37)	55 (0.37)	41 (0.32)	48 (0.33)	45 (0.32)	40 (0.37)	0.348
Ischemic heart disease	3090 (23.4)	4031 (22.2)	3086 (20.9)	2678 (20.7)	3089 (21.2)	2917 (20.8)	2296 (21.0)	<0.001
Ictus	3122 (23.6)	4574 (25.2)	3771 (25.6)	3301 (25.6)	3963 (27.2)	3887 (27.7)	3117 (28.5)	<0.001
Chronic kidney disease	4121 (31.2)	5778 (31.9)	4855 (32.9)	4535 (35.1)	5485 (37.6)	5381 (38.3)	4384 (40.1)	<0.001
Cirrhosis	425 (3.22)	616 (3.40)	495 (3.35)	420 (3.25)	506 (3.47)	398 (2.83)	313 (2.86)	0.010
Osteoporosis	1839 (13.9)	2823 (15.6)	2317 (15.7)	2163 (16.7)	2658 (18.2)	2604 (18.5)	2168 (19.8)	<0.001
Arthrosis	5556 (42.1)	7966 (43.9)	6863 (46.5)	6275 (48.6)	7452 (51.1)	7286 (51.9)	5922 (54.1)	<0.001
Arthritis	914 (6.92)	1352 (7.45)	1212 (8.21)	1219 (9.44)	1506 (10.3)	1474 (10.5)	1354 (12.4)	<0.001
Dementia	2050 (15.5)	3257 (18.0)	2761 (18.7)	2581 (20.0)	3172 (21.7)	3766 (26.8)	2905 (26.5)	<0.001

Abbreviations: GMA, adjusted morbidity groups (in Spanish "Grupos de Morbilidad Ajustados"); COPD, chronic obstructive pulmonary disease; HIV, human immunodeficiency virus. [a] Pearson's Chi-squared test with Yates' continuity correction.

3.3. *Evaluation of Demographic and Clinical Characteristics of CCPs and ACPs (2019)*

In 2019, the total number of CCP cases (prevalence) was 167,892, of which 98,676 were women and 69,216 were men. The prevalence of ACP was lower, with 19,741 individuals, of which 11,907 were women and 7834 were men. The distribution of these populations by age and gender is shown in Figure S1.

Regarding the distribution of CCPs and ACPs according to socioeconomic level and sex, the prevalence of both CCPs and ACPs progressively increased as the socioeconomic level decreased for both women and men, with an overall higher prevalence of both CCPs and ACPs in the low- and very low-income categories. The population of women with very low and low-income had the highest prevalence of CCPs and ACPs, respectively (Table 4).

Table 4. Prevalence rates of complex chronic and advanced chronic patients according to income level (year 2019), rate per 1000 people.

	CCP		ACP	
	Women	**Men**	**Women**	**Men**
Income level, €/year				
High (>100,000)	8.0	6.0	1.6	1.4
Intermediate (18,000–100,000)	12.0	13.1	2.1	2.3
Low (<18,000)	36.4	27.0	6.2	4.6
Very low (unemployed/receiving welfare support)	36.8	25.7	5.1	3.9

Abbreviations: ACP, advanced chronic patients; CCP, complex chronic patients.

The clinical characteristics of CCPs and ACPs were compared to those of adjusted non-CCP and non-ACP populations. According to the GMA stratification of the morbidity-associated risk, most CCP and ACP cases were at high risk, representing an increased proportion of patients in this risk level compared with their respective adjusted non-CCP and non-ACP populations (63% vs. 23% for CCP and 71% vs. 30% for ACP, respectively) (Figure 6).

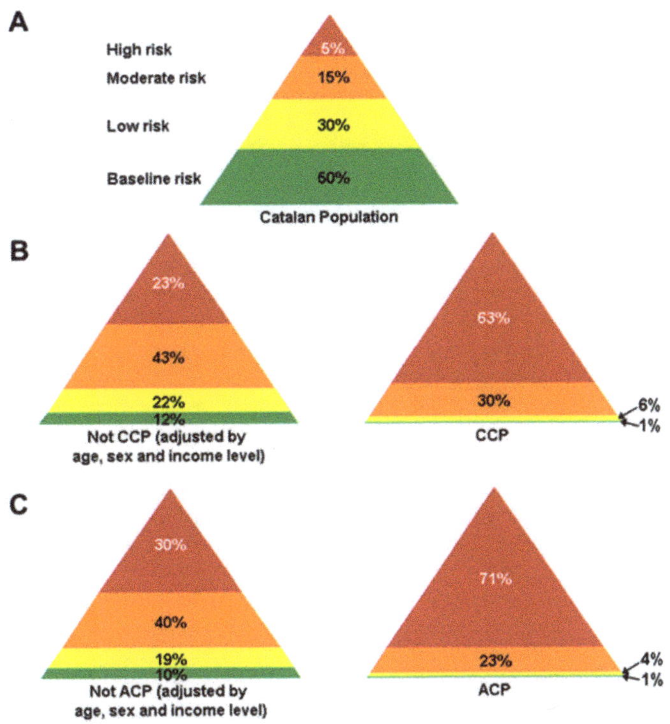

Figure 6. GMA stratification according to comorbidity risks of (**A**) total Catalan population, (**B**) CCPs, and (**C**) ACPs and their corresponding age-, sex-, and income level-adjusted non-CCP and non-ACP populations in 2019.

Accordingly, all comorbidities were present at significantly higher frequencies in CCPs and ACPs compared with their adjusted non-CCP and non-ACP populations (Table 5). The most frequent comorbidity in CCP patients was diabetes, followed, in this order, by chronic kidney disease, heart failure, cancer, and chronic obstructive pulmonary disease (COPD), whereas in ACPs, cancer was the most frequent comorbidity, followed by chronic kidney disease, heart failure, dementia, and diabetes. The distribution of morbidities by sex is shown in Figure S2 (Supplementary file).

Table 5. Main comorbidities in complex chronic and advanced chronic patients and their corresponding non-CCP and non-ACP populations adjusted by age, sex, and annual income (year 2019), %.

	CCP	Adjusted Non-CCP Population	*p*-Value [a]	ACP	Adjusted Non-ACP Population	*p*-Value [a]
Diabetes	43.8	24.4	<0.001	38.3	26.6	<0.001
Chronic kidney disease	41.5	24.2	<0.001	42.0	29.2	<0.001
Heart failure	39.0	13.8	<0.001	40.7	20.0	<0.001
Cancer	34.2	25.0	<0.001	48.5	26.4	<0.001
COPD	32.3	15.3	<0.001	32.5	18.0	<0.001
Dementia	30.4	13.9	<0.001	38.8	18.2	<0.001
Stroke	29.3	13.9	<0.001	31.2	17.0	<0.001
Ischemic heart disease	26.1	12.7	<0.001	24.1	15.3	<0.001
Arthritis	14.7	9.4	<0.001	13.0	10.1	<0.001
Asthma	12.8	7.2	<0.001	11.3	7.9	<0.001
Alcoholism	5.9	2.0	<0.001	5.9	2.2	<0.001
Atypical psychosis	4.6	1.6	<0.001	5.1	2.2	<0.001
Major depressive disorder	4.2	2.0	<0.001	3.4	2.1	<0.001
Cirrhosis	2.8	0.9	<0.001	3.4	1.0	<0.001
Schizophrenia	1.8	0.6	<0.001	1.3	0.7	<0.001
Bipolar disorder	1.4	0.6	<0.001	-	-	<0.001

Abbreviations: ACP, advanced chronic patients; CCP, complex chronic patients; COPD, chronic obstructive pulmonary disease. [a] Rate ratio by median-unbiased estimation (mid-p).

3.4. Evaluation of Health Service Utilization and Associated Expenditures of CCPs and ACPs (2019)

Table 6 summarizes the utilization of healthcare services by CCPs and ACPs and their associated expenditure. Compared with their adjusted non-CCP and non-ACP populations, CCPs and ACPs had significantly higher utilization of the different healthcare services, including primary care, outpatient care, emergency admissions, day hospital, and mental health, and were prescribed a higher number of drugs.

Admission rates in acute care hospitals, intermediate care hospitals, and psychiatric centers were also higher in CCPs and ACPs than in their respective non-CCP and non-ACP populations. Differences with their corresponding age, sex, and income level-adjusted non-CCP and non-ACP populations were particularly higher for utilization of mental health services and admission to psychiatric centers in CCPs and day hospital and intermediate care hospital admissions in ACPs. Expenditures derived from primary care, outpatient care, hospital admissions, emergency departments, mental health, intermediate care hospitals, prescribed drugs, and other healthcare services were also significantly higher for CCPs and ACPs compared to the non-CCP and non-ACP populations adjusted by age, sex, and income level, with the exception of ACPs' mental health services expenditure. Accordingly, total expenditures on healthcare services were substantially and significantly higher for CCPs and ACPs than for their corresponding adjusted non-CCP and non-ACP populations (Table 6). While hospital admissions and prescribed drugs were the main expenses in all patient groups (i.e., CCPs, ACPs, and the corresponding non-CCP and non-ACP populations), differences in expenditures associated with mental health and outpatient care between CCPs and ACPs and their corresponding non-CCP and non-ACP populations, respectively, were higher than those of other expenditures.

Table 6. Utilization of healthcare services by complex chronic (CCPs) and advanced chronic (ACPs) patients and their corresponding non-CCP and non-ACP populations adjusted by age, sex, and annual income, and their associated expenditures (year 2019).

	CCP	Adjusted Non-CCP Population	*p*-Value [a]	ACP	Adjusted Non-ACP Population	*p*-Value [a]
Healthcare services utilization						
Ambulatory healthcare services (visits or admissions per patient and year), mean						
Primary care	21.1	11.3	<0.001	22.2	12.8	<0.001
Outpatient care	4.3	2.6	<0.001	4.7	2.5	<0.001
Emergency department	1.3	0.6	<0.001	1.6	0.7	<0.001
Day hospital	0.7	0.2	<0.001	1.5	0.3	<0.001
Mental health	0.2	0.1	<0.001	0.1	0.1	<0.001
Prescribed drugs (number per patient and year)	12.6	8.0	<0.001	12.7	8.7	<0.001
Rate of admissions (institutionalizations), admissions per 100 patients and year						
Acute care hospital	64.4	27.1	<0.001	88.4	31.9	<0.001
Intermediate care hospital	17.0	5.7	<0.001	35.5	8.1	<0.001
Psychiatric center	0.5	0.1	<0.001	0.2	0.1	<0.001
Healthcare services expenditure (€ per person and year) (%) [b]			***p*-value** [c]			***p*-value** [c]
Primary care	653.5 (10.75)	367.8 (14.98)	<0.001	667.3 (8.35)	413.5 (14.59)	<0.001
Outpatient care	441.2 (7.26)	225.1 (9.17)	<0.001	618.1 (7.73)	221.9 (7.83)	<0.001
Hospital admissions	1713.6 (28.19)	698.8 (28.46)	<0.001	2385.9 (29.84)	821.1 (28.98)	<0.001
Emergency department	551.4 (9.07)	223.9 (9.12)	<0.001	696.3 (8.71)	286.0 (10.09)	<0.001
Mental health	30.8 (0.51)	10.4 (0.42)	<0.001	11.5 (0.14)	9.5 (0.34)	0.167
Intermediate care center	475.9 (7.83)	163.5 (6.66)	<0.001	774.4 (9.69)	224.4 (7.92)	<0.001
Prescribed drugs	1709.2 (28.12)	684.9 (27.90)	<0.001	2211.6 (27.66)	742.4 (26.20)	<0.001
Other healthcare services	502.8 (8.27)	80.9 (3.30)	<0.001	630.1 (7.88)	114.4 (4.04)	<0.001
Total healthcare costs	6078.3	2455.2	<0.001	7995.2	2833.3	<0.001

Abbreviations: ACPs, advanced chronic patients; CCPs, complex chronic patients. [a] Rate ratio test by median-unbiased estimation (mid-p); [b] Calculated over the total healthcare costs for each group; [c] Student's *t*-test.

3.5. Geographical Variability of CCPs and ACPs Incidence in Catalonia

CCPs' and ACPs' cumulative incidence for the 2017–2019 period in the different regions of Catalonia was adjusted to the population's age, sex, morbidity, and income level, revealing regions with increased and decreased case incidence compared to the expected rates (Figure 7). Overall, adjusted incidence indexes of CCPs and ACPs were similar in each individual region.

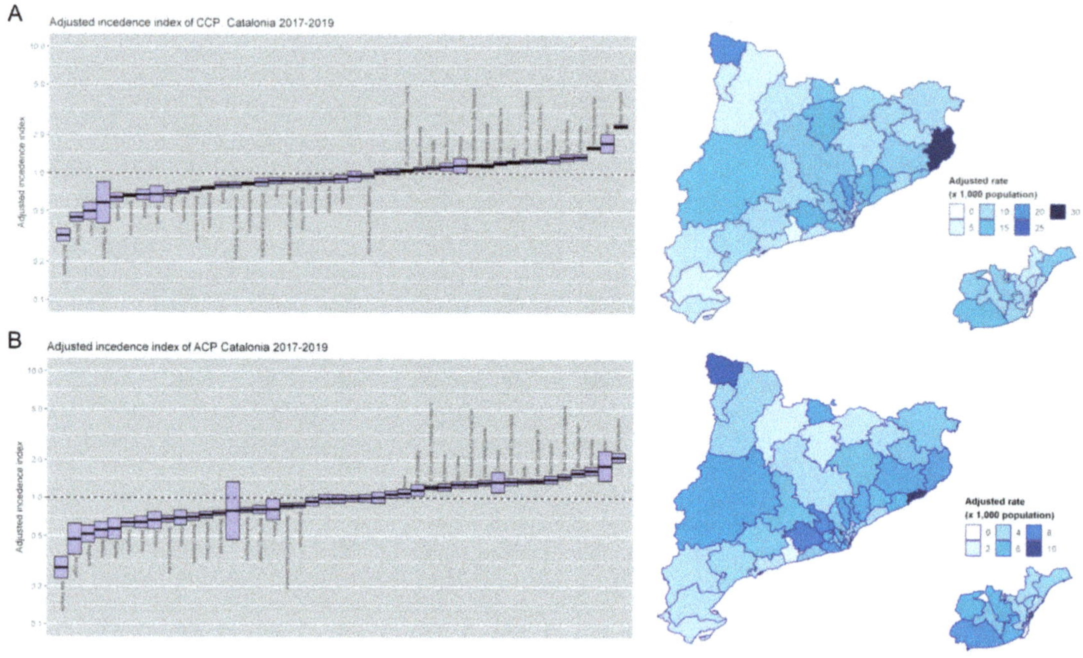

Figure 7. Incidence of (**A**) chronic complex patients (CCPs) and (**B**) advanced chronic patients (ACPs) adjusted by age, sex, comorbidities, and income level in the different regions of Catalonia in 2019.

4. Discussion

Owing to the healthcare needs and service utilization rates of individuals with chronic conditions and complex needs, the health plan of the Government of Catalonia for the 2011–2015 period prioritized this highly demanding population for the implementation of the chronic care program. To provide them with the best possible care, the Catalan Department of Health developed an integrated and individualized model of care structured in four stages, of which the first entails the screening and identification of CCPs and ACPs. In the healthcare system of Catalonia, the GMA stratification category and identification as CCP and ACP are registered on patients' medical records, enabling the use of this information. This observational, retrospective study, including all CCPs and ACPs from Catalonia identified between 2013 and 2019, assessed the first stage of this model during its initial implementation. During the study period, prevalence and incidence rates of CCPs and ACPs increased and decreased, respectively, while the probability of survival was significantly lower in ACPs compared to CCPs, and progressively decreased in both groups. The evolution of prevalence and incidence rates and survival probabilities are compatible with the progressive identification of CCPs and ACPs during the initial implementation of a novel, innovative care model. Likewise, the overall sociodemographic and clinical characteristics of CCPs and ACPs significantly changed. The prevalence of both CCPs and ACPs was higher in populations with low and very low income, and most cases were women. Compared to their respective age-, sex-, and income level-adjusted non-CCP and non-ACP populations, CCPs and ACPs were at higher morbidity-associated risk, had higher rates of all comorbidities, and higher utilization of healthcare services and associated health expenditure.

This observational study analyzed the implementation of a novel, innovative healthcare model centered on persons with complex conditions and care needs and advanced chronic diseases. Specifically, this study assessed the identification of the model's target population during the first years of implementation. The initial stages of the implementation of a novel model are likely associated with a learning curve for the identification of both CCPs and ACPs, and increased efforts to implement the program among physicians, which may explain the progressively decreased case incidence and increased case prevalence. In this regard, prevalence rates are expected to increase as the model is consolidated and reaches expected rates [23]. The incidence and prevalence rates found throughout the study revealed a faster identification of CCPs compared to ACPs. In this regard, given that patient identification relies on physicians' criteria and expertise applying the established perceptive criteria, effective identification is likely to be associated with a learning curve, even though GMAs have been published in the eHealth records since their introduction. Additionally, considering the time during visits required for identification, the substantial burden of primary care teams in our setting may have likely impacted the identification of CCPs and ACPs. Other shortcomings of this model are associated with ethical discussions during the first years, likely leading to a reluctance by some clinicians to identify these populations, particularly ACPs, given their palliative connotation. Ethical issues are particularly relevant in the case of ACPs, given the previously acknowledged prejudices and fears towards the identification and care of people needing palliative care, who may not receive appropriate care in case of a health crisis [36]. Furthermore, clinicians who need clear rules and guidelines may regard the need to combine objective and professional (subjective) perceptions of complexity as a barrier for identification, further contributing to increased reluctance. Future interventions from the Department of Health should focus on highlighting the benefits of early identification of the model's target population to overcome these shortcomings and implement an individual intervention plan shared among clinicians (second and third stage of the integrated care model), as depicted in Figure 2. In this respect, the patient identification information is accessible and visible to all the healthcare system, including acute and intermediate care centers, mental health centers, emergency services, and, in certain territories, social services. Despite the model's shortcomings, most CCP and ACP prevalent cases were identified during the first years. In this regard, in the first years of implementation of the model (2013–2017), identification was encouraged with economic incentives to clinicians, partly explaining the decreased identification after this period. The economic incentives were objectives introduced in the commissioning process and incorporated a variable pay based on the achievement of goals. Despite expediting identification, economic incentives raised clinical and ethical controversies among professionals, and, currently, quality (i.e., added value to the patient) of identification is prioritized over quantity. Case identification (i.e., incidence of CCPs and ACPs) peaked during the first years (in 2014 for both CCPs and ACPs) and decreased from 0.96% and 0.25% in 2014 to 0.39% and 0.16% in 2019, respectively. The higher incidence rates observed during the first years are compatible with the implementation of the first stage of the novel model and reflect the effective identification of CCPs and ACPs.

ACPs had a lower probability of survival, consistent with their identification as patients with low life expectancy prognosis, indirectly confirming the validity of ACP definition [16]. Regarding the prevalence of ACPs, previous cross-sectional studies using validated tools (i.e., NECPAL) to identify patients with chronic conditions in need of palliative care (similar to the ACP definition) reported rates of 1–1.5%, higher than the 0.26% found in this study [25,26]. However, whereas these previous studies aimed at prospectively identifying these patients, this study was conducted in a real-world setting and reflected the heterogeneity among clinicians, similar to the geographical variability, likely explaining the observed differences. As explained earlier, this heterogeneity may be related to ethical issues associated with the identification of ACPs. Furthermore, the reluctance of some clinicians to identify CCPs as ACPs may have additionally contributed to these discrepancies. In this regard, clinicians have shown increased reluctance to use the ACP identification, likely resulting in decreased identification of this subgroup of CCPs. Nevertheless, the decreased prevalence of ACPs compared to that estimated in previous cross-sectional studies (20% of estimated ACP prevalence) and the limited availability of international experiences in the field of chronicity focused on the proactive identification of ACPs warrant further research [25,26].

The use of the concepts of complexity and multimorbidity to define and identify CCPs and ACPs is novel and unique in this model. The concept of complexity in the context of healthcare lacks a precise definition and, in addition to clinical factors (i.e., chronic diseases), it encompasses other patient-related factors (i.e., socioeconomic), physician-related factors, including training, expertise, and experience, factors related to the organization of care, including decision-making, workflow, technology, and availability of time, team-related factors (i.e., leadership), contextual factors (physical and social), and organizational factors, including structures, politics, and procedures [21,37,38]. Unlike clinical variables, routine electronic clinical records do not systematically record most social factors and clinical fragmentation variables that determine the complexity and, overall, the availability of structured information regarding social variables is limited. Given the diversity of constructs that have been associated with complexity, an international consensus on its definition is needed to homogenize results from different studies and understand the care needs of complex patients [20,39]. In contrast, the availability of validated screening instruments enabled the identification of ACPs based on a robust construct. In our setting, the NECPAL tool, a validated instrument for the early identification of the need for palliative care among individuals with limited life expectancy, is routinely used [26,40,41]. Despite differences in the application of the concept of palliative care among countries (i.e., patients with oncologic conditions and in the last weeks or days of life vs. management of advanced chronic conditions), the definition of ACPs included in the Catalan healthcare plan used a robust method for identification, similar to other countries, potentially enabling comparisons among different countries/settings.

While the identification of CCPs depended on professionals' subjectivity (i.e., perception) regarding the concept of complexity, which was supported by information communication technology tools for stratification, the initial screening considered unique functional identifications related to patients' complexity status, regardless of the number of chronic conditions, using the automatic and hence, objective, GMA stratification system [16,32]. A complexity status detected by the initial GMA stratification system is likely to be associated with difficult management and decision-making. In this context, the GMA algorithm is a useful non-invasive support stratification tool for the initial identification by primary healthcare teams of people with potentially complex healthcare needs, candidates to be identified as CCPs, and to whom the integrated and individualized healthcare model developed by the Department of Health is applicable. This screening allows labeling patients and prioritizing them for their subsequent evaluation and identification as CCPs and ACPs. The results obtained regarding their demographic and clinical characteristics and their healthcare services utilization using the support stratification method (i.e., GMA) support the validity of the CCP and ACP constructs defined in the Integrated Chronic Care program to identify patients with specific care needs.

Analysis of the prevalence of comorbidities revealed that all of them were more frequent in CCPs and ACPs than in age-, sex-, and income level-adjusted non-CCP and non-ACP populations, showing an increased morbidity burden. Despite ranking in similar positions regarding frequency, cancer ranked fourth in CCPs (34.2%) and was the most frequent comorbidity in ACPs (48.5%) and, conversely, diabetes ranked as the most frequent comorbidity in CCPs (43.8%) and fifth in ACPs (38.3%), showing trends consistent with patients' end-of-life situation. In this regard, while most patients needing palliative care identified in previous studies using the NECPAL tool were in the dementia trajectory (55%), in this study, cancer was the most frequent comorbidity in patients identified as ACP. Even though patients with complex statuses included in this study may only have one chronic disease, considering the fact that most people with chronic disease have multimorbidity and the overall high prevalence of chronic conditions in CCPs and ACPs, most of this study's population likely had multimorbidity [42]. Regardless of the frequencies of comorbidities and the number of chronic diseases, CCPs and ACPs were at substantially higher morbidity-associated risk.

Despite potential differences in the definition of complexity, patients with multimorbidity and functional limitations have higher needs compared to multimorbid patients [43]. Accordingly, compared with age-, sex-, and income-level-adjusted non-CCP and non-ACP populations, CCPs and ACPs had substantially higher utilization of healthcare services. Primary care and acute care hospital admissions were the most frequently used. Furthermore, low- and very low-income population segments had a higher prevalence of CCPs and ACPs, indicating a relationship between socioeconomic and complex chronic statuses, similar to previous studies showing relationships between multimorbidity and income and educational levels [42,44–46]. In this regard, patients were classified according to income arbitrarily using data available from pharmacy records. These classification criteria used ad hoc precluded comparisons with other studies. The previously reported relationships between low income and more intensive use of primary care and high income and higher use of specialists and the higher prevalence of CCPs and ACPs in populations with lower income may explain their higher use of primary care services [47]. The substantial increase in the use of all healthcare resources and their associated expenditures in CCPs and ACPs compared to their corresponding adjusted non-CCP and non-ACP populations underscores the anticipated impact of complex chronic patients on the healthcare system.

The healthcare system of Catalonia uses a stratification algorithm (i.e., GMA) and specific identifiers for CCPs and ACPs, which are registered on the electronic health records, allowing us to monitor and use this information. To our knowledge, the availability of this information is unique to the Catalan healthcare system or is at least very rare in other settings. Furthermore, the tools (i.e., information system data and individual patient assessment) and criteria (i.e., clinical, context-related, and health and social care system-related) used for CCPs and ACPs identification are unique of the Chronic Care program precluding direct comparisons with previous reports. Previous studies have described similar populations using the high needs, high costs concept, corresponding to those patients who use the healthcare system the most. This criterion is typically used to define and identify patients with multiple chronic conditions [8]. A meta-analysis of studies evaluating patients with high needs, high costs showed that similar to this study, these patients had increased healthcare resource utilization, were more likely to die, and their most frequent comorbidities were similar to those of CCPs and ACPs [48]. Additionally, both social and material deprivation (similar to the context-related criteria considered in this model) were associated with higher costs [48]. However, the high needs, high costs patients identified in these previous studies using information system data were younger than the population identified in this study: half were younger than 65 years, whereas in this study, only 10.4% and 10.3% of CCPs and ACPs, respectively, were within this age range [48].

Regarding the prevalence of patients equivalent to CCPs, the previously reported prevalence rates of patients with multimorbidity differed across studies and settings [49]. Overall, prevalence rates in low- and middle-income countries were lower compared to those in high-income countries. Regarding ACPs, several previous models aiming to identify persons needing palliative care have been evaluated, but specific data regarding the prevalence of ACPs was not reported [50].

The results of this study should be interpreted in the context of some methodological limitations associated with its retrospective design and real-world setting, including variability in recorded data and the tools used for patient identification. In this regard, the GMA algorithm considers clinical variables of chronic diseases to measure the morbidity burden, and clinicians' subjective criteria are fundamental to evaluate other areas of complexity excluded from the GMA, including classification of chronic diseases according to severity and stages. In this regard, additional tools measuring the social- and healthcare system-related complexity dimensions are needed to gather structured, good-quality data on social variables, such as dependency, poverty, poor housing, and loneliness, beyond the management perspective of current algorithms. As structured variables become progressively available, they will be incorporated in the assessment of the different complexity dimensions. An additional limitation of this study is related to the identification of ACPs and their lack of validation on a case-by-case basis. Given the ethical issues associated with ACP identification, some clinicians may be hesitant to identify patients as ACP, potentially resulting in inaccurate ACP identifications of some cases. In this regard, future studies aimed at assessing healthcare service utilization and characteristics of the ACP population may require a prospective design to ensure data reliability. Nevertheless, the use of a large dataset including all people using the public healthcare system lacked selection bias and likely compensated for missing data and potential inaccuracies, at least partly, allowing us to capture the characteristics of chronic complex patients at the population level. The classification used in this study is unique to the Catalan Healthcare system. Therefore, the results of this study may not be applicable to other countries using other screening tools, identifiers, and other strategies to manage patients with multimorbidity and those with a short life prognosis [11,51,52]. Furthermore, the results from this model are unlikely to be applicable to healthcare systems of developing countries with poorly established primary healthcare systems.

Despite these limitations, this study assessed basic demographic and health indicators, allowing the characterization of the population with chronic complex conditions and describe their evolution in the context of the Chronic and Integrated Care Program. Future studies should focus on assessing trends in healthcare service utilization and expenditures to assess the impact of the Integrated Chronic Care Program in the management of these patients and their outcomes. Nevertheless, the results from this and future studies will be very useful to identify the challenges of implementing an integrated care model by the Department of Health. Future studies should assess other patient-related factors, such as patient experience outcome measures (PREMs), satisfaction, self-perceived health, and early access to palliative care, which may influence health-related quality of life, survival, healthcare costs, and end-of-life care, to ultimately improve the Integrated Care Program.

5. Conclusions

In the framework of the Chronic and Integrated Healthcare Program, the target populations of CCPs and ACPs were effectively identified, revealing their higher prevalence in low- and very low-income populations. CCPs and ACPs showed a higher frequency of multimorbidity, morbidity-associated risk, and utilization of healthcare services compared with the population of the same age, sex, and income level, reflecting their higher needs and expenditure. These results underscore the need to provide integrated care to complex chronic patients from the healthcare and social perspectives to improve and optimize their management. In the context of the increasing prevalence of people with complex chronic conditions, strategies, such as the Chronic Care Plan assessed in this study, which focus on this population of patients, should be implemented and assessed with the goal of decreasing their burden on the healthcare system.

Supplementary Materials: The following are available online at https://www.mdpi.com/article/10.3390/ijerph18189473/s1, Table S1: Types of complexity and criteria for the identification of complex chronic patients; Figure S1: Age and sex distribution of prevalent and cases of (A) CCPs and (B) ACPs in 2019; Figure S2: Main comorbidities of prevalent cases of (A) complex chronic (CCP) and (B) advanced chronic (ACP) patients by sex. Year 2019.

Author Contributions: Conceptualization, S.J.S., J.C.C., E.V., M.C., P.A., E.M.M.-A., E.G.-S., V.M. and J.A.-N.; Formal analysis, E.V. and M.C.; Investigation, S.J.S., J.C.C., E.V., M.C., P.A., E.M.M.-A., E.G.-S., V.M. and J.A.-N.; Methodology, S.J.S., J.C.C., E.V., M.C., P.A., E.M.M.-A., E.G.-S., V.M. and J.A.-N.; Software, E.V. and M.C.; Validation, E.V. and M.C.; Writing—original draft, S.J.S., J.C.C., E.V., M.C., P.A., E.M.M.-A., E.G.-S., V.M. and J.A.-N.; Writing—review & editing, S.J.S., J.C.C., E.V., M.C., P.A., E.M.M.-A., E.G.-S., V.M. and J.A.-N. All authors have read and agreed to the published version of the manuscript.

Funding: This research did not receive any funding from agencies in the public, commercial, or not-for-profit sectors.

Institutional Review Board: The study was conducted according to the guidelines of the Declaration of Helsinki, and was approved by the Research Ethics Committee of the Universitat de Vic—University of Vic-Central University of Catalonia (UVic-UCC) (ref. no. 63/2018 approved 3 December 2018).

Informed Consent Statement: Based on sample size and absence of personal information in the dataset used for analysis, the Ethics Committee determined that a written informed consent from each patient was not necessary.

Data Availability Statement: The datasets used and/or analyzed during the current study are available from the corresponding author upon reasonable request.

Acknowledgments: We would like to especially thank Albert Ledesma, Carles Blay, Assumció González-Mestre, Ester Sarquella, Laura Vila, and Xavier Gómez-Batiste for their role as promoters of the Chronic Care Program of Catalonia. Their vision and tenacity were essential to launching this pioneering program. The authors would like to thank i2e3 biomedical research institute for providing medical writing support during the preparation of the manuscript.

Conflicts of Interest: The authors declare that they have no conflict of interests.

References

1. Institut d'Estadística de Catalunya, Idescat. Taules de Vida (2015–2019). Available online: https://www.idescat.cat/pub/?id=iev (accessed on 12 March 2021).
2. Institut d'Estadística de Catalunya, Idescat. Fecunditat. Available online: https://www.idescat.cat/pub/?id=inddt&n=917 (accessed on 12 March 2021).
3. Institut d'Estadística de Catalunya, Idescat. Estructura per Edats, Envelliment i Dependència. Available online: https://www.idescat.cat/pub/?id=inddt&n=915 (accessed on 12 March 2021).
4. OECD. *Health at a Glance 2017*; OECD: Geneva, Switzerland, 2017.
5. Rechel, B.; Doyle, Y.; Grundy, E.; McKee, M. *How Can Health Systems Respond to Population Ageing?* World Health Organization: Geneva, Switzerland, 2009.
6. Anderson, G. *Chronic Care: Making the Case for Ongoing Care*; Robert Wood Johnson Foundation: Princeton, NJ, USA; The Johns Hopkins Bloomberg School of Public Health: Baltimore, MD, USA, 2010.
7. Alwan, A. *Global Status Report on Noncommunicable Diseases*; World Health Organization: Geneva, Switzerland, 2010; p. 176.
8. Blumenthal, D.; Chernof, B.; Fulmer, T.; Lumpkin, J.; Selberg, J. Caring for High-Need, High-Cost Patients—An Urgent Priority. *N. Engl. J. Med.* **2016**, *375*, 909–911. [CrossRef] [PubMed]
9. McPhail, S.M. Multimorbidity in chronic disease: Impact on health care resources and costs. *Risk Manag. Healthc. Policy* **2016**, *9*, 143–156. [CrossRef]
10. Lehnert, T.; Heider, D.; Leicht, H.; Heinrich, S.; Corrieri, S.; Luppa, M.; Riedel-Heller, S.; König, H.-H. Review: Health care utilization and costs of elderly persons with multiple chronic conditions. *Med. Care Res. Rev.* **2011**, *68*, 387–420. [CrossRef] [PubMed]
11. National Institute for Health and Care Excellence. *Multimorbidity: Assessment, Prioritisation and Management of Care for People with Commonly Occurring Multimorbidity*; NICE: London, UK, 2016.
12. Ham, C. *Integrated Care: What is it? Does it Work? What Does it Mean for the NHS?* The King's Fund: London, UK, 2010.
13. Nolte, E.; Knai, C.; Saltman, R.B. *Assessing Chronic Disease Management in European Health Systems*; European Observatory on Health Systems and Policies: Brussels, Belgium, 2014; ISBN 9789289050302.
14. Elwyn, G.; Vermunt, N.P.C.A. Goal-Based Shared Decision-Making: Developing an Integrated Model. *J. Patient Exp.* **2020**, *7*, 688–696. [CrossRef]
15. Mateo-Abad, M.; Fullaondo, A.; Merino, M.; Gris, S.; Marchet, F.; Avolio, F.; Graps, E.; Ravic, M.; Kovac, M.; Benkovic, V.; et al. Impact Assessment of an Innovative Integrated Care Model for Older Complex Patients with Multimorbidity: The CareWell Project. *Int. J. Integr. Care* **2020**, *20*, 8. [CrossRef]
16. Blay, C.; Limón, E.; Alburquerque, J.; Amblàs, J.; Ansa, X.; Ariño, S.; Benet, C. Bases per a un Model Català d'Atenció a les Persones amb Necessitats Complexes. *Programa Prev. i Atenció a La Cronicitat* **2017**, 1–66. Available online: http://hdl.handle.net/11351/3305 (accessed on 22 March 2021).
17. Tortajada, S.; Giménez-Campos, M.S.; Villar-López, J.; Faubel-Cava, R.; Donat-Castelló, L.; Valdivieso-Martínez, B.; Soriano-Melchor, E.; Bahamontes-Mulió, A.; García-Gómez, J.M. Case Management for Patients with Complex Multimorbidity: Development and Validation of a Coordinated Intervention between Primary and Hospital Care. *Int. J. Integr. Care* **2017**, *17*, 4. [CrossRef]
18. Ahgren, B. Chain of care development in Sweden: Results of a national study. *Int. J. Integr. Care* **2003**, 3, e01. [CrossRef]

19. Contel, J.C.; Ledesma, A.; Blay, C.; Mestre, A.G.; Cabezas, C.; Puigdollers, M.; Zara, C.; Amil, P.; Sarquella, E.; Constante, C. Chronic and integrated care in Catalonia. *Int. J. Integr. Care* **2015**, *15*, e025. [CrossRef]
20. Zullig, L.L.; Whitson, H.E.; Hastings, S.N.; Beadles, C.; Kravchenko, J.; Akushevich, I.; Maciejewski, M.L. A Systematic Review of Conceptual Frameworks of Medical Complexity and New Model Development. *J. Gen. Intern. Med.* **2016**, *31*, 329–337. [CrossRef]
21. Kuipers, P.; Kendall, E.; Ehrlich, C.; McIntyre, M.; Barber, L.; Amsters, D.; Kendall, M.; Kuipers, K.; Muenchberger, H.; Brownie, S. *Complexity and Health Care: Health Practicioner Workforce Services, Roles, Skills and Training, to Respond to Patients with Complex Needs*; Queensland Health: Brisbane, Australia, 2011; pp. 1–25.
22. Catalunya. Departament de La Presidència; Termcat, C.D.T. *Terminologia de L'atenció Integrada [en línia]*. Available online: http://www.termcat.cat/ca/Diccionaris_En_Linia/230/ (accessed on 17 May 2021).
23. Amblàs-Novellas, J.; Amil, P.; Bocanegra, I.; Bullich, Í.; Contel, J.C.; Gil, E.; Giráldez, G.; González, A.; Melendo, E.M.; Mir, V.; et al. *Bases Conceptuals i Model D'atenció per a Les Persones Fràgils, Amb Cronicitat Complexa (PCC) o Avançada (MACA)*; Generalitat de Catalunya (Government of Catalonia) Health Department: Barcelona, Spain, 2020.
24. Murray, S.A.; Kendall, M.; Mitchell, G.; Moine, S.; Amblàs-Novellas, J.; Boyd, K. Palliative care from diagnosis to death. *BMJ* **2017**, *356*, j878. [CrossRef] [PubMed]
25. Gómez-Batiste, X.; Martínez-Muñoz, M.; Blay, C.; Amblàs, J.; Vila, L.; Costa, X.; Villanueva, A.; Espaulella, J.; Espinosa, J.; Figuerola, M.; et al. Identifying patients with chronic conditions in need of palliative care in the general population: Development of the NECPAL tool and preliminary prevalence rates in Catalonia. *BMJ Support. Palliat. Care* **2013**, *3*, 300–308. [CrossRef]
26. Gómez-Batiste, X.; Martínez-Muñoz, M.; Blay, C.; Amblàs, J.; Vila, L.; Costa, X.; Espaulella, J.; Espinosa, J.; Constante, C.; Mitchell, G.K. Prevalence and characteristics of patients with advanced chronic conditions in need of palliative care in the general population: A cross-sectional study. *Palliat. Med.* **2014**, *28*, 302–311. [CrossRef]
27. Amblàs-Novellas, J.; Espaulella, J.; Rexach, L.; Fontecha, B.; Inzitari, M.; Blay, C.; Gómez-Batiste, X. Frailty, severity, progression and shared decision-making: A pragmatic framework for the challenge of clinical complexity at the end of life. *Eur. Geriatr. Med.* **2015**, *6*, 189–194. [CrossRef]
28. Kumpunen, S.; Edwards, N.; Georghiou, T.; Hughes, G. Why do evaluations of integrated care not produce the results we expect? *Int. J. Care Coord.* **2020**, *23*, 9–13. [CrossRef]
29. Cancio, J.M.; Vela, E.; Santaeugènia, S.; Clèries, M.; Inzitari, M.; Ruiz, D. Influence of demographic and clinical characteristics of elderly patients with a hip fracture on mortality: A retrospective, total cohort study in North-East Spain. *Bone* **2018**, *117*, 123–129. [CrossRef] [PubMed]
30. Cancio, J.M.; Vela, E.; Santaeugènia, S.; Clèries, M.; Inzitari, M.; Ruiz, D. Long-term Impact of Hip Fracture on the Use of Healthcare Resources: A Population-Based Study. *J. Am. Med. Dir. Assoc.* **2019**, *20*, 456–461. [CrossRef]
31. Amblàs-Novellas, J.; Santaeugènia, S.J.; Vela, E.; Clèries, M.; Contel, J.C. What lies beneath: A retrospective, population-based cohort study investigating clinical and resource-use characteristics of institutionalized older people in Catalonia. *BMC Geriatr.* **2020**, *20*, 187. [CrossRef]
32. Monterde, D.; Vela, E.; Clèries, M. Adjusted morbidity groups: A new multiple morbidity measurement of use in Primary Care. *Aten. Primaria* **2016**, *48*, 674–682. [CrossRef]
33. Monterde, D.; Vela, E.; Clèries, M.; Garcia-Eroles, L.; Roca, J.; Pérez-Sust, P. Multimorbidity as a predictor of health service utilization in primary care: A registry-based study of the Catalan population. *BMC Fam. Pract.* **2020**, *21*, 39. [CrossRef] [PubMed]
34. Dueñas-Espín, I.; Vela, E.; Pauws, S.; Bescos, C.; Cano, I.; Cleries, M.; Contel, J.C.; de Manuel Keenoy, E.; Garcia-Aymerich, J.; Gomez-Cabrero, D.; et al. Proposals for enhanced health risk assessment and stratification in an integrated care scenario. *BMJ Open* **2016**, in press. [CrossRef]
35. Vela, E.; Clèries, M.; Vella, V.A.; Adroher, C.; García-Altés, A. Population-based analysis of the Healthcare expenditure in Catalonia (Spain): What and who consumes more resources? *Gac. Sanit.* **2019**, *33*, 24–31. [CrossRef] [PubMed]
36. Gómez-Batiste, X.; Blay, C.; Broggi, M.A.; Lasmarias, C.; Vila, L.; Amblàs, J.; Espaulella, J.; Costa, X.; Martínez-Muñoz, M.; Robles, B.; et al. Ethical Challenges of Early Identification of Advanced Chronic Patients in Need of Palliative Care. *J. Palliat. Care* **2018**, *33*, 247–251. [CrossRef]
37. Nardi, R.; Scanelli, G.; Corrao, S.; Iori, I.; Mathieu, G.; Cataldi Amatrian, R. Co-morbidity does not reflect complexity in internal medicine patients. *Eur. J. Intern. Med.* **2007**, *18*, 359–368. [CrossRef]
38. Loeb, D.F.; Binswanger, I.A.; Candrian, C.; Bayliss, E.A. Primary care physician insights into a typology of the complex patient in primary care. *Ann. Fam. Med.* **2015**, *13*, 451–455. [CrossRef] [PubMed]
39. Valderas, J.M.; Starfield, B.; Sibbald, B.; Salisbury, C.; Roland, M. Defining Comorbidity: Implications for Understanding Health and Health Services. *Ann. Fam. Med.* **2007**, *5*, 395–402. [CrossRef] [PubMed]
40. Gómez-Batiste, X.; Turrillas, P.; Tebé, C.; Calsina-Berna, A.; Amblàs-Novellas, J. NECPAL tool prognostication in advanced chronic illness: A rapid review and expert consensus. *BMJ Support. Palliat. Care* **2020**. [CrossRef]
41. Central Catalonia Chronicity Research Group Utility of NECPAL. Available online: https://en.c3rg.com/necpal (accessed on 29 November 2020).
42. Barnett, K.; Mercer, S.W.; Norbury, M.; Watt, G.; Wyke, S.; Guthrie, B. Epidemiology of multimorbidity and implications for health care, research, and medical education: A cross-sectional study. *Lancet* **2012**, *380*, 37–43. [CrossRef]

43. Hayes, S.L.; Salzberg, C.A.; McCarthy, D.; Radley, D.C.; Abrams, M.K.; Shah, T.; Anderson, G.F. High-Need, High-Cost Patients: Who Are They and How Do They Use Health Care? A Population-Based Comparison of Demographics, Health Care Use, and Expenditures. *Issue Brief* **2016**, *26*, 1–14.
44. Agborsangaya, C.B.; Lau, D.; Lahtinen, M.; Cooke, T.; Johnson, J.A. Multimorbidity prevalence and patterns across socioeconomic determinants: A cross-sectional survey. *BMC Public Health* **2012**, *12*, 201. [CrossRef]
45. McLean, G.; Gunn, J.; Wyke, S.; Guthrie, B.; Watt, G.C.M.; Blane, D.N.; Mercer, S.W. The influence of socioeconomic deprivation on multimorbidity at different ages: A cross-sectional study. *Br. J. Gen. Pract.* **2014**, *64*, e440–e447. [CrossRef] [PubMed]
46. Schiøtz, M.L.; Stockmarr, A.; Høst, D.; Glümer, C.; Frølich, A. Social disparities in the prevalence of multimorbidity—A register-based population study. *BMC Public Health* **2017**, *17*, 422. [CrossRef]
47. Mbuya-Bienge, C.; Simard, M.; Gaulin, M.; Candas, B.; Sirois, C. Does socio-economic status influence the effect of multimorbidity on the frequent use of ambulatory care services in a universal healthcare system? A population-based cohort study. *BMC Health Serv. Res.* **2021**, *21*, 202. [CrossRef] [PubMed]
48. Wammes, J.J.G.; van der Wees, P.J.; Tanke, M.A.C.; Westert, G.P.; Jeurissen, P.P.T. Systematic review of high-cost patients' characteristics and healthcare utilisation. *BMC Open* **2018**, *8*, e023113. [CrossRef] [PubMed]
49. Nguyen, H.; Manolova, G.; Daskalopoulou, C.; Vitoratou, S.; Prince, M.; Prina, A.M. Prevalence of multimorbidity in community settings: A systematic review and meta-analysis of observational studies. *J. Comorb.* **2019**, *9*. [CrossRef]
50. Finucane, A.M.; Davydaitis, D.; Horseman, Z.; Carduff, E.; Baughan, P.; Tapsfield, J.; Spiller, J.A.; Meade, R.; Lydon, B.; Thompson, I.M.; et al. Electronic care coordination systems for people with advanced progressive illness: A mixed-methods evaluation in Scottish primary care. *Br. J. Gen. Pract.* **2020**, *70*, e20–e28. [CrossRef]
51. Gold Standards Framework. Proactive Identification Guidance. 2017. Available online: https://goldstandardsframework.org.uk/pig (accessed on 22 March 2021).
52. Tools to Aid Clinical Identification of End of Life: An Evidence Check Rapid Review Brokered by the Sax Institute for the Agency for Clinical Innovation. 2017. Available online: https://www.saxinstitute.org.au/wp-content/uploads/Tools-to-aid-clinical-identification-of-end-of-life.pdf (accessed on 22 March 2021).

International Journal of
Environmental Research and Public Health

MDPI

Article

Primary Care Professionals' Self-Efficacy Surrounding Advance Care Planning and Its Link to Sociodemographics, Background and Perceptions: A Cross-Sectional Study

Cristina Lasmarías [1,2], Amor Aradilla-Herrero [3,*], Cristina Esquinas [4,5], Sebastià Santaeugènia [2,6], Francisco Cegri [7,8], Esther Limón [9,10] and Mireia Subirana-Casacuberta [11,12]

1 Department of Education and Training, Catalan Institute of Oncology, 08908 Barcelona, Spain; clasmarias@iconcologia.net
2 Catalonia Chronic Care Research Group, University of Vic-Central University of Catalonia, 08500 Vic, Spain; sebastia.santaeugenia@gencat.cat
3 School of Nursing, Escoles Universitàries Gimbernat (Universitat Autònoma de Barcelona), 08174 Sant Cugat del Vallès, Spain
4 Department of Pneumology, Hospital Universitari Vall d'Hebron, 08031 Barcelona, Spain; crise4@hotmail.com
5 Public Health, Mental, Maternal and Child Health Nursing Departament, Faculty of Medicine and Health Sciences, University of Barcelona, 08031 Barcelona, Spain
6 Chronic Care Program, Health Department, Generalitat de Catalunya, 08028 Barcelona, Spain
7 Responsible for Strategy and Innovation of the Association of Family and Community Nursing of Catalonia (AIFiCC), 08001 Barcelona, Spain; pacocegri@gmail.com
8 Primary Care Centre Sant Martí de Provençals, Institut Català de la Salut, 08001 Barcelona, Spain
9 Permanent Board of the Catalan Society of Family and Community Medicine (CAMFiC), 08019 Barcelona, Spain; elimonramirez@gmail.com
10 Primary Care Centre Mataró-7, Institut Català de la Salut, 08019 Mataró, Spain
11 Nursing Department, Parc Taulí Hospital Universitari, Institut d'Investigació i Innovació Parc TaulíI3PT, Universitat Autònoma de Barcelona, Consorci Sanitari Parc Taulí, 08208 Sabadell, Spain; msubiranac@tauli.cat
12 Research Group on Methodology, Methods, Models, and Health and Social Outcomes, Faculty of Health Science and Welfare, Centre for Health and Social Care Research, University of Vic-Central University of Catalonia, 08500 Vic, Spain
* Correspondence: amor.aradilla@eug.es; Tel.: +34-935-893-727

Citation: Lasmarías, C.; Aradilla-Herrero, A.; Esquinas, C.; Santaeugènia, S.; Cegri, F.; Limón, E.; Subirana-Casacuberta, M. Primary Care Professionals' Self-Efficacy Surrounding Advance Care Planning and Its Link to Sociodemographics, Background and Perceptions: A Cross-Sectional Study. *Int. J. Environ. Res. Public Health* **2021**, *18*, 9034. https://doi.org/10.3390/ijerph18179034

Academic Editor: Paul B. Tchounwou

Received: 27 July 2021
Accepted: 24 August 2021
Published: 27 August 2021

Abstract: Primary care (PC) professionals have been considered the most appropriate practitioners for leading Advance care planning (ACP) processes with advanced chronic patients. Aim: To explore how PC doctors' and nurses' self-efficacy surrounding ACP is linked to their sociodemographic characteristics, background and perceptions of ACP practices. Methods: A cross-sectional study was performed. Sociodemographics, background and perceptions about ACP in practice were collected using an online survey. The Advance Care Planning Self-Efficacy Spanish (ACP-SEs) scale was used for the self-efficacy measurement. Statistical analysis: Bivariate, multivariate and backward stepwise logistic regression analyses were performed to identify variables independently related to a higher score on the ACP-SEs. Results: N = 465 participants, 70.04% doctors, 81.47% female. The participants had a mean age of 46.45 years and 66.16% had spent >15 years in their current practice. The logistic regression model showed that scoring ≤ 75 on the ACP-SEs was related to a higher score on feeling sufficiently trained, having participated in ACP processes, perceiving that ACP facilitates knowledge of preferences and values, and perceiving that ACP improves patients' quality of life. Conclusion: Professionals with previous background and those who have a positive perception of ACP are more likely to feel able to carry out ACP processes with patients.

Keywords: advance care planning; primary care; self-efficacy

1. Introduction

Advance care planning (ACP) enables individuals to define goals, values, and preferences for future medical treatment and care, to discuss these matters with their family and care providers, and to record and review preferences as necessary [1]. In recent years, several international initiatives have been developed to improve ACP implementation in clinical practice. Initiatives have explored, for instance, the factors that make ACP processes feasible in daily practice [2], the main benefits for patients and families [3] and the health outcomes expected from an ACP process [4,5]. In this sense, proposals about how to integrate models of ACP into health systems [6–8] and how to adapt ACP content to individuals' health condition [9] and to specific settings [10,11] have been studied.

A wide range of studies define the professional profile that should lead the ACP process. A multidisciplinary approach seems to be the most appropriate to ensure a broad, multidimensional and individualised ACP process [12]. Specifically, primary care (PC) professionals have been recognized as the most suited to promote ACP processes, since the long-term nature of the relationships between PC professionals and patients is unique compared to other fields within the public health system [13–15]. This characteristic might help PC professionals build trust with patients, and it could explain the positive attitudes expressed to ACP process and its value on the improvement of the end-of life process [16]. However, PC professionals express difficulties in initiating ACP processes, such as lack of time and skills (skills that are common in other disciplines [6]) and lack of knowledge about how to choose the best moment to initiate an ACP [17]. These obstacles are especially important in the PC setting, where professionals take care of people with diverse disease trajectories [18], and some differences have been found in the approach of physicians and allied health professionals to ACP process [15].

In Spain, recent initiatives have emerged to promote the development of ACP. For example, the ACP model of Spain's Catalonia region was established within the framework of the 2011–2015 health plan and started in 2014 [19]. Some key aspects of this project included a consensus of experts about the concept of ACP, a practical guide to ACP for clinical practice, and a training programme, including a 10 h online course and face-to-face workshops lasting from 4 to 8 h [19]. In addition, a dedicated section for documenting the ACP process was added to the medical record of complex chronic patients and advanced chronic patients.

In Catalonia, as in other places, advanced chronic patients are considered to benefit when the ACP process begins early [20,21]. These patients are coded as advanced chronic patients using the NECPAL©(Palliative Needs) tool (Chair of Palliative Care, Vic, Spain, 2012), which includes the question, "Would I be surprised if this patient died in the next twelve months?" [22,23]. PC professionals are responsible both for coding these patients and recording their ACP process. Documentation of the process is visible to all health professionals in the public health system through a shared clinical record, which is accessible throughout the Catalan public health system [24]. It is understood that having a specific section in the medical record for documenting the ACP is a quality indicator of the implementation of this process [1]. Considering the importance of recording the ACP process, the Catalan model of ACP defined the ACP as a recurring continuity improvement cycle, divided into six phases. One phase was to record preferences and other medical decisions explored in an ACP process as part of the medical record [19]. However, the low quality and accuracy of this documentation in Catalonia are currently areas for improvement, especially considering the ACP record's potential impact on patient safety [25].

Self-efficacy refers to the individual's belief in his/her ability to carry out a particular task [26]. This topic has frequently been recommended as an indicator for measuring the impact of ACP programs [1,27] and measuring how training programmes improve self-efficacy in ACP [28]. Bandura stated that a person is more self-efficacious when the activity to be carried out (or the learning to be integrated) makes sense and the person also feels prepared to carry it out [26]. In a previous work, the Advance Care Planning

Self-Efficacy Spanish scale (ACP-SEs) has been validated to evaluate the self-efficacy of PC, palliative specialists and geriatricians surrounding ACP [29]. These professionals are frequently responsible for carrying out ACP processes with advanced chronic patients. The ACP-SEs scale is composed of 19 items and shows adequate psychometric properties (Cronbach's alpha = 0.95) to be used with these professionals [29]. Having a validated scale to measure self-efficacy in ACP will facilitate an understanding of this item in our context in order to design and develop public health programs of ACP.

Several projects have been carried out to evaluate the impact of ACP programmes in PC settings with good results [20,30]. However, to our knowledge, the link between, on the one hand, self-efficacy surrounding ACP and, on the other hand, the demographic characteristics, background, and perception of PC doctors and nurses has not been described. Understanding the relationships among these variables in PC in a public health system could help not only the development of public ACP programs but also the integration of ACP in models of care for people with advanced chronic illness.

We aim to explore the relationship between, on the one hand, PC doctors' and nurses' self-efficacy in ACP and, on the other, their sociodemographic characteristics, training and experience, and perception of ACP and to analyse differences between the two disciplines.

2. Materials and Methods

2.1. Study Design and Selection of Study Subjects

A cross-sectional study was performed. The sample was composed of doctors and nurses from the Catalan society of family and community medicine (CAMFiC) and the Association of family and community nursing of Catalonia (AIFiCC), which represents a large amount of doctors and nurses that work in primary care in Catalonia.

2.2. Procedure and Measurements

We created a survey on the *RedCap* platform (https://www.project-redcap.org/, accessed on 12 November 2020). All members of the two scientific societies were invited to participate via a link sent by the presidents of both societies in October 2018. We did not have access to the email addresses of participants. By responding to the survey, participants demonstrated their consent to take part in the study, which was voluntary and anonymous. Three reminders were sent (final reminder in December 2018). The estimated time required to respond to the survey was between 15 and 20 min.

2.3. Measurement Instruments

The survey was composed of four sections: (1) sociodemographic variables: age, gender, profession and years in current practice; (2) 9 variables about background: knowledge about ACP, previous training in ACP and practical experience in conducting ACP processes; (3) 12 items were created ad-hoc to measure perception, including applicability, of ACP practices (scored from a minimum of 1 to a maximum of 10); (4) the ACP-SEs scale, in its Spanish version [29], containing 19 items rated on a 5-point Likert-type scale (1 = not at all capable and 5 = completely capable).

2.4. Institutional Review Board Statement

All research procedures used in this study were established in accordance with the Declaration of Helsinki. The ethics committee at the University of Vic reviewed and approved the study protocol (code RS005_S). Moreover, we designed the study in accordance with the ethics criteria established by Spanish Organic Law 3/2018 of 5 December on personal data protection and the guarantee of digital rights, following the General Regulation (EU) 2016/679 of 27 April 2016 on data protection.

2.5. Statistical Analysis

Categorical variables were described with absolute frequencies and percentages. Quantitative variables were described using the mean and standard deviation (SD). The Kolmogorov-Smirnov test was used to assess the normality of distributions.

We analysed the variables related to sociodemographics, background and perception according to field (medicine vs. nursing). The total score on the ACP-SEs was calculated as the sum of scores for the items, which was rescaled from 0 (minimum) to 100 (maximum). In the case of quantitative variables, the Student's *t*-test (Mann-Whitney U-test if normality was not assumed) or ANOVA tests (in the case of variables with more than two categories) were carried out. The Chi-squared test (Fisher test for frequencies < 5) was used for the comparison of categorical variables.

To identify the variables related to a high score on the ACP-SEs, we transformed the total score of the scale into a binary variable using the third quartile as a cut-off point (≤ and >75 points). For the bivariate analysis, the variables related to sociodemographics and ACP background were included. From the group about perceptions' variables, eight were also included: we also transformed these quantitative items into binary variables (< and ≥8 points/10) and included those with a score ≥8 points.

In the multivariate analysis, we performed a backward stepwise logistic regression analysis to identify variables independently related to ACP > 75 points. Variables with a *p*-value < 0.2 in the bivariate analysis were included as independent factors. The results were described with odds ratios (OR), 95% confidence interval (CI) and *p*-values. The combination of predictors from the final model was used to calculate the probabilities of ACP-SEs > 75 points. A receiver operating characteristic (ROC) analysis and Hosmer-Lemeshow goodness-of-fit test were performed to assess the overall fit of the model [31]. For all the tests, *p*-values < 0.05 were considered statistically significant. We used the statistics package R Studio (V2.5.1) (The R Project for Statistical Computing, Viena, Austria), for the analysis.

3. Results

A total of 465 professionals participated in the study, of whom 70.04% were doctors and 29.96% were nurses (one social worker was excluded from the analysis based on field). The mean age was 46.45 years, 81.47% were women, and 66.16% had more than 15 years of professional experience (Table 1). Table 2 shows the results for ACP background (knowledge, training, and experience in ACP). A total percentage of 70.26% had completed training in the subject, of which 30.58% (n = 100) had completed more than 8 h. A sum of 52.89% of the participants had carried out an ACP with patients, of which 30.39% (n = 141) stated that they had experienced difficulties. The professionals frequently carried out ACPs with people with advanced chronic disease (71.02%), advanced cancer (64.49%) and frailty (62.04%).

Table 1. Sociodemographic characteristics and field.

		Total n = 464 *	Medicine n = 325	Nursing n = 139
Age	Mean (SD)	46.45 (10.17)	46.38 (10.23)	46.61 (10.04)
Gender	Female	379 (81.47%)	253 (77.85%)	125 (89.93%)
	Male	86 (18.53%)	72 (22.15%)	14 (10.07%)
Years in current practice	<1 year	1 (0.22%)	1 (0.31%)	0
	1–5 years	60 (12.93%)	47 (14.46%)	13 (9.35%)
	6–10 years	45 (9.7%)	35 (10.77%)	10 (7.19%)
	11–15 years	45 (9.7%)	34 (10.46%)	11 (7.91%)
	>15 years	307 (66.16%)	204 (62.77%)	103 (74.1%)
	No active	6 (1.29%)	4 (1.23%)	2 (1.44%)

* The social worker has been excluded.

Table 2. Training and professional experience in ACP [1] according to field.

		Total n = 464	Medicine n = 325	Nursing n = 139
Have you heard of ACP?	No	12 (2.59%)	9 (2.77%)	3 (2.16%)
	Yes	452 (97.41%)	316 (97.23%)	136 (97.84%)
Have you completed training in ACP?	No	138 (29.74%)	108 (33.23%)	30 (21.58%)
	Yes	326 (70.26%)	217 (66.77%)	109 (78.42%)
Number of hours of training completed (n = 326)	<1 h	9 (2,75%)	7 (3.23%)	2 (1.83%)
	1 h–2 h	66 (26.61%)	50(23.04%)	16 (18.35%)
	2 h–4 h	87 (19.88%)	55 (25.35%)	32 (14.68%)
	4 h–8 h	64 (20.18%)	44 (20.28%)	20 (29.36%)
	>8 h	100(30.58%)	61 (28.11%)	39 (35.78%)
Do you consider yourself to be sufficiently trained to carry out ACP processes?	Value: 1 to 10 $\overline{x}$ (SD)	5.54 (2.29)	5.44 (2.26)	5.78 (2.37)
Have you participated in an ACP process with a patient?	No	219 (47.2%)	149 (45.85%)	70 (50.36%)
	Yes	245 (52.89%)	176 (54.15%)	69 (49.64%)
Number of ACP processes per month (n = 245)	None	81 (33.06%)	59 (33.52%)	22 (31.88%)
	1–5	155 (63.27%)	112 (63.64%)	43 (62.32%)
	6–10	6 (2.45%)	3 (1.7%)	3 (4.35%)
	11–20	3 (1.22%)	2 (1.14%)	1 (1.45%)
Time since the last ACP process	Less than one week	34 (13.88%)	19 (10.8%)	15 (21.74%)
	Between one week and one month	70 (28.57%)	54 (30.68%)	16 (23.19%)
	More than one month	141 (57.55%)	103 (58.52%)	38 (55.07%)
Have you had any difficulties in carrying out ACP processes?	No	104 (22.41%)	75 (23.08%)	29 (20.86%)
	Yes	141 (30.39%)	101 (31.08%)	40 (28.78%)
Main disease of the patients with whom you carried out ACP processes (more than one option was possible)	Advanced organ failure disease	174 (71.02%)	126 (71.59%)	48 (69.57%)
	Advanced cancer	158 (64.49%)	116 (65.91%)	42 (60.87%)
	Advanced dementia	99 (40.41%)	67 (38.07%)	32 (46.38%)
	Advanced neurological disease	77 (31.43%)	53 (30.11%)	24 (34.78%)
	Frailty	152 (62.04%)	108 (61.36%)	44 (63.77%)
	Another chronic disease	88 (35.92%)	65 (36.93%)	23 (33.33%)

[1] ACP = advance care planning.

The mean score for the item "Do you consider yourself to be sufficiently trained to carry out ACP processes?" was 5.54 out of 10.

Table 3 describes the professionals' perception of ACP in practice according to field. Of the 12 items explored, 8 show a mean greater than 8 on a scale from 1 to 10. In the set of overall means, the item with the highest score, with a mean of 9.08 (SD = 1.19), is "The ACP process facilitates the expression of wishes and preferences to be taken into consideration when the patient is not able to express by him or herself" and is followed by "ACP is important for complex chronic patients and advanced chronic patients", with a mean of 8.99 (SD = 1.19). The two items with the lowest score are "ACP is important for healthy people" (mean = 6.77; SD = 2.47) and "The ACP process is feasible in my professional setting" (mean = 7.11; SD = 2.17).

Table 4 shows the results in relation to the 19 ACP-SEs items by field. The ACP-SEs scale shows an overall mean of 65.90 (SD = 16.01) out of 100, with no statistically significant differences between doctors and nurses.

Table 3. Perception of ACP [1] practices by field.

	Total $n = 464$ (Mean, SD)	Medicine $n = 325$ (Mean, SD)	Nurse $n = 139$ (Mean, SD)
The ACP process facilitates the expression of wishes and preferences to take into account when the patient is not able to express him or herself	9.08 (1.19)	9.01 (1.22)	9.26 (1.1)
ACP is important for complex chronic patients and advanced chronic patients	8.99 (1.19)	8.91 (1.21)	9.18 (1.14)
ACP facilitates knowledge of patients' values and preference	8.98 (1.11)	8.91 (1.12)	9.13 (1.08)
ACP makes it possible to identify the patient's personal representative	8.78 (1.46)	8.76 (1.44)	8.82 (1.5)
ACP enables the patient to die at the place he/she wishes	8.40 (1.66)	8.33 (1.63)	8.58 (1.72)
ACP process makes it possible to adapt treatments to realistic therapeutic options	8.26 (1.58)	8.15 (1.57)	8.5 (1.59)
ACP is important for patients with a chronic disease even they are not identified as complex chronic patients or advanced chronic patients	8.24 (1.71)	8.23 (1.68)	8.28 (1.79)
The ACP process contributes to improving patients' quality of life	8.12 (1.71)	8.05 (1.7)	8.27 (1.71)
ACP helps me to coordinate with other professionals	7.97 (1.85)	7.78 (1.92)	8.41 (1.61)
ACP gives me confidence as a professional that I'm caring for patients properly	7.92 (1.88)	7.9 (1.85)	7.96 (1.96)
The ACP process is feasible in my professional setting	7.11 (2.17)	6.95 (2.23)	7.46 (2.01)
ACP is important to healthy people	6.77 (2.47)	6.74 (2.37)	6.83 (2.69)

[1] ACP = advance care planning; Correlation between the ACP-SEs and the Groups of Interest.

Table 4. Relation between the ACP-SEs [1] and field.

	Total $n = 464$ (Mean, SD)	Medicine $n = 325$ (Mean, SD)	Nurse $n = 139$ (Mean, SD)	p-Value
1. Find the time to discuss the patient's prognosis preferences and care plan with the patient	3.48 (0.98)	3.49 (0.98)	3.45 (0.97)	0.60
2. Determine how much the patient wants to know about the prognosis	3.68 (0.82)	3.73 (0.78)	3.58 (0.89)	0.10
3. Determine the level of involvement the patient wants in decision-making	3.65 (0.81)	3.69 (0.78)	3.54 (0.86)	0.07
4. Determine who else (e.g., family members) the patient would like to be involved in decision-making	3.98 (0.75)	4.03 (0.73)	3.88 (0.8)	0.07
5. Provide the desired level of information and guidance needed to help the patient in decision-making	3.69 (0.81)	3.71 (0.77)	3.64 (0.88)	0.40
6. Describe the pros and cons of different life-sustaining treatments	3.5 (0.92)	3.56 (0.87)	3.37 (1.02)	**0.049**
7. Determine the patient's specific wishes for types of medical treatment	3.53 (0.85)	3.58 (0.84)	3.41 (0.86)	**0.049**
8. Discuss and negotiate individualised treatment goals and plans with patient	3.55 (0.89)	3.6 (0.86)	3.42 (0.93)	**0.049**
9. Ensure that patient's treatment preferences will be honoured at your facility	3.98 (0.83)	4.01 (0.81)	3.93 (0.87)	0.40
10. Ensure that patient's treatment preferences will be honoured at a hospital if patient is hospitalised	2.77 (1.13)	2.73 (1.1)	2.87 (1.2)	0.20
11. Discuss how to complete a living will with the patient	3.5 (1.1)	3.46 (1.11)	3.6 (1.09)	0.20
12. Determine when there should be a shift in care goals	3.4 (0.98)	3.36 (0.96)	3.5 (1.04)	0.20
13. Reassess the patient's wishes when a shift in care goals is needed	3.59 (0.95)	3.57 (0.96)	3.65 (0.92)	0.40
14. Openly discuss uncertainty with patient when it exists	3.81 (0.85)	3.87 (0.82)	3.67 (0.9)	**0.02**
15. Educate patient and clarify any misperceptions about the disease or prognosis	3.8 (0.77)	3.86 (0.72)	3.63 (0.86)	**0.006**
16. Respond empathetically to patient's and family's concerns	4.11 (0.72)	4.18 (0.67)	3.95 (0.81)	**0.003**
17. Communicate "bad news" to patients and their families	3.67 (0.84)	3.84 (0.71)	3.29 (0.99)	**<0.001**
18. Engage patients in ACP conversations	3.69 (0.88)	3.71 (0.82)	3.64 (1.01)	0.50
19. Correctly register the decisions and care plan agreed to over the course of the ACP	3.69 (1.06)	3.62 (1.08)	3.83 (1.01)	**0.049**
TOTAL SCORE Mean (SD) (re-scaled to 100)	65.90 (16.01)	66.59 (15.40)	64.28 (17.30)	0.200

[1] ACP-SEs = Advance Care Planning Self-Efficacy Spanish. p-values < 0.05; Statistically significant differences have been marked in bold.

Of the 19 items, 7 show significant differences between doctors and nurses (6, 7, 8, 14, 15, 16, 17, 19). Doctors scored higher on all of these except item 19.

Table 5 shows the bivariate analysis of the main variables of Tables 1 and 2 and the high-scoring variables from Table 3 (≥8 points), categorised as binary variables, and the score obtained in the ACP-SEs with the cut-off point of the third quartile (≤ and >75 points). A total of 24.52% (n = 114) of the participants scored >75 points out of 100. Years in current practice, hours of training in ACP, having previously participated in ACP processes, and believing that ACP offers greater knowledge of patients' values and preferences, among others, showed statistically significant differences (Table 5).

Table 5. Comparison of the total ACP score categorized according to Q3 cut-off of 75 points on the APC-SEs scale and sample characteristics and variables of interest in the questionnaire. Bivariate analysis.

		N = 465 [1] (100%)	Total Score ACPS-SEs [2] (100 Points)		
			<= 75	>75	*p*-Value
Age	**<50 years**	273 (58.71%)	208 (59.26%)	65 (57.02%)	0.754
	≥50 years	192 (41.29%)	143 (40.74%)	49 (42.98%)	
Gender	Female	379 (81.51%)	291 (82.91%)	88 (77.19%)	0.220
	Male	86 (18.49%)	60 (17.09%)	26 (22.81%)	
Field	Medicine	325 (69.89%)	243 (69.43%)	82 (71.93%)	0.698
	Nursing	139 (29.89%)	107 (30.57%)	32 (28.07%)	
Years in current practice	≤15 years	151 (32.90%)	120 (34.38%)	31 (28.18%)	**<0.001**
	>15 years	308 (67.10%)	229 (65.62%)	79 (71.82%)	
Have you completed training in ACP?	No	138 (29.68%)	111 (31.62%)	27 (23.68%)	0.135
	Yes	327 (70.32%)	240 (68.38%)	87 (76.32%)	
Number of hours of training completed	≤4 h	162 (49.54%)	131 (54.58%)	31 (35.63%)	**0.004**
	>4 h	165 (50.46%)	109 (45.42%)	56 (64.37%)	
Do you consider yourself to be sufficiently trained to carry out ACP processes?	<8 points	361 (77.63%)	302 (86.04%)	59 (51.75%)	**0.001**
	≥8 points	104 (22.37%)	49 (13.96%)	55 (48.25%)	
Have you participated in an ACP process with a patient?	No	219 (47.1%)	186 (52.99%)	33 (28.95%)	**<0.001**
	Yes	246 (52.9%)	165 (47.01%)	81 (71.05%)	
Number of ACP processes per month	None	81 (32.93%)	56 (33.94%)	25 (30.86%)	**<0.001**
	≥1	156 (63.42%)	109 (66.06%)	56 (69.14%)	
Have you had any difficult in carrying out ACP processes?	No	105 (22.58%)	56 (33.94%)	49 (60.49%)	**<0.001**
	Yes	141 (30.32%)	109 (66.06%)	32 (39.51%)	
ACP is important for complex chronic patients and advanced chronic patients	≤8 points	135 (29.03%)	119 (33.90%)	16 (14.04%)	**0.003**
	>8 points	330 (70.97%)	232 (66.10%)	98 (85.96%)	
ACP is important for patients with a chronic disease even they are not identified as complex chronic patients or advanced chronic patients	≤8 points	237 (50.97%)	197 (56.13%)	40 (35.09%)	0.632
	>8 points	228 (49.03%)	154 (43.87%)	74 (64.91%)	
The ACP process facilitates expression of wishes and preferences to taken into account when the patient is not able to express him or herself	≤8 points	114 (24.52%)	97 (27.64%)	17 (14.91%)	0.097
	>8 points	351 (75.48%)	254 (72.36%)	97 (85.09%)	
ACP makes it possible to identify the patient's personal representative	≤8 points	153 (32.90%)	130 (37.04%)	23 (20.18%)	**0.013**
	>8 points	312 (67.10%)	221 (62.96%)	91 (79.82%)	
The ACP process contributes to improving patients' quality of life	≤8 points	241 (51.83%)	203 (57.83%)	38 (33.36%)	**<0.001**
	>8 points	224 (48.17%)	148 (42.17%)	76 (66.67%)	
ACP enables the patient to die at the place he/she wishes	≤8 points	202 (43.44%)	171 (48.72%)	31 (27.19%)	**<0.001**
	>8 points	263 (56.56%)	180 (51.28%)	83 (72.81%)	
The ACP process makes it possible to adapt the treatments to realistic therapeutic options	≤8 points	120 (25.81%)	97 (27.64%)	23 (20.18%)	**<0.001**
	>8 points	345 (74.19%)	254 (72.36%)	91 (79.82%)	
ACP facilitates knowledge of patients' values and preferences	≤8 points	46 (9.89%)	42 (11.97%)	4 (3.51%)	**<0.001**
	>8 points	419 (90.11%)	309 (88.03%)	110 (96.49%)	

[1] Including the social worker. [2] ACP-SEs = Advance Care Planning Self Efficacy Spanish p-values < 0.05; Statistically significant differences have been marked in bold

The result of the logistic regression analysis shows four variables independently related to the increase in the probability of scoring >75 points on the ACP-SEs: having previously participated in ACP processes (OR = 1.70; p-value = 0.043), perceiving that ACP contributes to improving people's quality of life (OR = 1.93; p-value = 0.013), perceiving that ACP facilitates the knowledge about patients' values and preferences (OR = 2.24; p-value = 0.028) and considering oneself to be sufficiently trained in ACP (OR = 3.98; p-value < 0.001) (Table S1).

The probabilities for scoring above 75 on the ACP-SEs were obtained by the following formula: Exp(β)/(1 + Exp(β)), where β = −2.838 + 1381 (in case of consider yourself to be sufficiently trained to carry out ACP processes ≥8 points) + 0.528 (previous participation in an ACP process with a patient) + 0.656 (ACP process contributes to improving the patients' quality of life >8 points) + 0.807 (ACP facilitates knowledge of patients' values and preferences >8 points). The probability of having >75 points on the ACP-SEs increased with the number of predictors, from 5% when no factor was present to 68% for patients having all four variables (Table S2). The model is well calibrated with a Hosmer-Lemeshow p = 0.875. The predictive power of the final model was AUC = 0.739 ((0.688–0.79)) (Figure 1).

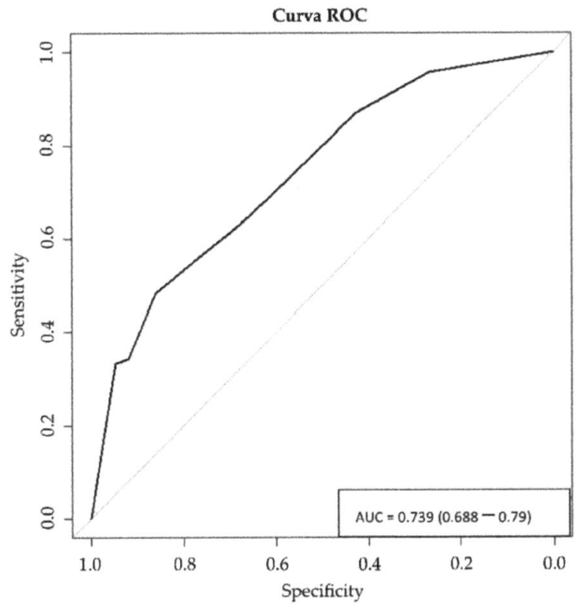

Figure 1. ROC curve [1]. Predictive value of the model. AUC = 0.739 (0.688–0.79) [2]. ([1] ROC = Receiver operating characteristic curve; [2] AUC = Area Under Curve).

4. Discussion

This study describes the relationship between, on the one hand, PC doctors' and nurses' self-efficacy in ACP and, on the other, variables related to their sociodemographics, background and perception of ACP. We have shown that, for the most part, professionals are familiar with ACP, have received prior training, and positively value the ACP process for use with people with advanced chronic diseases. Additionally, the factors related to greater self-efficacy in ACP are associated with considering oneself to be sufficiently trained in ACP, having previously participated in ACP processes, believing that ACP can contribute to improving the quality of life of the people cared for and believing that ACP makes it easier to know patients' values and preferences.

Although we observed no statistical differences across fields in the overall data of the ACP-SEs, there are significant differences in certain items of the scale. These differences are related to the fact that certain aspects explored by the ACP-SEs have to do with medical processes in which nurses are less comfortable to participate, coinciding with findings from international settings [32]. However, nurses have the competencies and skills to lead ACP processes [16,33] and to be able to talk with patients about all aspects of the ACP. Conversations about ACP should focus on promoting people's autonomy and participation in shared decision-making about their health, their illness, their concerns and preferences, and their values [34] and not exclusively about medical decisions. This holistic conception of the ACP could contribute to strengthen the involvement of all health fields in the process.

One of the most important aspects of our study is that training and implementation of ACP processes as part of the clinical routine are linked to greater self-efficacy in ACP. In this sense, training can improve self-efficacy in ACP [28,35], because it is key to promoting the development of safe practices in professionals who feel insecure or describe barriers when carrying out these processes.

Although, to our knowledge, there are no standard training models, specific short training sessions can stimulate the implementation of ACP processes among professionals [36], which is consistent with our results. The self-efficacy measurement can be useful for exploring the design and impact of training programmes [27] and recognising gold standard professionals who have a greater predisposition to lead ACP processes. This will allow the identification of ACP experts or facilitators within teams, especially in nursing [37].

In light of the data from our study, we suggest that ACP training include awareness of the importance of the ACP process for patients, reflections about barriers in practice and how to manage them, and the specification of aspects to be discussed with the sick person. The content of the training should promote a multidisciplinary approach that integrates all the dimensions of the individual, beyond discussion about medical treatments [33]. Finally, it could be useful to use simulation practices and reflective diaries about daily practice, which facilitate a learning process focussed on the professional's real environment. Education about ACP should proceed gradually, training professionals in how to discuss ACP with patients firstly with a low degree of complexity and later with patients with more complex needs, according to the experience of the professional. This would allow the professional responsible for driving the ACP to gain confidence and greater self-efficacy in facing more complex ACP processes that necessitate wide-ranging reflection.

Another result to be highlighted is related to the feasibility of implementing ACP, since this category received one of the lowest scores. Further analysis could examine the specific relationships between feasibility and other variables. Previous studies have shown that one of the barriers to ACP is the lack of time and the unavailability of spaces that facilitate respectful dialogue [14]. Policymakers and administrators who develop ACP programmes should consider not only the motivation and training of professionals, but also the improvement of ACP processes, their incorporation into team planning, the creation of meeting spaces to carry out the process, the development of incentives for professionals and the recording of the ACP process in medical records in a way that is both accessible and visible [25].

Finally, our participants gave high scores to the importance of ACP in patients with advanced and complex diseases, mainly cancer, organ disease, and frailty. Although a PC professional is best positioned to conduct the PCA process [38], given the high prevalence of people with chronic diseases and palliative care needs in the community [22], future studies should analyse facilitators and barriers to the implementation of ACP within an integrated health system [18,20], including the participation of professionals from multiple disciplines and different settings (acute care hospitals, intermediate care hospitals, mental health facilities and palliative home care teams). Such an approach would facilitate decision-making in patients with diverse trajectories and diseases.

ACP has the potential to promote shared decision-making and it is therefore essential that professionals feel prepared and motivated to lead the process. After the health threat caused by the SARS-COVID-19 pandemic, it is especially crucial that institutions, healthcare professionals in general and PC in particular, strengthen their efforts to improve healthcare quality by continuing to integrate practices of ACP in their daily routine [4,39].

4.1. Strengths

The results are based on a large sample of doctors and nurses from all over Catalonia who care for people with advanced chronic diseases and who work in a public health system that adopted an ACP process more than 5 years ago. Another strength is that we included both doctors and nurses and broke down the results by field, allowing us to see differences in the link between ACP self-efficacy and the other variables across the two fields. Finally, we used a validated self-efficacy scale, improving the accuracy of the results.

4.2. Limitations of Study

Since participation in the study was voluntary, the survey responses may be biased by participants' motivation towards and interest in the subject. This could explain the high scores in most of the survey items. Additionally, because only PC professionals were surveyed, we must be cautious when extrapolating the data to other healthcare settings. Such settings could be examined in future studies. The variables related to ACP were measured with a numerical scale that was not accompanied by qualitative descriptors, making it somewhat difficult to interpret participants' responses. The cut-off point of 8 was decided by the team and may have been influenced by the team members' own interpretation of the numerical scale. Finally, the data obtained cannot be extrapolated or applied to healthcare systems that are very different from the Catalan system.

4.3. Implications for Practice

The data from this study can contribute to improving the design and implementation of ACP programmes in PC settings, especially in relation to the definition of possible competencies and levels of responsibility in the implementation of ACP processes by the professionals involved. The study makes it possible to identify professionals that are most suited to act as facilitators of ACP processes, especially for complex cases.

5. Conclusions

To our knowledge, this is the first study that analyses in detail the relationship between self-efficacy and the perception of PC professionals about ACP practices in the context of an integrated ACP model within a public health system. We have identified that considering oneself to be sufficiently trained in ACP, having previously participated in ACP processes, believing that ACP can contribute to improving the quality of life of the people cared for and believing that ACP makes it easier to know patients' values and preferences are factors linked to PC professionals feeling more able to carry out ACP processes with patients. Strategies to improve the feasibility of ACP in practice could include systematically integrating ACP into care models, improving recording systems in medical records, and identifying and promoting predictive and facilitating factors for the implementation of ACP in care for people with advanced chronic diseases. This article provides important information about the role of self-efficacy in ACP practices in interdisciplinary PC teams, and it may help policy makers and administrators to promote the ACP process within public health systems.

Supplementary Materials: The following are available online at https://www.mdpi.com/article/10.3390/ijerph18179034/s1, Table S1: Supplemental table. Multivariate analysis. Logistic regression model (dependent variable ACP-SEs > 75 points), Table S2: Supplemental table. Statistical probability of showing a condition and scoring the ACP-SEs > 75 points.

Author Contributions: C.L., A.A.-H., F.C., E.L., S.S. and M.S.-C. conceptualized and designed the study. C.L., F.C., E.L. and A.A.-H. were involved in methodology and data collection, C.E. performed the statistical analysis, and C.L., F.C., E.L., A.A.-H. and M.S.-C. contributed to data analysis and interpretation of the findings (formal analysis, investigation, and data curation). C.L., A.A.-H. and M.S.-C. wrote the draft of the article. C.L., A.A.-H., C.E., F.C., E.L., S.S. and M.S.-C. reviewed and approved the final article. All authors meet conditions of the International Committee of Medical Journal Editors regarding authorship. All authors have read and agreed to the published version of the manuscript.

Funding: Publication of this article is supported by the grant awarded by Barcelona's Official School of Nursing (Project Code: PR-363/2019).

Institutional Review Board Statement: All research procedures used in this study were established in accordance with the Declaration of Helsinki. The ethics committee at the University of Vic reviewed and approved the study protocol (code RS005_S). Moreover, we designed the study in accordance with the ethics criteria established by Spanish Organic Law 3/2018 of 5 December on personal data protection and the guarantee of digital rights, following the General Regulation (EU) 2016/679 of 27 April 2016 on data protection.

Informed Consent Statement: Informed consent was obtained from all subjects involved in the study.

Data Availability Statement: Data sharing not applicable.

Acknowledgments: We would like to thank the people and organisations that have supported this study.

Conflicts of Interest: The authors declare no conflict of interest. The funders had no role in the design of the study; in the collection, analyses, or interpretation of data; in the writing of the manuscript, or in the decision to publish the results.

References

1. Rietjens, J.A.C.; Sudore, R.L.; Connolly, M.; van Delden, J.J.; Drickamer, M.A.; Droger, M.; van der Heide, A.; Heyland, D.K.; Houttekier, D.; Janssen, D.J.A.; et al. Definition and Recommendations for Advance Care Planning: An International Consensus Supported by the European Association for Palliative Care. *Lancet Oncol.* **2017**, *18*, e543–e551. Available online: https://linkinghub.elsevier.com/retrieve/pii/S147020451730582X (accessed on 24 July 2021). [CrossRef]
2. Gilissen, J.; Pivodic, L.; Smets, T.; Gastmans, C.; Stichele, R.V.; Deliens, L.; Block, L.V.D. Preconditions for Successful Advance Care Planning in Nursing Homes: A Systematic Review. *Int. J. Nurs. Stud.* **2017**, *66*, 47–59. Available online: https://linkinghub.elsevier.com/retrieve/pii/S0020748916302401 (accessed on 24 July 2021). [CrossRef] [PubMed]
3. Brinkman-Stoppelenburg, A.; Rietjens, J.A.; van der Heide, A. The effects of advance care planning on end-of-life care: A systematic review. *Palliat. Med.* **2014**, *28*, 1000–1025. [CrossRef] [PubMed]
4. Sable-Smith, A.; Arnett, K.R.; Nowels, M.A.; Colborn, K.; Lum, H.D.; Nowels, D. Interactions with the healthcare system influence advance care planning activities: Results from a representative survey in 11 developed countries. *Fam. Pract.* **2018**, *35*, 307–311. [CrossRef]
5. McMahan, R.D.; Tellez, I.; Sudore, R.L. Deconstructing the Complexities of Advance Care Planning Outcomes: What Do We Know and Where Do We Go? A Scoping Review. *J. Am. Geriatr. Soc.* **2021**, *69*, 234–244. [CrossRef]
6. Lally, K.M.; Ducharme, C.M.; Roach, R.L.; Towey, C.; Filinson, R.; Tuya Fulton, A. Interprofessional training: Geriatrics and palliative care principles for primary care teams in an ACO. *Gerontol. Geriatr. Educ.* **2019**, *40*, 121–131. [CrossRef]
7. Pecanac, K.E.; Repenshek, M.F.; Tennenbaum, D.; Hammes, B.J. Respecting Choices® and Advance Directives in a Diverse Community. *J. Palliat. Med.* **2014**, *17*, 282–287. Available online: http://www.ncbi.nlm.nih.gov/pubmed/24325558 (accessed on 24 July 2021).
8. Reidy, J.; Halvorson, J.; Makowski, S.; Katz, D.; Weinstein, B.; McCluskey, C.; Doering, A.; DeCarli, K.; Tjia, J. Health System Advance Care Planning Culture Change for High-Risk Patients: The Promise and Challenges of Engaging Providers, Patients, and Families in Systematic Advance Care Planning. *J. Palliat. Med.* **2017**, *20*, 388–394. [CrossRef]
9. Korfage, I.J.; Carreras, G.; Christensen, C.M.A.; Billekens, P.; Bramley, L.; Briggs, L.; Bulli, F.; Caswell, G.; Červ, B.; Van Delden, J.J.M.; et al. Advance care planning in patients with advanced cancer: A 6-country, cluster-randomised clinical trial. *PLoS Med.* **2020**, *17*, e1003422. [CrossRef]
10. Lakin, J.R.; Neal, B.J.; Maloney, F.L.; Paladino, J.; Vogeli, C.; Tumblin, J.; Vienneau, M.; Fromme, E.; Cunningham, R.; Block, S.D.; et al. A systematic intervention to improve serious illness communication in primary care: Effect on expenses at the end of life. *Healthcare* **2020**, *8*, 26–31. [CrossRef]
11. Der Schmitten, J.I.; Rothärmel, S.; Mellert, C.; Rixen, S.; Hammes, B.J.; Briggs, L.; Wegscheider, K.; Marckmann, G. A complex regional intervention to implement advance care planning in one town's nursing homes: Protocol of a controlled inter-regional

study. *BMC Health Serv. Res.* **2011**, *11*, 14. Available online: http://www.ncbi.nlm.nih.gov/pubmed/21261952 (accessed on 24 July 2021).

12. Arnett, K.; Sudore, R.L.; Nowels, D.; Feng, C.X.; Levy, C.R.; Lum, H.D. Advance Care Planning: Understanding Clinical Routines and Experiences of Interprofessional Team Members in Diverse Health Care Settings. *Am. J. Hosp. Palliat. Med.* **2017**, *34*, 946–953. [CrossRef]
13. Nelson-Brantley, H.; Buller, C.; Befort, C.; Ellerbeck, E.; Shifter, A.; Ellis, S. Using Implementation Science to Further the Adoption and Implementation of Advance Care Planning in Rural Primary Care. *J. Nurs. Scholarsh.* **2020**, *52*, 55–64. [CrossRef] [PubMed]
14. Howard, M.; Bernard, C.; Klein, D.; Elston, D.; Tan, A.; Slaven, M.; Barwich, D.; You, J.J.; Heyland, D.K. Survey of health care providers Barriers to and enablers of advance care planning with patients in primary care. *Can. Fam. Physician Médecin Fam. Can.* **2018**, *64*, 190–198. Available online: https://www-cfp-ca.proxy.queensu.ca/content/cfp/64/4/e190.full.pdf (accessed on 24 July 2021).
15. Hafid, A.; Howard, M.; Guenter, D.; Elston, D.; Fikree, S.; Gallagher, E.; Winemaker, S.; Waters, H. Advance care planning conversations in primary care: A quality improvement project using the Serious Illness Care Program. *BMC Palliat. Care* **2021**, *20*, 122. [CrossRef]
16. Fan, E.; Rhee, J.J. A self-reported survey on the confidence levels and motivation of New South Wales practice nurses on conducting advance-care planning (ACP) initiatives in the general-practice setting. *Aust. J. Prim. Health* **2017**, *23*, 80–86. [CrossRef] [PubMed]
17. De Vleminck, A.; Pardon, K.; Beernaert, K.; Deschepper, R.; Houttekier, D.; Van Audenhove, C.; Deliens, L.; Stichele, R.V. Barriers to advance care planning in cancer, heart failure and dementia patients: A focus group study on general practitioners' views and experiences. *PLoS ONE* **2014**, *9*, e84905. Available online: http://www.ncbi.nlm.nih.gov/pubmed/24465450 (accessed on 22 June 2021).
18. Howard, M.; Day, A.G.; Bernard, C.; Tan, A.; You, J.; Klein, D.; Heyland, D.K. Development and Psychometric Properties of a Survey to Assess Barriers to Implementing Advance Care Planning in Primary Care. *J. Pain Symptom Manag.* **2018**, *55*, 12–21. [CrossRef]
19. Lasmarías, C.; Aradilla-Herrero, A.; Santaeugènia, S.; Blay, C.; Delgado, S.; Ela, S.; Terribas, N.; Gómez-Batiste, X. Development and implementation of an advance care planning program in Catalonia, Spain. *Palliat. Support. Care* **2019**, *17*, 415–424. [CrossRef] [PubMed]
20. De Vleminck, A.; Houttekier, D.; Deliens, L.; Vander Stichele, R.; Pardon, K. Development of a complex intervention to support the initiation of advance care planning by general practitioners in patients at risk of deteriorating or dying: A phase 0–1 study. *BMC Palliat. Care* **2016**, *15*, 17. Available online: http://www.ncbi.nlm.nih.gov/pubmed/26868650 (accessed on 22 June 2021).
21. Miller, H.; Tan, J.; Clayton, J.M.; Meller, A.; Hermiz, O.; Zwar, N.; Rhee, J. Patient experiences of nurse-facilitated advance care planning in a general practice setting: A qualitative study. *BMC Palliat. Care* **2019**, *18*, 25. [CrossRef] [PubMed]
22. Gómez-Batiste, X.; Martínez-Muñoz, M.; Blay, C.; Amblàs, J.; Vila, L.; Costa, X.; Espaulella, J.; Espinosa, J.; Constante, C.; Mitchell, G.K. Prevalence and characteristics of patients with advanced chronic conditions in need of palliative care in the general population: A cross-sectional study. *Palliat. Med.* **2014**, *28*, 302–311. Available online: http://www.ncbi.nlm.nih.gov/pubmed/24403380 (accessed on 20 January 2021).
23. Lakin, J.R.; Robinson, M.G.; Obermeyer, Z.; Powers, B.W.; Block, S.D.; Cunningham, R.; Tumblin, J.M.; Vogeli, C.; Bernacki, R.E. Prioritizing Primary Care Patients for a Communication Intervention Using the "Surprise Question": A Prospective Cohort Study. *J. Gen. Intern. Med.* **2019**, *34*, 1467–1474. [CrossRef]
24. Contel, J.C.; Ledesma, A.; Blay, C.; González -Mestre, A.; Cabezas, C.; Puigdollers, M.; Zara, C.; Amil, P.; Sarquella, E.; Constante, C. Chronic and integrated care in Catalonia. *Int. J. Integr. Care* **2015**, *15*, 1–13. Available online: http://www.ijic.org/article/10.5334/ijic.2205/ (accessed on 20 August 2021). [CrossRef]
25. Walker, E.; McMahan, R.; Barnes, D.; Katen, M.; Lamas, D.; Sudore, R. Advance Care Planning Documentation Practices and Accessibility in the Electronic Health Record: Implications for Patient Safety. *J. Pain Symptom Manag.* **2018**, *55*, 256–264. [CrossRef]
26. Bandura, A. Self-efficacy: Toward a unifying theory of behavioral change. *Psychol. Rev.* **1977**, *84*, 191–215. [CrossRef]
27. Baughman, K.; Ludwick, R.; Fischbein, R.; McCormick, K.; Meeker, J.; Hewit, M.; Drost, J.; Kropp, D. Development of a Scale to Assess Physician Advance Care Planning Self-Efficacy. *Am. J. Hosp. Palliat. Care* **2016**, *34*, 435–441. [CrossRef] [PubMed]
28. Detering, K.; Silvester, W.; Corke, C.; Milnes, S.; Fullam, R.; Lewis, V.; Renton, J. Teaching general practitioners and doctors-in-training to discuss advance care planning: Evaluation of a brief multimodality education programme. *BMJ Support. Palliat. Care* **2014**, *4*, 313–321. Available online: http://www-ncbi-nlm-nih-gov.sire.ub.edu/pubmed/24844586 (accessed on 24 July 2021).
29. Lasmarías, C.; Subirana-Casacuberta, M.; Mancho, N.; Aradilla-Herrero, A. Spanish Cross-Cultural Adaptation and Psychometric Properties of the Advance Care Planning Self-Efficacy: A Cross-Sectional Study. *J. Palliat. Med.* **2021**. Available online: http://www.ncbi.nlm.nih.gov/pubmed/34143670 (accessed on 24 July 2021).
30. Rose, B.L.; Leung, S.; Gustin, J.; Childers, J. Initiating Advance Care Planning in Primary Care: A Model for Success. *J. Palliat. Med.* **2019**, *22*, 427–431. [CrossRef] [PubMed]
31. Hosmer, D.W.; Lemeshow, S.; Sturdivant, R.X. *Applied Logistic Regression*; Wiley Series in Probability and Statistics; John Wiley & Sons: Hoboken, NJ, USA, 2013. [CrossRef]

32. Howard, M.; Langevin, J.; Bernard, C.; Tan, A.; Klein, D.; Slaven, M.; Barwich, D.; Elston, D.; Arora, N.; Heyland, D.K. Primary care clinicians' confidence, willingness participation and perceptions of roles in advance care planning discussions with patients: A multi-site survey. *Fam. Pract.* **2020**, *37*, 219–226. [CrossRef] [PubMed]
33. Seymour, J.; Almack, K.; Kennedy, S. Implementing advance care planning: A qualitative study of community nurses' views and experiences. *BMC Palliat. Care* **2010**, *9*, 4. Available online: http://www.ncbi.nlm.nih.gov/pubmed/20377876 (accessed on 24 July 2021).
34. Doukas, D.J.; McCullough, L.B. The Values History: The evaluation of the patient's values and advance directives. *J. Fam. Pract.* **1991**, *32*, 145–153.
35. Pan, H.-H.; Wu, L.-F.; Chang, L.-F.; Hung, Y.-C.; Lin, C.; Ho, C.-L. Effects of Dispositional Resilience and Self-Efficacy on Practice in Advanced Care Planning of Terminally Ill Patients among Taiwanese Nurses: A Study Using Path Modeling. *Int. J. Environ. Res. Public Health* **2021**, *18*, 1236. Available online: https://www.mdpi.com/1660-4601/18/3/1236 (accessed on 20 August 2021). [CrossRef]
36. Lakin, J.R.; Koritsanszky, L.A.; Cunningham, R.; Maloney, F.L.; Neal, B.J.; Paladino, J.; Palmor, M.C.; Vogeli, C.; Ferris, T.G.; Block, S.D.; et al. A Systematic Intervention to Improve Serious Illness Communication In Primary Care. *Health Aff.* **2017**, *36*, 1258–1264. [CrossRef]
37. Risk, J.; Mohammadi, L.; Rhee, J.; Walters, L.; Ward, P.R. Barriers, enablers and initiatives for uptake of advance care planning in general practice: A systematic review and critical interpretive synthesis. *BMJ Open* **2019**, *9*, e030275. [CrossRef]
38. De Vleminck, A.; Pardon, K.; Beernaert, K.; Houttekier, D.; Vander Stichele, R.; Deliens, L. How Do General Practitioners Conceptualise Advance Care Planning in Their Practice? A Qualitative Study. *PLoS ONE* **2016**, *11*, e0153747. Available online: http://www.ncbi.nlm.nih.gov/pubmed/27096846 (accessed on 24 July 2021).
39. Block, B.L.; Smith, A.K.; Sudore, R.L. During COVID-19, Outpatient Advance Care Planning is Imperative: We need All Hands on Deck. *J. Am. Geriatr. Soc.* **2020**, *68*, 1–8. [CrossRef]

International Journal of
Environmental Research and Public Health

MDPI

Review

Differentiating between Seronegative Elderly-Onset Rheumatoid Arthritis and Polymyalgia Rheumatica: A Qualitative Synthesis of Narrative Reviews

Ryuichi Ohta [1,*] and Chiaki Sano [2]

1 Community Care, Unnan City Hospital, Unnan 699-1221, Japan
2 Department of Community Medicine Management, Faculty of Medicine, Shimane University, Izumo 690-0823, Japan
* Correspondence: ryuichiohta0120@gmail.com; Tel.: +81-90-5060-5330

Abstract: Elderly-onset rheumatoid arthritis (EORA) is prevalent among older patients, and its incidence is increasing due to aging societies. However, differentiating between EORA and polymyalgia rheumatica (PMR) is challenging for clinicians and hinders the initiation of effective treatment for rheumatoid arthritis among older generations, thereby allowing its progression. Therefore, we conducted a qualitative synthesis of narrative reviews via meta-ethnography regarding seronegative EORA diagnosis to clarify the methods to differentiate seronegative EORA from PMR. Three databases (PubMed, EMBASE, and Web of Science) were searched for relevant reviews published between January 2011 and October 2022. The extracted articles were synthesized using meta-ethnography, and 185 studies were selected following the protocol. Seven reviews were analyzed, and four themes and nine concepts were identified. The four themes included difficulty in differentiation, mandatory follow-up, and factors favoring rheumatoid arthritis and those favoring PMR. Factors favoring seronegative EORA and PMR should be considered for effective diagnosis and prompt initiation of disease-modifying anti-rheumatic drugs. Mandatory and long follow-ups of suspected patients are essential for differentiating the two diseases. The attitude of rheumatologists toward tentatively diagnosing seronegative EORA and flexibly modifying their hypotheses based on new or altered symptoms can aid in effective management and avoiding misdiagnosis.

Keywords: seronegative; elderly; aged; older; rheumatoid arthritis; polymyalgia rheumatica; follow; differentiation

Citation: Ohta, R.; Sano, C. Differentiating between Seronegative Elderly-Onset Rheumatoid Arthritis and Polymyalgia Rheumatica: A Qualitative Synthesis of Narrative Reviews. *Int. J. Environ. Res. Public Health* **2023**, *20*, 1789. https://doi.org/10.3390/ijerph20031789

Academic Editors: Francisco José Tarazona Santabalbina, Sebastià Josep Santaeugènia Gonzàlez, José Augusto García Navarro and José Viña

Received: 21 December 2022
Revised: 11 January 2023
Accepted: 16 January 2023
Published: 18 January 2023

1. Introduction

Early diagnosis and treatment of rheumatoid arthritis is essential to improving the quality of life of patients. Rheumatoid arthritis is prevalent in 0.5–1% of the total population, primarily occurs in middle-aged women [1], and commonly exists as peripheral arthritis with a progressive clinical course [2]. Rheumatoid arthritis is diagnosed based on classification criteria that comprise clinical findings, inflammatory markers, and autoantibodies [3]. Early treatment of rheumatoid arthritis is crucial to preventing progressive joint destruction [4,5]. Early initiation of disease-modifying anti-rheumatic drugs (DMARDs) can mitigate disease progression and preserve the quality of life of patients [4,5].

As society ages, the number of older patients suffering from rheumatoid arthritis increases [6]. Elderly-onset rheumatoid arthritis (EORA) is prevalent among older patients and differs in presentation from rheumatoid arthritis that affects younger generations. Additionally, rheumatoid arthritis in patients aged over 65 years has a different presentation than in younger patients [7]. The onset of EORA is more drastic than young-onset rheumatoid arthritis (YORA) [8]. Joint involvements may differ between EORA and YORA. EORA can involve proximal joints, such as the shoulder, neck, and femoral joints [9]. Furthermore,

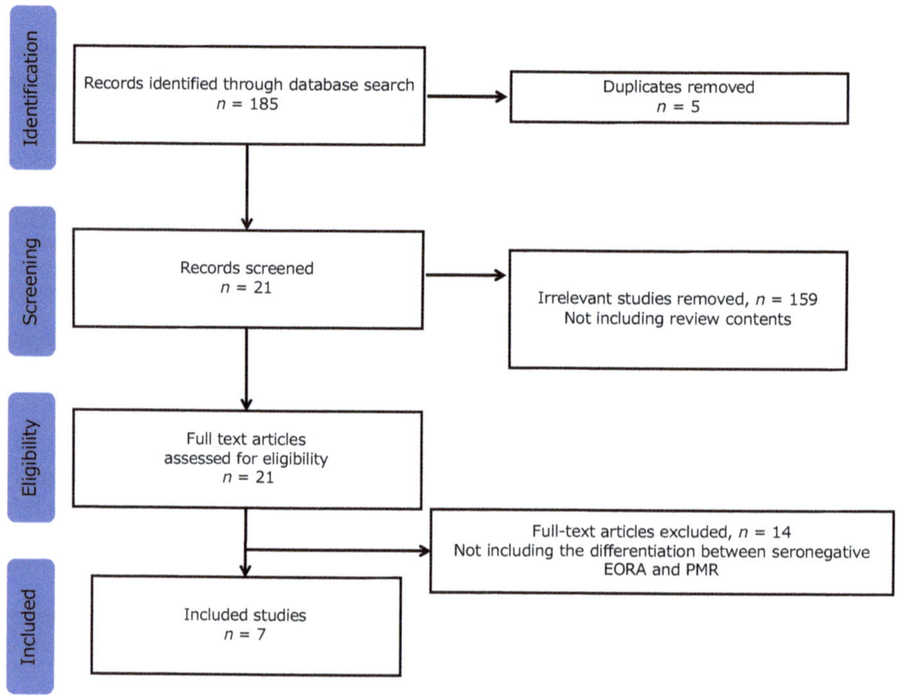

Figure 1. Selection flow. Note: EORA, elderly-onset rheumatoid arthritis; PMR, polymyalgia rheumatica.

Table 2. Studies included in the review.

Year	Title	Purpose	Included Codes
2009 [8]	Elderly Onset Rheumatoid Arthritis Differential Diagnosis and Choice of First Line and Subsequent Therapy	To review the EORA subset of patients with regard to demographic and clinical features, therapeutic options, outcomes, and differential diagnosis of EORA from other elderly rheumatological conditions	Duration (in months and years) Asymmetrical small joint involvement
2014 [17]	Polymyalgia rheumatica—diagnosis and classification	To differentiate PMR from other diseases	Symptomatic similarity at initial presentation Duration (in months and years)
2016 [18]	Targeting Low Disease Activity in Elderly-Onset Rheumatoid Arthritis: Current and Future Roles of Biological Disease-Modifying Antirheumatic Drugs	To review the clinical features of EORA and obstacles that prevent rheumatologists from providing standard treatment to patients with EORA	Symptom similarity at initial presentation Duration (in months and years)
2018 [19]	An autumn tale: geriatric rheumatoid arthritis	To review the clinical characteristics, prognosis, and treatment principles of EORA	Clinical picture of EORA with PMR phenotype Follow-up for peripheral lesions leading to EORA diagnosis Initial radiographic change Peripheral arthritis

Table 2. *Cont.*

Year	Title	Purpose	Included Codes
2018 [20]	Morning Stiffness in Elderly Patients with Rheumatoid Arthritis: What is Known About the Effect of Biological and Targeted Agents?	To review the impact of morning stiffness in patients with rheumatoid arthritis and summarize the efficacy of the biologic and targeted synthetic disease-modifying anti-rheumatic drugs in the alleviation of morning stiffness	Clinical picture of EORA with PMR phenotype Duration (in months and years)
2019 [21]	Polymyalgia Rheumatica and Seronegative Elderly-Onset Rheumatoid Arthritis: Two Different Diseases with Many Similarities	To highlight the main differences and similarities between seronegative EORA, PMR, and PMR-like EORA	Symptom similarity at initial presentation Duration (in months and years) Initial radiographic change Peripheral arthritis
2022 [22]	Treatment strategies for elderly-onset rheumatoid arthritis in the new era	To review effective differential diagnosis and treatment for EORA	Peripheral pain occurring in EORA and PMR Hip joint pain favoring PMR Extracapsular inflammation of peripheral lesions suggesting PMR Follow-up of peripheral lesions leading to EORA diagnosis Initial radiographic change Peripheral arthritis

Note: EORA, elderly-onset rheumatoid arthritis; PMR, polymyalgia rheumatica.

3.1. *Meta-Ethnography Results*

Through meta-ethnography, four themes and nine concepts were identified (Table 3). The four themes included difficulty in differentiation, mandatory follow-up, and factors favoring rheumatoid arthritis and PMR. Each theme was described after reviewing the quotes and concepts appearing in the included articles.

Table 3. Thematic analysis results in meta-ethnography.

Theme	Concept
Difficulty in differentiation	Symptom similarity at initial presentation Clinical picture of EORA with PMR phenotype
Mandatory follow-up	Duration (in months and years) Peripheral lesions follow-up leading to EORA diagnosis
Favoring rheumatoid arthritis	Initial radiographic change Peripheral arthritis
Favoring PMR	Hip joint pain Extracapsular inflammation of peripheral lesions Asymmetrical small joint involvement

EORA, elderly-onset rheumatoid arthritis; PMR, polymyalgia rheumatica.

3.2. *Difficulty in Differentiation*

Differentiating between seronegative EORA and PMR presents various difficulties, primarily due to two disease traits: symptom similarity at initial presentation and the clinical picture of EORA with PMR phenotype.

3.2.1. Symptom Similarity at Initial Presentation

The presentation of EORA and PMR may be similar at the time of initial presentation. EORA may present with proximal joint and muscular symptoms that are similar to those of PMR. One review stated the difficulty in distinguishing between patients with PMR and those with early-stage EORA and PMR-like presentation [17]. It also mentioned that PMR symptoms are the principal initial manifestation in 25% of patients with EORA, of whom 10% eventually develop characteristic rheumatoid arthritis features [17].

This difficulty is caused by the presence of arthralgia due to joint pain in PMR. The presence of joint pain does not rule out the possibility of PMR. Accordingly, another review revealed not only the similarity between the two at the initial presentation but also the continuity from PMR to seronegative EORA, which may hinder diagnosis [21]. Furthermore, shoulder pain and arthritis can appear in both diseases; therefore, shoulder symptoms are not used for their differentiation. One review demonstrated that 13–23% of patients with early EORA have an explosive onset of shoulder arthritis, which resembles PMR symptoms [18].

3.2.2. Clinical Picture of EORA with PMR Phenotype

One of the EORA presentations may include PMR symptoms throughout its clinical course. Differentiation from PMR can be achieved in the presence of autoimmune antibodies. Some types of seronegative EORA have symptoms that mirror those of PMR. One review suggested that one of the forms of EORA presentation is a PMR-like pattern that involves proximal limb joints and more acute disease onset with lower RF positivity and erosive disease [20]. Moreover, EORA displays a large joint involvement with the possibility of small joint involvement as well. Several reviews have suggested that there are three distinct clinical patterns within EORA's heterogeneity. A PMR-like and usually RF-negative form has an acute onset, non-eroding joints, and a good prognosis. A quarter of PMR patients may also have asymmetric non-erosive polyarthritis, which highlights the need for a good differential diagnosis [19]. Another review reported that >20% of patients with PMR were later diagnosed with rheumatoid arthritis, which lent to the theory that PMR and EORA are components of a single disease process [21]. Some types of seronegative EORA involve PMR presentation; therefore, the differentiation from PMR can be challenging because this disease is a part of the EORA clinical disease picture.

3.3. *Mandatory Follow-Up*

For an effective diagnosis, mandatory follow-up for a duration of months and years is essential.

3.3.1. Duration of Months and Years

Similar initial presentation and the clinical picture of EORA with PMR symptoms demand symptomatic follow-up of patients suspected to have seronegative EORA. One review suggested that follow-up is required in cases in which differentiating between PMR and seronegative EORA with PMR-like presentation is a challenge [18]. Moreover, the same review stated that patients who initially present with polymyalgia may evolve into a clinical situation closer to seronegative rheumatoid arthritis [18]. However, the duration of follow-up remains unclear. One review suggested, "A follow-up of several months may be required to make a definite distinction between PMR and EORA" [17]. In another, "EORA pattern is characterized by clinical and prognostic similarity to PMR. It is characterized by sudden onset, wrist tenosynovitis, common pitting edema in the hands, and spontaneous remission within 3–18 months" [19]. Another review reported, "The least common pattern is characterized by a sudden onset of symptoms, wrist tenosynovitis, pitting edema in the hands, and spontaneous remission within 3–18 months. Every attempt must be made to rule out other differential diagnoses (such as PMR, polyarticular gout, systemic vasculitis, and paraneoplastic manifestations) in older patients presenting with joint symptoms" [20].

Thus, year-long observations can help differentiate seronegative EORA from PMR because PMR can undergo remission in this duration.

3.3.2. Peripheral Lesions Follow-Up Leading to EORA Diagnosis

During follow-up, peripheral joint pain and arthritis appear in patients with seronegative EORA. Therefore, a follow-up to detect peripheral lesions is useful for diagnosing seronegative EORA. One review suggested that approximately 50% of patients with PMR with peripheral lesions were diagnosed with EORA 1 year post-treatment [22]. The observation of patients suspected of having seronegative EORA can focus on the appearance of peripheral lesions to effectively diagnose the disease.

3.4. Factors Favoring Rheumatoid Arthritis

In the differentiation between seronegative EORA and PMR, initial radiographic change and the presence of peripheral arthritis are factors that favor seronegative EORA.

3.4.1. Initial Radiographic Change

Rheumatoid arthritis is an inflammatory disease of the joints that progressively destroys them. Signs of deformities and joint destruction on radiography indicate the diagnosis of seronegative rheumatoid arthritis. One review suggested that "ACPA-positive EORA or ACPA-seronegative EORA with bone erosion at baseline is clearly different from PMR in terms of the progression of joint destruction, while seronegative EORA without bone erosion might have a benign course in terms of radiological joint destruction. Progression of bone erosion is essential for precise differential diagnosis, but rheumatoid factor- or anti-citrullinated protein antibody-negative early EORA with a PMR phenotype would not progress to erosive arthritis if initially treated with csDMARDs with or without GC therapy. These findings suggest a phenotypic overlap of PMR and seronegative early EORA with a PMR phenotype" [22]. To diagnose seronegative EORA at initial presentation, performing radiography is crucial for detecting the presence of deformities and erosions.

3.4.2. Peripheral Arthritis

The presence of peripheral arthritis may indicate seronegative rheumatoid arthritis and differentiate it from PMR at initial presentation. Arthritis of various peripheral joints suggests seronegative EORA. One review suggested, "An erosive arthritis or the symmetrical involvement of metacarpophalangeal and/or proximal interphalangeal joints can help to diagnose seronegative EORA" [21]. Another reported, "The presence of metacarpophalangeal (MCP)/proximal interphalangeal (PIP) joint arthritis with proximal limb joint involvement is considered a predictive factor for seronegative EORA" [19]. Detecting peripheral arthritis such as MCP and PIP joint arthritis can aid in the effective diagnosis of seronegative EORA.

3.5. Factors Favoring Polymyalgia Rheumatica

In the differentiation between seronegative EORA and PMR, hip joint pain, extracapsular inflammation of peripheral lesions, and asymmetrical small joint involvement are factors that favor polymyalgia rheumatica.

3.5.1. Hip Joint Pain

To diagnose seronegative EORA, the characteristics of PMR should be considered. Compared to EORA, PMR tends to involve the proximal joints with muscle pain with morning stiffness. One review suggested that in 50% and 20% of patients with PMR and EORA with shoulder lesions, respectively, pain and limited range of motion of the hip joints were reported; therefore, both shoulder and hip lesions are PMR phenotypic features [22]. The dominant symptoms such as hip and shoulder pain and limited range of motion of the joints indicate the possibility of PMR.

3.5.2. Extracapsular Inflammation of Peripheral Lesions

In addition to joint and muscle pain, surrounding tissues are involved in PMR pathophysiology. Even if peripheral pain occurs, whether the pain originates from the joints should be examined. As one review suggested, "Extracapsular inflammation of peripheral lesions with tenosynovitis and surrounding pitting edema is reportedly characteristic of PMR" [22]. Examining the joints and extracapsular tissues is essential for the effective diagnosis of seronegative EORA. PMR should be suspected in the presence of extracapsular tissue inflammation.

3.5.3. Asymmetrical Small Joint Involvement

The distribution of joint pain should be considered to differentiate seronegative EORA from PMR. When patients have peripheral symptoms, the laterality of the symptoms should be taken into account. One review showed that approximately 25% of patients with PMR present with peripheral synovitis that is frequently asymmetrical and non-erosive [8]. The peripheral joint examination should include an observation of the distribution of positive physical findings. When laterality exists in the findings, PMR should be considered as a possible diagnosis.

4. Discussion

This meta-ethnography of narrative reviews regarding the differentiation between seronegative EORA and PMR clarified four themes: difficulty in differentiation, mandatory follow-up, and factors favoring rheumatoid arthritis and those favoring PMR. Although the clear differentiation of the two diseases is challenging, factors favoring seronegative EORA and PMR should be considered for an effective diagnosis. The fundamental method for differentiating between the two diseases is the mandatory follow-up of suspected patients for months and years. The attitude of rheumatologists toward tentatively diagnosing seronegative EORA and flexibly changing their hypothesis based on the appearance or new or changed symptoms can be critical.

Differential diagnosis of seronegative EORA and PMR is difficult; therefore, a definitive diagnosis at the initial stage may be challenging. As this article showed, this difficulty mainly stems from symptom similarity at initial presentation and the clinical picture of EORA with a PMR phenotype. Therefore, clinicians may not be able to differentiate the two diseases in the initial phases [23,24]. Both diseases can have an acute onset with systemic symptoms rather than only joint pain in seropositive rheumatoid arthritis or YORA [25]. In addition, seronegative EORA can present with symptoms of PMR [23,24,26]. Thus, rheumatologists and family physicians who deal with these diseases should be careful when proposing a definitive diagnosis to suspected patients. For effective management, meticulous follow-up of symptoms is required.

Continuous follow-up can help effectively diagnose seronegative EORA because symptoms can change during the clinical course and reach the spectrum of either seronegative EORA or PMR. This study showed that mandatory follow-up involves durations of months and years in addition to follow-up of peripheral lesions for EORA diagnosis. The duration of follow-up varied depending on the review [8,17–20]. The clinical courses of seronegative EORA and PMR are acute and affect the activities of daily living in older patients [27,28]. Steroids can be initiated to mitigate the acute symptoms of seronegative EORA and PMR after ruling out other diseases such as bacteremia, sepsis, and other systemic inflammations [29]. While tapering steroids, several symptoms such as joint and muscular pain may appear [21]. PMR symptoms may not reappear for months and years after tapering steroids [17,21,30]; however, those of seronegative EORA may recur [26,31], which can be a sign of seronegative rheumatoid arthritis [26,31]. Furthermore, the appearance of peripheral arthritis during the follow-up period can be used to diagnose seronegative EORA. As this study showed, peripheral arthritis can appear during follow-up with steroid tapering. Thus, for an effective diagnosis, extensive follow-up is essential during steroid tapering, during which detecting recurrences and the appearance of peripheral arthritis is useful.

Investigating the factors that favor rheumatoid arthritis may aid in the diagnosis of seronegative rheumatoid arthritis after ruling out PMR and contribute to the smooth initiation of DMARDs for rheumatoid arthritis. The factors that favor rheumatoid arthritis are initial radiographic changes and the presence of peripheral arthritis. Radiographic changes in peripheral joints such as the PIP and MP joints suggest the possibility of seronegative EORA rather than PMR [32,33]. During the initial diagnosis of patients suspected to have seronegative EORA, peripheral radiography is essential for an effective diagnosis. In addition, initial examinations should focus on detecting peripheral arthritis to differentiate EORA and PMR because the presence of peripheral arthritis can indicate the possibility of seronegative EORA [31]. Effective and comprehensive physical joint examination is crucial to diagnose seronegative EORA. Regarding the treatment of seronegative EORA, prompt initiation of DMARDs is necessary to prevent joint destruction and deformities [4,5]. Based on these results, DMARDs should be prescribed for older patients suspected to have EORA with initial radiographic changes of the peripheral joints or physical findings of peripheral arthritis.

Contrary to the factors that favor seronegative EORA, PMR-favoring factors such as hip joint pain, extracapsular inflammation of peripheral lesions, and asymmetrical small joint involvement should be considered when predicting the clinical course. PMR can be a self-limiting condition in some cases; therefore, discontinuation of steroids without the use of DMARDs may be possible after steroid intake for a period ranging from months to 2 years [17,30]. Patients with hip pain, extracapsular inflammation, and asymmetrical small joint involvement should be followed up after tapering steroids without prescribing DMARDs.

This study had some limitations. Only a few review articles were found that investigated the differentiation between seronegative EORA and PMR. Rheumatology is more commonly studied among young to middle-aged patients; therefore, EORA and PMR may be differentiated according to the different clinical courses among older patients, but the research available is limited. Future studies should prospectively investigate ways to differentiate between these two diseases. In addition, this meta-ethnography did not include original articles to confirm the evidence present in the review articles. The inclusion criterion used in this study may have excluded some grey articles by experienced practitioners. Furthermore, due to accessibility limitations, this review may have missed relevant articles published in languages other than English. To overcome this limitation, we employed global search engines. Future studies can define the findings of this review as clinical factors and analyze older patients diagnosed with seronegative EORA and PMR to identify the sensitivity and specificity of each factor.

5. Conclusions

This meta-ethnography of narrative reviews regarding the differentiation between seronegative EORA and PMR clarified four themes: difficulty in differentiation, mandatory follow-up, and factors favoring rheumatoid arthritis and those favoring PMR. The favoring factors for both conditions should be checked for an effective diagnosis and a prompt initiation of DMARDs. Additionally, mandatory follow-up of suspected patients for months and years is essential for differentiating between the two diseases. The attitude of rheumatologists toward tentatively diagnosing seronegative EORA and flexibly changing their hypothesis based on the appearance of new or altered symptoms can be important. With the rise of aging societies globally, differentiating between EORA and PMR is becoming increasingly important for physicians.

Author Contributions: Conceptualization, R.O.; Methodology, R.O.; Software, R.O.; Validation, R.O. and C.S.; Formal Analysis, R.O. and C.S.; Investigation, R.O.; Resources, R.O.; Data Curation, R.O. and C.S.; Writing—Original Draft Preparation, R.O. and C.S.; Writing—Review and Editing, R.O. and C.S.; Visualization, R.O. and C.S.; Supervision, C.S.; Project Administration, R.O. All authors have read and agreed to the published version of the manuscript.

Funding: This research received no external funding.

Institutional Review Board Statement: Not applicable.

Informed Consent Statement: Not applicable.

Data Availability Statement: All relevant datasets used in this study are presented in the manuscript.

Acknowledgments: We would like to thank all the participants in this research.

Conflicts of Interest: The authors declare no conflict of interest.

References

1. Aletaha, D.; Smolen, J.S. Diagnosis and management of rheumatoid arthritis: A review. *JAMA* **2018**, *320*, 1360–1372. [CrossRef] [PubMed]
2. Liu, X.; Barber, C.E.H.; Katz, S.; Homik, J.; Bertazzon, S.; Patel, A.B.; Robert, J.; Smith, C.; Mosher, D.; Marshall, D.A. Geographic variation in the prevalence of rheumatoid arthritis in Alberta, Canada. *ACR Open Rheumatol.* **2021**, *3*, 324–332. [CrossRef] [PubMed]
3. Aletaha, D.; Neogi, T.; Silman, A.J.; Funovits, J.; Felson, D.T.; Bingham, C.O.; Birnbaum, N.S.; Burmester, G.R.; Bykerk, V.P.; Cohen, M.D.; et al. Rheumatoid arthritis classification criteria: An american college of rheumatology/european league against rheumatism collaborative initiative. *Arthritis Rheum.* **2010**, *62*, 2569–2581. [CrossRef] [PubMed]
4. Upchurch, K.S.; Kay, J. Evolution of treatment for rheumatoid arthritis. *Rheumatology* **2012**, *51*, vi28–vi36. [CrossRef]
5. Negoescu, A.; Ostör, A.J. Early recognition improves prognosis in elderly onset RA. *Practitioner* **2014**, *258*, 11–14.
6. Bajocchi, G.; La Corte, R.; Locaputo, A.; Govoni, M.; Trotta, F. Elderly onset rheumatoid arthritis: Clinical aspects. *Clin. Exp. Rheumatol.* **2000**, *18*, S49.
7. Gonzalez-Gay, M.A.; Hajeer, A.H.; Dababneh, A.; Makki, R.; Garcia-Porrua, C.; Thomson, W.; Ollier, W. Seronegative rheumatoid arthritis in elderly and polymyalgia rheumatica have similar patterns of HLA association. *J. Rheumatol.* **2001**, *28*, 122–125.
8. Villa-Blanco, J.I.; Calvo-Alén, J. Elderly onset rheumatoid arthritis: Differential diagnosis and choice of first-line and subsequent therapy. *Drugs Aging* **2009**, *26*, 739–750. [CrossRef]
9. Olivieri, I.; Pipitone, N.; D' Angelo, S.; Padula, A.; Salvarani, C. Late-onset rheumatoid arthritis and late-onset spondyloarthritis. *Clin. Exp. Rheumatol.* **2009**, *27*, S139–S145.
10. Healey, L.A.; Sheets, P.K. The relation of polymyalgia rheumatica to rheumatoid arthritis. *J. Rheumatol.* **1988**, *15*, 750–752.
11. Bitik, B.; Mercan, R.; Tufan, A.; Tezcan, E.; Küçük, H.; İlhan, M.; Öztürk, M.A.; Haznedaroğlu, S.; Göker, B. Differential diagnosis of elevated erythrocyte sedimentation rate and C-reactive protein levels: A rheumatology perspective. *Eur. J. Rheumatol.* **2015**, *2*, 131–134. [CrossRef] [PubMed]
12. Wiebe, E.; Huscher, D.; Schaumburg, D.; Palmowski, A.; Hermann, S.; Buttgereit, T.; Biesen, R.; Burmester, G.R.; Palmowski, Y.; Boers, M.; et al. Optimising both disease control and glucocorticoid dosing is essential for bone protection in patients with rheumatic disease. *Ann. Rheum. Dis.* **2022**, *81*, 1313–1322. [CrossRef] [PubMed]
13. Nikiphorou, E.; Santos, E.J.F.; Marques, A.; Böhm, P.; Bijlsma, J.W.; Daien, C.I.; Esbensen, B.A.; Ferreira, R.J.O.; Fragoulis, G.E.; Holmes, P.; et al. 2021 EULAR recommendations for the implementation of self-management strategies in patients with inflammatory arthritis. *Ann. Rheum. Dis.* **2021**, *80*, 1278–1285. [CrossRef] [PubMed]
14. Sepriano, A.; Kerschbaumer, A.; Smolen, J.S.; van der Heijde, D.; Dougados, M.; van Vollenhoven, R.; McInnes, I.B.; Bijlsma, J.W.; Burmester, G.R.; de Wit, M.; et al. Safety of synthetic and biological DMARDs: A systematic literature review informing the 2019 update of the EULAR recommendations for the management of rheumatoid arthritis. *Ann. Rheum. Dis.* **2020**, *79*, 760–770. [CrossRef] [PubMed]
15. Cahill, M.; Robinson, K.; Pettigrew, J.; Galvin, R.; Stanley, M. Qualitative synthesis: A guide to conducting a meta-ethnography. *Br. J. Occup. Ther.* **2018**, *81*, 129–137. [CrossRef]
16. France, E.F.; Uny, I.; Ring, N.; Turley, R.L.; Maxwell, M.; Duncan, E.A.S.; Jepson, R.G.; Roberts, R.J.; Noyes, J. A methodological systematic review of meta-ethnography conduct to articulate the complex analytical phases. *BMC Med. Res. Methodol.* **2019**, *19*, 35. [CrossRef] [PubMed]
17. Nesher, G. Polymyalgia rheumatica—Diagnosis and classification. *J. Autoimmun.* **2014**, *48*, 76–78. [CrossRef]
18. Sugihara, T.; Harigai, M. Targeting low disease activity in elderly-onset rheumatoid arthritis: Current and future roles of biological disease-modifying antirheumatic drugs. *Drugs Aging* **2016**, *33*, 97–107. [CrossRef]
19. Kobak, S.; Bes, C. An autumn tale: Geriatric rheumatoid arthritis. *Ther. Adv. Musculoskelet. Dis.* **2018**, *10*, 3–11. [CrossRef]
20. Mok, C.C. Morning stiffness in elderly patients with rheumatoid arthritis: What is known about the effect of biological and targeted agents? *Drugs Aging* **2018**, *35*, 477–483. [CrossRef]

21. Manzo, C.; Emamifar, A. Polymyalgia Rheumatica and Seronegative Elderly-Onset Rheumatoid Arthritis: Two Different Diseases with Many Similarities. *Eur. Med. J. Rheumatol.* **2019**, *4*, 111–119. [CrossRef]
22. Sugihara, T. Treatment strategies for elderly-onset rheumatoid arthritis in the new era. *Mod. Rheumatol.* **2022**, *32*, 493–499. [CrossRef] [PubMed]
23. Paalanen, K.; Puolakka, K.; Nikiphorou, E.; Hannonen, P.; Sokka, T. Is seronegative rheumatoid arthritis true rheumatoid arthritis? A nationwide cohort study. *Rheumatology* **2021**, *60*, 2391–2395. [CrossRef] [PubMed]
24. Maassen, J.M.; Bergstra, S.A.; Chopra, A.; Govind, N.; Murphy, E.A.; Vega-Morales, D.; Huizinga, T.W.J.; Allaart, C.F. Phenotype and treatment of elderly onset compared with younger onset rheumatoid arthritis patients in international daily practice. *Rheumatology* **2021**, *60*, 4801–4810. [CrossRef] [PubMed]
25. Choi, S.; Lee, K.H. Clinical management of seronegative and seropositive rheumatoid arthritis: A comparative study. *PLoS ONE* **2018**, *13*, e0195550. [CrossRef]
26. Paalanen, K.; Rannio, K.; Rannio, T.; Asikainen, J.; Hannonen, P.; Sokka, T. Does early seronegative arthritis develop into rheumatoid arthritis? A 10-year observational study. *Clin. Exp. Rheumatol.* **2019**, *37*, 37–43.
27. Manzo, C.; Castagna, A. Subjective sleep disturbances at the time of diagnosis in patients with polymyalgia rheumatica and in patients with seronegative elderly-onset rheumatoid arthritis. A pilot study. *Reumatologia* **2020**, *58*, 196–201. [CrossRef]
28. Korkmaz, C.; Yıldız, P. Giant cell arteritis, polymyalgia rheumatica, and late-onset rheumatoid arthritis: Can they be components of a single disease process in elderly patients? *Eur. J. Rheumatol.* **2017**, *4*, 157–160. [CrossRef]
29. Earwood, J.S.; Walker, T.R.; Sue, G.J.C. Septic arthritis: Diagnosis and treatment. *Am. Fam. Physician* **2021**, *104*, 589–597.
30. Li, H.; Altman, R.D.; Yao, Q. RS3PE: Clinical and research development. *Curr. Rheumatol. Rep.* **2015**, *17*, 49. [CrossRef]
31. Caporali, R.; Montecucco, C.; Epis, O.; Bobbio-Pallavicini, F.; Maio, T.; Cimmino, M.A. Presenting features of polymyalgia rheumatica (PMR) and rheumatoid arthritis with PMR-like onset: A prospective study. *Ann. Rheum. Dis.* **2001**, *60*, 1021–1024. [CrossRef] [PubMed]
32. Sudoł-Szopińska, I.; Jans, L.; Teh, J. Rheumatoid arthritis: What do MRI and ultrasound show. *J. Ultrason.* **2017**, *17*, 5–16. [CrossRef] [PubMed]
33. Jang, S.; Kwon, E.J.; Lee, J.J. Rheumatoid arthritis: Pathogenic roles of diverse immune cells. *Int. J. Mol. Sci.* **2022**, *23*, 905. [CrossRef] [PubMed]

International Journal of
Environmental Research and Public Health

MDPI

Review

Is Frailty Diagnosis Important in Patients with COPD? A Narrative Review of the Literature

Francisco José Tarazona-Santabalbina [1,2,3,*], Elsa Naval [4], Juan María De la Cámara-de las Heras [5], Cristina Cunha-Pérez [6] and José Viña [2,7]

1 Geriatric Medicine Department, Hospital Universitario de la Ribera, Carretera de Corbera km. 1, 46600 Alzira, Spain
2 Centro de Investigación Biomédica en Red Fragilidad y Envejecimiento Saludable (CIBERFES), 28029 Madrid, Spain
3 Medical School, Universidad Católica de Valencia San Vicente Mártir, 46001 Valencia, Spain
4 Department of Pneumology, Hospital Universitari la Ribera, 46600 Alzira, Spain
5 Library Service, Hospital Universitario la Ribera, 46600 Alzira, Spain
6 School of Doctorate, Universidad Católica de Valencia San Vicente Mártir, 46001 Valencia, Spain
7 Departament of Physiology, Universitat de Valencia, 46010 Valencia, Spain
* Correspondence: fjtarazonas@gmail.com; Tel.: +34-962542800

Abstract: Frailty is prevalent in older adults and is related to a worsening functionality, quality of life, and health outcomes. Though there is an increasing interest in this field, the relationship between frailty and worsening COPD outcomes remains unknown. A narrative review of the literature with studies published between 2018 and 2022 was carried out to address three questions: the prevalence of frailty and other geriatric syndromes in COPD patients, the link between frailty and worsening health outcomes in COPD patients, and the non-pharmacological interventions performed in order to reverse frailty in these patients. A total of 25 articles were selected. Frailty prevalence ranged from 6% and 85.9%, depending on the COPD severity and the frailty measurement tool used. Frailty in COPD patients was related to a high prevalence of geriatric syndromes and to a high incidence of adverse events such as exacerbations, admissions, readmissions, and mortality. One study showed improvements in functionality after physical intervention. In conclusion, the prevalence of frailty is associated with a high incidence of geriatric syndromes and adverse events in COPD patients. The use of frailty screenings and a comprehensive geriatric assessment of COPD patients is advisable in order to detect associated problems and to establish individualized approaches for better outcomes.

Keywords: COPD; frailty; comprehensive geriatric assessment; outcomes; interventions; geriatric syndromes

Citation: Tarazona-Santabalbina, F.J.; Naval, E.; De la Cámara-de las Heras, J.M.; Cunha-Pérez, C.; Viña, J. Is Frailty Diagnosis Important in Patients with COPD? A Narrative Review of the Literature. *Int. J. Environ. Res. Public Health* **2023**, *20*, 1678. https://doi.org/10.3390/ijerph20031678

Academic Editor: Paul B. Tchounwou

Received: 28 November 2022
Revised: 11 January 2023
Accepted: 15 January 2023
Published: 17 January 2023

1. Introduction

The generally agreed-upon definition of frailty states that frailty pertains to "a medical syndrome with multiple causes and contributors that is characterized by diminished strength, endurance, and reduced physiological function that increases an individual's vulnerability for developing increased dependency and/or death" [1]. From a clinical perspective, frailty implies a decrease in physiological reserves and an increased vulnerability to stressors, leading to greater overall vulnerability [2]. This vulnerability increases the risk of adverse health outcomes, such as disability, hospitalizations, and mortality [2]. Frailty is dynamic in nature, and it is possible to identify positive and negative trajectories among the robust, pre-frailty, and frailty categories, both spontaneously and after interventions [3]. Despite having previously mentioned the generally accepted definition of frailty, there are a large number of operational definitions and measurement tools [4], including the phenotypic model established by Linda Fried (FFP, short for Fried frailty phenotype), one of the most used tools [2]. This model is defined by the presence of at least three of the following criteria: weight loss, self-reported exhaustion, weakness, slowed gait speed, and lower energy expenditure. The pre-frailty condition is defined by the presence of one or two of these criteria, and a fit adult is defined as someone who does not meet any of these five criteria.

Frailty has been misleadingly associated with aging, comorbidity, or disability, although it does increase the risks of increased morbidity and lower survival. In addition, frailty is bidirectionally associated with chronic diseases. The presence of chronic conditions increases the risk of frailty [5], and the presence of frailty increases the risk of morbidity and mortality by accelerating the physical impairment of patients [6]. Therefore, a frailty assessment is of great importance in older adults, regardless of the main diagnosis [2].

Frailty, as a geriatric syndrome, is traditionally underdiagnosed outside the field of geriatric medicine. However, an increasing interest in inquiring about the relationship between chronic obstructive pulmonary disease (COPD) and frailty has been recently observed. COPD is one of the most prevalent and disabling chronic diseases in older adults, and it leads to a relevant decrease in survival [7]. The prevalence of frailty in older adults with COPD is high, even in younger populations (under 65 years old), ranging between 4% and 59% [8], although the use of the FFP decreases the prevalence down to 9.9% [8].

A significant increase in scientific works about the relationship between frailty and COPD has been observed in the last few years, in which the important prognostic role of a frailty diagnosis in the progression of COPD has been highlighted [9,10].

Likewise, COPD is the third leading cause of death worldwide. Many factors contribute to the development of COPD, including genetic factors (alpha1-antitrypsin deficiency), pollution, cigarette smoking, and occupational exposure to various chemicals. COPD manifests as an inflammatory disease that affects the airways, lung parenchyma, and pulmonary vasculature. Inhalation exposure can trigger an inflammatory response, leading to a decrease in forced expiratory volume and tissue destruction and eventually to airflow limitation. Airflow limitation is the main pathophysiological feature of COPD [11]. This relationship between the two may be because COPD and frailty share some risk factors, including aging, smoking, and inflammation [12] as well as clinical manifestations, such as fatigue, anorexia, muscle weakness, and slowed walking speed [13]. In this sense, Lahousse et al. suggested that both pathologies might have a single physiopathology in common [14].

Finally, it should not be forgotten that several common risk factors shared by COPD and frailty have been identified, including alcohol consumption, cigarette smoking, lower physical activity, poor diet, and older age [15]. All are related to functional and cognitive declines. For these reasons, an early diagnosis of frailty is needed to provide a specific and individualized care plan for the recovery and mitigation of the deleterious effects of frailty in these patients. Clinicians should actively screen frailty in COPD patients to improve their outcomes [15,16].

A systematic review and meta-analysis recently showed that the risk of frailty among patients with COPD diagnoses was twice as high as those of adults of the same age without COPD [17]. Both COPD and frailty share some risk factors, such as less favorable aging trajectories, nicotinism, and inflammation [18], along with clinical signs such as anorexia, fatigue, muscular weakness and slower gait speed (the latter three are the phenotypic criteria of frailty syndrome) [18], which could be due to their shared physiopathology [13].

However, the choice of the frailty assessment is one of the most common problems in frailty studies in different clinical settings. Given the increase in scientific literature on this topic, the authors of this paper considered it important to perform a narrative review of the literature in order to understand the actual prevalence of frailty in patients with COPD, the association of frailty with other geriatric syndromes in these patients, the influence of frailty on patient clinical evolution and survival, and the therapeutic approaches to reverse or reduce frailty in COPD patients.

2. Materials and Methods

The present review was carried out by conducting an electronic search in OVID® (Ovid Technologies, Inc., Wolters Kluwer Health, NY, USA) (Medline and Embase) and PubMed® (National Library of Medicine, Bethesda, MD, USA), combining the following MeSH keywords: "pulmonary disease, chronic obstructive", and "frailty". The search was completed on 30 September 2022 and limited to publications posted in the last 5 years; in English and Spanish; and in human subjects aged 60 or older. A total of 575 articles were obtained, of which 25 were finally selected. Details of the evaluation and selection process of the items are shown in Figure 1.

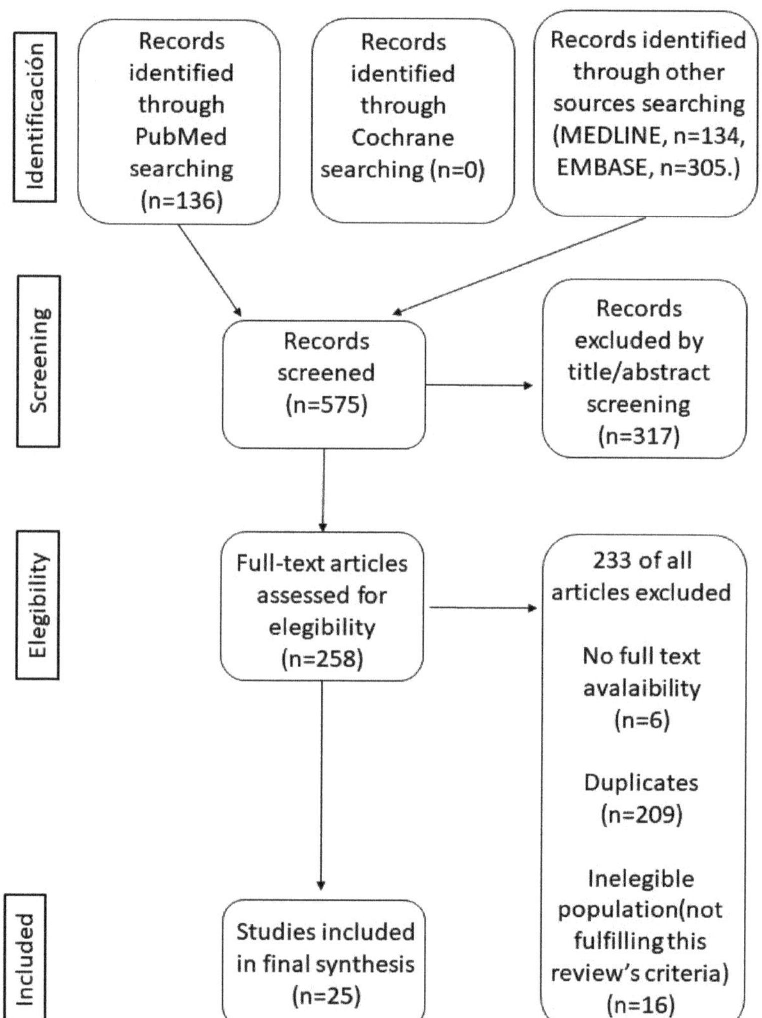

Figure 1. PRISMA flow diagram for database search and study selection process. Based on Moher et al. [19].

The articles were selected by four researchers based on the following eligibility criteria: The articles included in the criteria were meta-analyses, randomized clinical trials, cohort studies, case-control studies, observational studies, and before-and-after studies; the population needed for the review were patients 65 years old or older with an established COPD diagnosis; and prevalence and incidence studies using reported frailty assessment tools and reporting frailty prevalence were included, along with interventions to reverse frailty. It was stated in the exclusion criteria that letters-to-the-editor; case reports, narrative and systematic reviews without meta-analyses; articles with no available abstracts or those with only the abstract published; and studies that met the inclusion criteria but less than 50% of their study sample was under the age of 65 (i.e., predominantly non-geriatric), which indicated these were not in compliance with the aim of this review.

The review authors re-evaluated all articles, and the final inclusion was restricted to those of high enough quality with information relevant to the objectives of this review. The outcome measures examined were mortality, length of hospital stay, functional status, medical complications, destination after discharge, functional recovery, frailty reversion, readmissions, and survival. The authors used an Excel® spreadsheet to record all the bibliographic references in order to identify and exclude duplicate papers. The Excel spreadsheet was also used to work with the selected articles and proceed to the selection of the articles. For this process, a bibliographic manager was used.

3. Results

The 25 selected articles for this review were classified according to their type: 1 meta-analysis, 1 cohort study, 1 retrospective cohort study, 8 longitudinal studies, 8 cross-sectional studies, 2 case-control studies and 4 retrospective studies. Most studies (22) reported frailty prevalence, which ranged from 6% to 85.9% and was partly related to the variability of the recruited patients' age, ranging from 61.2 to 75.9 years, and most of the sample being male. There was a considerable variation in the frailty measurement tools, as up to 13 tools used in the studies included in the review were found: the hospital frailty risk score (HFRS) in two studies; the Fried frailty phenotype (FFP) in eight studies; the frailty index (FI) in four studies; the Kihon checklist (KCL) in two studies; the fatigue, resistance, ambulation, illnesses, and loss of weight (FRAIL) in two studies; the reported Edmonton frail scale (REFS), the Chinese–Canadian study of health and aging clinical frailty scale, the evaluative frailty index for physical activity (EFIP), the clinical frailty scale (CFS), the short physical performance test (SPPB), and the frail non-disabled questionnaire (FiND), in one study each; and the timed up and go (TUG) and the FI-lab in one study.

3.1. Frailty Prevalence

Out of the 25 selected studies, 22 of them included frailty prevalence data in the studied sample, which, as discussed in the previous section, ranged from 6% to 85.9%. The frailty prevalence reported by each study included in this narrative review are reported in Table 1.

Table 1. Summary of studies with reported prevalence of frailty in COPD patients.

Article	Type	Sample	Age (Years)	Tool Used	Frailty Prevalence	Spirometric COPD Confirmation
Ushida, K. et al. [20]	Retrospective cohort	3396 COPD patients	75.9 (SD 11.2)	HFRS	14%	No (ICD-10 codes: J41–J44)
Neo, H.Y. et al. [21]	Prospective, propensity score-match study	100 matched pairs	73.9 (SD 8.2)	HFRS	57%	No
Kennedy, C. C. et al. [22]	Retrospective cohort	902 COPD patients	67 [IQR 63–70]	FFP	6%	Yes

Table 1. *Cont.*

Article	Type	Sample	Age (Years)	Tool Used	Frailty Prevalence	Spirometric COPD Confirmation
Witt, L. J. et al. [23]	Observational study	70 patients admitted due to COPD exacerbation	63.5 (SD 58.1, 71.3)	FFP	67%	Yes
Luo, J. et al. [24]	Cross-sectional study	309 COPD patients	86 [IQR 80–90]	FFP	49.8%	Yes
Yee, N. et al. [25]	Prospective cohort	280 COPD patients	68.6 (SD 9.2)	FFP	23%	Yes
Bernabeu-Mora, R. et al. [26]	Prospective study	119 COPD patients	66.9 (SD 7.9)	FFP	7.6%	Yes
Naval, E. et al. [27]	Cross-sectional study	127 COPD patients	66.5 (SD 7.9)	FFP	24.4%	Yes
Hanlon, P. et al. [28]	Observational study	3132 COPD patients	61.9 (SD 5.9)	FFP FI	17% FFP 32% FI	Yes
Zhang, D. et al. [29]	Prospective study	302 COPD patients	86 [IQR 80–90]	FFP CFS FI-CD SPPB	FFP 51% CFS 64% FI 58.6% SPPB 59.6%	Yes
Soni, N. et al. [30]	Case-control study	150 COPD 150 Controls	65.98 (SD 5.43) 65.72 (SD 5.65)	FRAIL	25.3%	Yes
Dias, L. S. et al. [31]	Cross-sectional study	150 COPD patients	67.0 (SD 61.0–71.5)	FRAIL	50.3%	Yes
Gale, N. S. et al. [32]	Case-control study	520 COPD 150 controls	(66.1 (SD 7.6)) (65 (SD 7.4))	FI	28%	Yes
Albarrati, A. M. et al. [33]	Case-control study	520 COPD 120 controls	66.1 (SD 7.6) (65 (SD 7.4))	FI	76%	Yes
Gu, J. J. et al. [34]	Observational retrospective study	154 COPD patients	79.73 (SD 8.38)	FI-lab	75.3%	Yes
Takahashi, S. et al. [35]	Cross-sectional study	40 COPD patients	70.6 (SD 8.21)	KCL	50%	Yes
Oishi, K. et al. [36]	Observational study	128 COPD patients	73 [IQR 69–78]	KCL	37.5%	Yes
Witt, L. J. et al. [37]	Cross-sectional study	322 COPD patients	69.6 (SD 7.4)	modified frailty	16%	No
Bernabeu-Mora, R. et al. [38]	Prospective cohort	103 hospitalized COPD patients	71 (SD 9.1)	REFS	35.9% moderate or severe frailty	Yes
Chen, P. J. et al. [39]	Cross-sectional study	125 COPD patients	77.36 (SD 10.26)	Chinese Canadian study of health and aging clinical frailty scale	85.9% dyspnea group 26.7% non-dyspnea group	Yes
Ter Beek, L. et al. [40]	Cross-sectional study	57 COPD patients	61.2 (SD 8.7)	EFIP	83%	Yes

Table 1. *Cont.*

Article	Type	Sample	Age (Years)	Tool Used	Frailty Prevalence	Spirometric COPD Confirmation
Chin, M. et al. [41]	Prospective study	46 patients admitted due to COPD exacerbation Mild frailty Moderate frailty Severe frailty	72 (SD 9) 72 (SD 10) 76 (SD 12)	CFS	54%	No described
Ierodiakonou, D. et al. [42]	Cross-sectional study	257 COPD patients	(65 (SD 12.3))	FiND (frail non-disabled)	82%	Yes

Legend: COPD = chronic obstructive pulmonary disease; HFRS = hospital frailty risk score; EFIP = evaluative frailty index for physical activity; FFP = Fried frailty phenotype; FI = frailty index; REFS = reported Edmonton frail scale; FRAIL = the fatigue, resistance, ambulation, illnesses, and loss of weight; KCL = Kihon checklist; Chinese–Canadian Study of health and aging clinical frailty scale; FI-lab = frailty index based on deficits in laboratory test; CFS = clinical frailty scale; SPPB = short physical performance test; FiND = the frail non-disabled questionnaire; SD = standard deviation; IQR = interquartile range.

Two studies, including a retrospective cohort study [20] and a prospective propensity score-match study [21] that used the HFRS as a frailty screening tool, showed a disparate prevalence of 14% and 57%, respectively, in samples of similar age. The former was conducted in hospitalized patients, with 68.7% of the sample having the highest degree of dyspnea (score 4 and 5) estimated with the Hugh–Jones dyspnea scale score while the latter had no data on disease severity.

However, the most widely used tool for the diagnosis of frailty was Linda Fried's phenotypic criteria, as it was used in eight COPD studies [22–29]. Thus, when using the FFP, the prevalence of frailty was 6% [23] and 67% [24], respectively. The ages of the participants in these two studies were similar. In the first study, which used FFP to detect frailty, patients had to have moderate or severe COPD while the second study was performed in patients admitted for an acute exacerbation of COPD. Of the 8 studies that used FFP as a screening tool for frailty, the second with the highest prevalence (49.8%) was completed in patients almost 20 years older than the 2 previous studies and who were diagnosed with COPD based on the 2017 Global Initiative for COPD (GOLD) guidelines. They had respiratory symptoms or risk factors and a post-bronchodilator ratio of forced expiratory volume in 1 s (FEV1) to forced vital capacity (FVC) < 0.70. The study by Kennedy et al. [22] reported an estimated incidence of frailty of 6.4 per 100 human years and the presence of frailty being associated with a lower quality of life. The fourth study that used FFP for the diagnosis of frailty had patients of an age similar to the first described with these criteria and showed a prevalence of frailty of 23% in patients with a diagnosis of COPD (post-bronchodilator FEV1/FVC < 80% and FEV1% predicted < 80%) and who also had been stable for at least the last 4 weeks [25]. A prospective study conducted in 119 patients with stable COPD during a 2-year follow-up assessed transitions between non-frail, pre-frail, and frail conditions using the FFP. A frailty prevalence of 7.6% and a pre-frailty prevalence of 73.1% were reported at the beginning of the study. After 2 years, 11.7% of patients worsened, 17.6% improved, and 70.5% remained the same. The variables associated with deterioration were dyspnea, disability, and lower handgrip strength, while those who improved presented greater handgrip and quadricep strength. The authors concluded that frailty is a dynamic condition, and therefore, the transitions between the states of frailty are related to significant changes in the clinical outcomes [26]. Furthermore, the authors of this narrative review designed a cross-sectional study in a prospective cohort of 127 patients with stable COPD (diagnosis based on the GOLD 2017 guideline) and classified 24.4% of the recruited sample as frail. The variables with independent associations with frailty included the mMRC,

HAD-DEP, and age [27]. The last two articles that used FFP as a frailty screening tool compared this measure to the IF [28] and to the IF, SPPB, and CFS [29], respectively. The first analyzed data extracted from the UK Biobank of patients with COPD, who had been identified by linked primary care data, using a previously validated list of diagnostic codes and a severity assessment of the disease that was measured with spirometry data (20% of patients with $FEV_1 < 50\%$) and then compared two frailty measurements, the FFP and the FI, during an 8-year follow-up period. Frailty prevalence was high regardless of the frailty assessment tool used (17% frail according to the FFP and 28% moderate and 4% severely frail according to the FI), but only the FFP was associated with a lower FEV_1 [28] while the second, a prospective study with a 2.18-year [IQR 1.56–2.62] follow-up of 302 COPD patients (diagnosis based on the GOLD 2017 guidelines), assessed the predictive capacity of 4 frailty scales: FFP, CFS, FI, and SPPB. Frailty prevalence was lower when using the FI (18%) and higher when using the other three tools (FFP, 51%; SPPB, 58.6%; CFS, 64.2%) [29].

Two papers measured frailty prevalence with FRAIL [30,31]. The first one, a case-control study [30], found a frailty prevalence of 25.3% in COPD patients (FEV1 to forced vital capacity of <0.7) assessed by the outpatient department and measured with the FRAIL scale, and these were linked to a higher prevalence of geriatric syndromes in COPD patients. The second one, an observational cross-sectional study in stable COPD patients (FEV1/FVC ratio of <0.7), found a greater risk of frailty in the GOLD groups B and D. Frailty was significantly correlated to the COPD assessment test score (CAT) and the MRC. The authors reported that the combination of CAT/MRC $\geq$ 5.5 was associated with the presence of frailty (odds ratio (OR) 6.7; 95% CI: 3.2–13.9) [31]. Two articles from the same research group used the IF as a diagnostic tool for frailty [32,33]. The first one, a cross-sectional study conducted in 520 community-based patients with COPD (confirmed by spirometry at entry), showed that patients with COPD presented a higher likelihood in the FI (0.16 (SD 0.08)) than the controls (0.05 (SD 0.03)), and the frailty prediction factors were the distance covered in a 6-minute walk distance (6MWD), the number of comorbidities, handgrip strength, and the number of acute exacerbations [32]. The second one, a case-control study, assessed the possible usefulness of the timed-up-and-go (TUG) tool for frailty detection. COPD patients presented a higher FI and a higher time using TUG (11.55 (SD 4.03 s)) than the controls (9.2 (SD 1.6 s)) [33]. This difference remained after adjusting for age and pulmonary function. The frailty predictive capacity using TUG (a cut-off point of 0.09 for the FI and 8 s for the TUG) was OR 2.67 (95% CI: 1.5–4.6) [33].

A retrospective study assessed the possible usefulness of a frailty evaluation tool (FI-lab) in COPD patients (primary clinical diagnosis of AECOPD at hospital admission) distributed between survivors and deceased. The FI-lab values were classified into 4 groups: <0.2, 0.2–0.4, 0.4–0.6, and >0.6. FI-lab < 0.4 values were seen in 88.3% of survivors while 75.3% of the deceased presented those values. The difference in the FI-lab was statistically significant between survivors (0.5 (SD 0.1)) and deceased (0.3 (SD 0.1)), $p < 0.001$ [34]. Two papers used KLC as a frailty detection tool in COPD patients (GOLD 2021) [35,36]. The first one, a cross-sectional study, found a frailty prevalence of 50% and reported negative effects of COPD on the central nervous system, along with depressive symptoms [36]. At the same time, a cross-sectional study conducted in 128 COPD patients found a frailty prevalence of 37.5% and estimated the relationship between the patient-reported outcome measures for dyspnea-related behavior, activity limitation (PROMs-D, which was consistent between the activity-limiting dyspnea scale (ADS) and the self-limiting dyspnea scale (SDS)), and frailty. Both ADS and SDS presented a high predictive capacity for frailty, although it was the PROMs-D (the sum of ADS and SDS) that was the most effective measure to classify frailty. For this reason, the authors found that the PROMs-D could be used as a frailty screening measure in patients with COPD [36].

A cross-sectional study conducted in COPD patients that lived in the community (posing the question, "Has a medical doctor ever told you that you have any of the following conditions: emphysema, chronic bronchitis, or chronic obstructive lung disease?") analyzed the prevalence of geriatric syndromes in the recruited sample, including frailty (at 16%, evaluated according to the modified frailty scale) [37]. A prospective observational study conducted in hospitalized patients with acute exacerbations of COPD (AECOPD) reported a prevalence of moderate or severe frailty of 35.9% (18.4% severe frailty) using the REFS tool within 48–96 h of hospital admission [38]. At the same time, a cross-sectional study recruited 125 patients with COPD and assessed frailty using the Chinese–Canadian Study of health and aging clinical frailty scale and dyspnea using the modified medical research council questionnaire. The patients were divided into two groups: dyspnea and non-dyspnea. The prevalence of frailty in the non-dyspnea group was 26.7%, and it was 85.9% in the dyspnea group. The predicting factor of frailty in the dyspnea group was the number of prescribed drugs, and both the polypharmacy and the CAT were positively correlated with the conversion time from fit to frail in both groups [39].

A cross-sectional study of 57 patients with COPD (GOLD 1–4) had the highest prevalence of frailty (EFIP) detected in this narrative review at 83%, which was associated with malnutrition and physical frailty [40]. On the other hand, the prevalence of slight-to-moderate frailty assessed using the CFS was 54% in 46 patients admitted to the hospital due to AECOPD [41]. Finally, similar results were reported in a study in primary care. Thus, a cross-sectional study conducted in 257 COPD patients in primary care assessed the frailty determinants of these patients. A frailty prevalence of 82% was reported using the FiND questionnaire. The risk of presenting frailty was associated with age, hypertension, uncontrolled disease (CAT $\geq$ 10), and mMRC $\geq$ 2, or the presence of $\geq$2 AECOPD and GOLD status (B and D vs. A and C groups). The authors concluded that the severity of COPD increased frailty prevalence [42].

This frailty prevalence was consistent with the one published in a meta-analysis (27 studies: 23 were cross-sectional, 3 were longitudinal, and 1 was mixed (both cross-sectional and longitudinal)). These established a collective prevalence of frailty of 20% in COPD patients (95% CI: 15–24%; I^2 = 94.4%) while pre-frailty prevalence was 56% (95% CI: 52–60%; I^2 = 80.8%). Patients with COPD showed an OR of presenting frailty of 1.97 (95% CI: 1.53–2.53) [17].

The Prevalence of Geriatric Syndromes Linked to Frailty in COPD Patients

The frailty prevalence in COPD patients was linked to a high prevalence of geriatric syndromes, as four of the studies included in this narrative review showed. Thus, malnutrition and frailty coexisted in 40% of the COPD patients, malnutrition and physical frailty coexisted in 18%, and malnutrition and disability coexisted in 21% of the cases in a cross-sectional study [40]. This overlap of geriatric patients indicates the need for nutritional intervention in COPD patients, especially before starting a rehabilitation program [40]. The EFIP and PG-SGA scores were significantly correlated (r = 0.43, p = 0.001), as well as the Fried's criteria and the PG-SGA score (r = 0.37, p = 0.005). A nutritional intervention should be delivered by health care professionals in COPD patients before starting a rehabilitation program [40]. A case-control study [30] assessed the prevalence of geriatric syndromes in 150 COPD patients. Frailty was linked to a high prevalence of geriatric syndromes, such as impairment in instrumental activities of daily living (IADL), 37.3%; cognitive impairment, 35.3%; urinary incontinence, 20.7%; and malnutrition (20.7%). Two variables increased the risk of the prevalence of geriatric syndromes: dyspnea ($\geq$2 mMRC grade) and low socioeconomic status [30]. The authors emphasized the importance of frail COPD patient performance in the comprehensive geriatric assessment (CGA). Frailty has been associated with mood disorders in COPD patients. A cross-sectional study of 40 COPD patients reported a low perception of quality of life. In addition, patients classified as frail using the KCL score presented lower left and right hippocampal, subiculum, and presubiculum volume, as compared to non-frail patients. The authors concluded by pointing to the impact of frailty on the hippocampal volume and its combined associations with a poor quality of life [35]. Finally, a cross-sectional study conducted in 3222 older adults (69.6 (SD 7.4) years) with COPD who lived in the community analyzed the prevalence of geriatric syndromes in the recruited sample, including frailty, functional disability, physical function impairment, lower physical activity, falls, polypharmacy, loneliness, depression, cognitive impairment, urinary incontinence, and comorbidity [37]. The prevalence of geriatric syndromes was higher in COPD patients than in non-COPD patients: 16%, frailty; 53.9%, urinary incontinence; 57.7%, loneliness perception; 58.1%, functional disability; 12.9%, moderate cognitive impairment; 32%, depressive symptoms; and 37.5%, severe polypharmacy (>10 prescribed drugs). As compared to non-COPD patients, the frailty-adjusted OR was 6.3 (95% CI: 3.0, 13.0). This meant that the COPD patients had a risk of frailty more than six times higher than the controls. Likewise, the adjusted OR for functional disability was 1.4 (95% CI: 1.01, 2.0); for impaired physical function, the adjusted OR was 2.1 (95% CI: 1.1, 3.7); for extreme low physical activity, the adjusted OR was 2.3 (95% CI: 1.5, 3.5); for polypharmacy (≥10 medications), the adjusted OR was 2.9 (95% CI: 2.0, 4.2); and for depression, the adjusted OR was 1.9 (95% CI: 1.4, 2.7) [37].

3.2. Frailty as a Predictive Factor of Poor Outcomes

Out of the 25 articles selected for this review, 12 associated the presence of frailty with worsening outcomes, such as AECOPD, hospitalization, length of stay, greater difficulty in returning home, hospital readmissions, and mortality. Table 2 summarizes the association between frailty and poor outcomes in COPD patients, as reported by the studies included in this narrative review.

Table 2. Relationship between health outcomes and frailty in COPD patients.

Article	Type	Sample	Poor Outcomes Associated with Frailty
Ushida, K. et al. [20]	Retrospective cohort	3396 COPD patients	Hospital admissions (32.9% vs. 17.5%) In-hospital mortality (16.4% vs. 12.5%) Greater difficulty in returning home (34.6% vs. 22.9%)
Kennedy, C. C. et al. [22]	Retrospective cohort	902	Increased rate of hospitalization: Adjusted HR, 1.6 (95% CI: 1.1–2.5) Increase in hospital use of 8.0 days: (95% CI: 4.4–11.6) Higher mortality rate: Adjusted HR 1.4 (95% CI: 0.97–2.0); $p = 0.07$
Witt, L. J. et al. [23]	Observational study	70 patients admitted due to COPD exacerbation	30-day readmissions: OR 11.2 (95% CI: 1.3–93.2)
Luo, J. et al. [24]	Cross-sectional study	309	AECOPD: IRR = 1.75 (95% CI: 1.09–2.82) All-cause hospitalizations: IRR = 1.4 (95% CI: 1.0–1.9) All-cause mortality risk: HR = 2.5 (95% CI: 1.0–6.4)
Yee, N. et al. [25]	Cohort study	280	Handgrip strength increased AECOPD risk: IRR 1.46 (95% CI: 1.09–1.97)
Naval, E. et al. [26]	Cross-sectional study	127	AECOPD: Frail COPD patients 2.2 (SD 1.7) vs. fit COPD patients 1.0 (SD 1.0)

Table 2. *Cont.*

Article	Type	Sample	Poor Outcomes Associated with Frailty
Hanlon, P. et al. [27]	Observational study	3132	FFP Mortality risk: HR 2.3 (95% CI: 1.8–3.0) MACE: HR 2.7; 95% CI: 1.7–4.5 Hospital admissions HR 3.4 (95% CI: 2.8–4.1) AECOPD hospital admissions: HR 5.2; 95% CI: 3.8–7.1 Community exacerbations: HR 2.1 (95% CI: 1.8–2.5) FI Mortality HR 2.6 (95% CI: 1.7–4.0) MACE HR 6.8 (95% CI: 2.7–17.0) Hospital admission HR 3.7 (95% CI: 2.5–5.4) AECOP hospital admissions HR 4.3; 95% CI: 2.4–7.7 Community exacerbations HR 2.4 (95% CI: 1.7–3.3)
Zhang, D. et al. [28]	Prospective study	302	1-year mortality risk FFP: HR = 3.11 (95% CI: 1.30–7.44) CFS: HR = 3.68 (95% CI: 1.03–13.16) SPPB: HR = 3.74 (95% CI: 1.39–10.06)
Gu, J. J. et al. [34]	Observational retrospective study	154	FI-lab increased AECOPD and mortality: OR 8.705 (95% CI: 3.646–20.782)
Bernabeu-Mora, R. et al. [38]	Prospective cohort	103 hospitalized COPD patients	Hospital readmission (45% vs. 18%) 90-day readmission (OR = 5.19; 95% CI: 1.26–21.50)
Chin, M. et al. [41]	Prospective study	46	Severe frailty vs. managing well and vulnerable: Total length of stay: 11 days [IQR 10–12] vs. 4 [IQR 2–7] Total cost CAD 14,109 [IQR 13,182–15,037] vs. 4366 [IQR 2490–7094] Previous hospitalization in the last 2 years, 6 [IQR 6–6] vs. 1 [IQR 0–2]
Patino-Hernandez, D. et al. [42]	Longitudinal study	2706 patients (76.4 years)	3-year mortality risk HR 1.95 (95% CI: 1.18–3.2)

Legend: COPD = chronic obstructive pulmonary disease; OR = odds ratio; HR = hazard ratio; AECOPD = acute exacerbation chronic obstructive pulmonary disease; IRR = incidence rate ratio; IQR = interquartile rank; CAD = Canadian dollars; MACE = major adverse cardiovascular events; FFP = Fried frailty phenotype; FI = frailty index; CFS = clinical frailty scale; SPPB = short physical performance battery.

3.3. Frailty as a Predictive Factor of Poor Outcomes

Out of the 25 articles selected for this review, 12 of them associated the presence of frailty with worsening outcomes such as AECOPD, hospitalization, length of stay, greater difficulty in returning home, hospital readmissions, and mortality. Thus, in a retrospective cohort study [20], frailty was associated with an increase in hospitalization (32.9% vs. 17.5%), in-hospital mortality (16.4% vs. 12.5%), greater difficulty in returning home (34.6% vs. 22.9%), and a poorer quality of life at discharge (8.7% vs. 12.4%) [20]. The higher HFRS score was independently associated with a prolonged hospital admission (length of stay $\geq$ 30 days) (OR 2.0; 95% CI: 1.4–2.9) [20]. The ability of the HFRS to predict prolonged hospitalization was slightly higher than that of the Charlson comorbidity index (CCI) [21]. In a similar way, the severely frail patients were also much more likely to be readmitted than the non-frail patients (45% vs. 18%), and after adjusting for age and relevant disease-related factors, severe frailty remained an independent risk factor for 90-day readmission (OR = 5.19; 95% CI: 1.26–21.50) [38]. Frail participants also had an increased rate of hospitalization (adjusted hazard ratio (HR), 1.6; 95% CI: 1.1–2.5; p = 0.02), an adjusted increase in hospital use of 8.0 days (95% CI: 4.4–11.6; $p = 0.001$), and a higher mortality rate (adjusted HR 1.4; 95% CI: 0.97–2.0; $p = 0.07$) [22]. Likewise, after adjustment, frailty increased the incidence of AECOPD (IRR = 1.75, 95% CI: 1.09–2.82) and all-cause hospital admissions (IRR = 1.39, 95% CI: 1.04–1.87). All-cause mortality risk also increased during the 1-year follow-up, which was higher in frail patients (HR = 2.54, 95% CI: 1.01–6.36) [24].

However, in another study, the global frailty rate was not associated with the incidence of AECOPD, although the estimated weakness component was associated with a higher risk of AECOPD, which was assessed by handgrip strength (IRR 1.46, 95% CI: 1.09–1.97); moreover, frailty was associated with the risk of having non-COPD hospital admissions [25]. In contrast, the FI-lab scores statistically increased the risk of AECOPD and mortality [41], and greater frailty severity was associated with an increase in care costs, a longer hospital stay, more previous hospitalizations, and subsequent hospitalizations with an "alternate level of care" (ALC) [41].

It was previously mentioned that weakness, established by handgrip strength performance, was a predictor of AECOPD. This indicator was also present in another study that had a 30-day readmission predictive capacity (OR 11.2; 95% CI: 1.3–93.2) [23]. Frailty assessed by the FFP was associated with a higher risk of mortality (HR 2.3; 95% CI: 1.8–3.0), major adverse cardiovascular events (MACE) (HR 2.7; 95% CI: 1.7–4.5), hospital admissions (HR 3.4; 95% CI: 2.8–4.1), hospital exacerbations (HR 5.2; 95% CI: 3.8–7.1), and community exacerbations (HR 2.1; 95% CI: 1.8–2.5). The FI (severe frailty compared to the robust condition) obtained similar outcomes as to mortality (HR 2.6; 95% CI: 1.7–4.0), MACE (HR 6.8; 95% CI: 2.7–17.0), hospital admissions (HR 3.7; 95% CI: 2.5–5.4), hospital exacerbations (HR 4.3; 95% CI: 2.4–7.7), and community exacerbations (HR 2.4; 95% CI: 1.7–3.3) [28]. Similarly, in a different study, all assessed scales (FFP, FI, CFS, and SPPB) were associated with an increase in 1-year mortality [29]. During the follow-up, all tools, except for the FI, were associated with mortality in a multivariate analysis: FFP, HR = 3.11 (95% CI: 1.30–7.44); CFS, HR = 3.68 (95% CI: 1.03–13.16); and SPPB, HR = 3.74 (95% CI: 1.39–10.06). The FFP was associated with AECOPD, and the FFP, the CFS, and the FI-CD were associated with the number of hospital admissions [29]. The CFS and the FI showed a sensitivity of 96% for predicting all-cause 1-year mortality in COPD patients, but all scales showed low specificity that ranged from 39% to 44% [30]. The receiver-operating characteristic (ROC) curves for the four frailty scales each had a similar capacity for predicting mortality, presenting no statistically significant differences. When associating variables, such as age, sex, medication, CCI: GOLD severity, and CAT, the tools improved their capacity to predict mortality [29].

The last two papers included in this section showed that frailty was associated with more COPD exacerbations, being 2.2 (SD 1.7) in frail patients and 1.0 (SD 1.0) in fit COPD patients [27]; at the same time, a longitudinal study conducted in 2706 patients (76.4 years) assessed the link between the presence of frailty, which was estimated with the FI, and the presence of COPD, and highlighted that the presence of both conditions increased 3-year mortality risk up to a 95% (HR 1.95; 95% CI: 1.18–3.2) [43].

3.4. Interventions in Frailty COPD Patients

Only one article focusing on frailty intervention was found in this narrative review. The intervention consisted of an integrative multidisciplinary approach that included evidence-based pharmacological and non-pharmacological interventions, such as palliative, respiratory, and rehabilitative therapies. The physical intervention consisted of physiotherapy and occupational therapy 4–5 times a week, each session lasting from 30 to 45 min, in which limb strengthening and aerobic activities were performed and tailored to the individual baseline function. In this study, 100 older-adult-matched pairs were recruited (73.9 (SD 8.2) years). High HFRS was present in 57%, and 71% had overlapping respiratory diagnoses. In this sample, integrated care for advanced respiratory disorders was associated with a further reduction in the length of hospital stay, down to 9.1 (SD 19.9) days; fewer admission days, 0.8 (SD 1.9); and fewer ED visits, 0.6 (SD 2.2). The 6MWD and the MBI scores improved. Greater improvement was observed in patients with lower baselines in their 6MWD and MBI scores. Prescriptions of slow-release opioids rose from 9% to 49%, and treatments for anxiety and depression rose from 5% to 19% [21].

4. Discussion

In the present review, the authors observed a significant variability in the frailty measurement tools used, in the associations among the presence of frailty, a high prevalence of geriatric syndromes and COPD-related aspects, in the close relationship between the presence of frailty and poorer health outcomes, and finally, in a single study that explored the potential benefits of a combined intervention (pharmacological and non-pharmacological, including rehabilitation and a nutritional approach), which showed improvements in functional parameters.

Despite having a generally accepted definition of frailty [1], the number of operational frailty definitions was high [44], along with the number of frailty assessment tools, which presented a significant heterogeneity for the validation in different settings [4]. This exact situation was observed in the review, in which up to 13 different measurement tools used for the assessment of frailty in COPD patients were identified. Only one study [29] compared frailty prevalence and its prognostic capacity for adverse health outcomes according to four scales (FFP, CFS, FI, and SPPB). The FI presented the lowest frailty prevalence, but the rest (FFP, CFS, and SPPB) presented similar outcomes. The FI is an index based on the accumulation of deficiencies and detects more advanced severity than the FFP. In fact, as the authors of a cross-sectional study highlighted, different frailty instruments may capture overlapping, albeit distinct, parameters, and thus, they should not be used interchangeably [45]. Therefore, the prevalence of frailty was lower when using the FFP (17%), as compared to the FI [28]. Although when using four tools (FFP, 51%; SPPB, 58.6%; FI-CD 59.6%; CFS, 64.2%), the authors reported a moderate-to-substantial agreement between the instruments [29]. Despite the concordance reported in a previous study [30], the heterogeneity in the assessed domains of the different frailty tools [4], and the severity of COPD, the prevalence of frailty was inconsistent among the studies included in this review, regardless of the stability of the COPD, the setting where the patients were recruited, the age range of the sample, or the instrument (even within the same instrument). Nevertheless, the prognostic capacity of frailty in patients with COPD persisted despite the variance in ages, instruments, and the degrees of severity of COPD, confirming the importance of frailty as an indicator of disease progression in these patients.

The only meta-analysis included in this review [17] highlighted that the risk of frailty in COPD patients was 97%, and it had important clinical implications, which required the use of the CGA in patients with positive frailty screenings to order to establish appropriate programs for reversing or reducing frailty. The authors of the aforementioned meta-analysis emphasized the common mechanisms and risk factors of COPD and frailty as well as the increased risks of mortality and hospital readmissions in the acute exacerbation of COPD in frail patients. However, the relationship among frailty, geriatric syndromes, and COPD were not assessed in the meta-analysis, as reflected in this narrative review, nor were other types of adverse events associated with frail COPD patients reported, as described in this narrative review.

Likewise, frailty has been related to the severity of COPD [20]. Since frailty prevalence is high in COPD patients, regular assessments of frailty in clinical practices with COPD patients should be performed to prevent or mitigate the progression of the disease [42]. Since the CGA is time-consuming, the TUG could be used as a screening tool [33] due to its ability to predict frailty in COPD patients.

Frailty prevalence increases with each decade of life. In fact, the only meta-analysis included in the study stated that age was one of the variables that increased the risk of frailty. However, most of the prevalence risks reported in these studies were higher, as compared to those measured in populations of similar age without COPD. Some published studies have linked the higher prevalence of frailty to the severity level of COPD and found a direct association using such measurements as mMRC [27,30,31,42], CAT [31,39,42], GOLD [31,42], FEV_1 [28], PROMs-D (ADS plus SDS) [36].

Just as the progression of diseases such as HIV/AIDS and higher rates of inflammatory activity increased frailty prevalence, as compared to patients of the same age without these diseases [46], the prevalence of frailty in COPD patients was also higher, which was likely due to higher inflammatory activity. Therefore, it would be interesting to explore whether pathophysiological factors (Figure 2) or possible adverse effects of drugs given for active disease management, such as corticosteroids, may accelerate the onset of frailty and other geriatric syndromes such as sarcopenia, which is an important element in the functional performance in older adults. As a recent review described, beyond chronic inflammation and reduced physical activity, factors that decrease muscle strength and endurance in COPD patients include oxidative stress, inactivity, hypoxemia, hormone abnormality, lack of nutrients such as protein and vitamin D, and the use of corticosteroids [15].

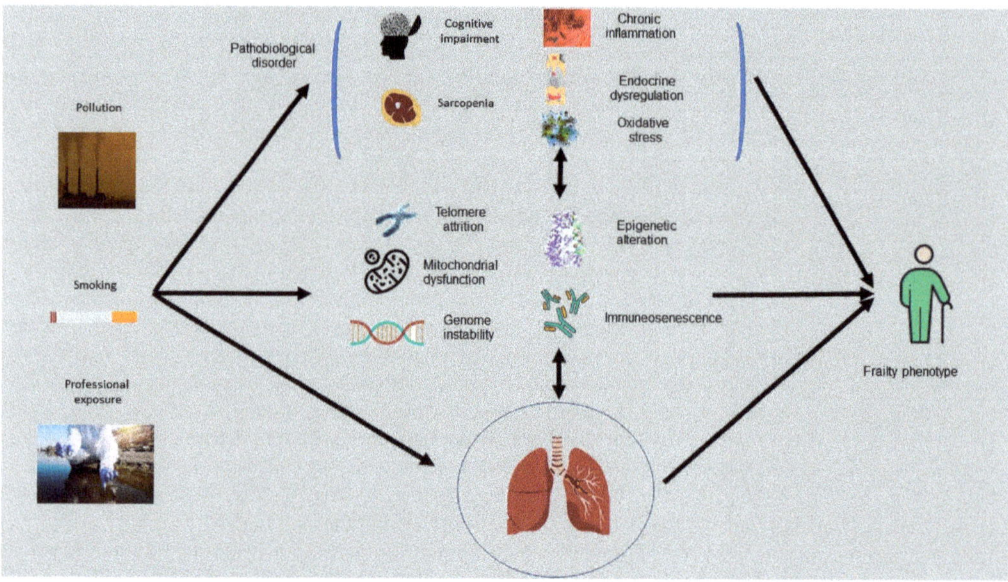

Figure 2. Graphic description of the link between environmental factors, pathobiological disorders, genetic factors, immunosenescence, and frailty.

Despite the high prevalence of sarcopenia in COPD patients (15.5%) and its relation to the severity of pulmonary disease, the FEV_1, poor exercise tolerance, and poor quality of life [47], none of the studies included in the review assessed the presence of sarcopenia in their samples.

The presence of frailty has been related to the high prevalence of geriatric syndromes in COPD patients. Malnutrition, cognitive impairment, polypharmacy, disability, and depressive symptomatology have been frequently reported in these patients. Consequently, various studies have highlighted the importance of early detection and using the CGA for the global assessment of these patients, with the aim of defining individualized care programs for the improvement of their clinical and functional outcomes. Thus, six of the included articles linked the presence of frailty in COPD patients with a high prevalence of other geriatric syndromes, such as malnutrition [30,40], IADL impairment [30], cognitive impairment [22,30,35,36], urinary incontinence [28,30], physical disabilities [22,26,28], depression [27,28,35], polypharmacy [28], falls [28], loneliness [28], and poor quality of life [22,35]. Given the increased presence of geriatric syndromes in older adult patients with COPD, the authors of various studies have highlighted the importance of the regular use of the CGA [22,32] for a complete multidimensional assessment that detects potentially correctable impairments and comorbidities, with the aim of reducing mortality rates [32]. An approach "beyond the lung" [37] is necessary in the care of these patients and should be focused on the management of geriatric syndromes and conditions, as their improvement enhances patient quality of life and clinical COPD outcomes.

Frailty and comorbidity are predictors of a worsening progression in patients under different clinical conditions [48]. A study in the present review stated that frailty was a better predictor of adverse events than the CCI [20]. Frailty, as a recent review highlighted, presents limited evidence regarding the increase in morbidity and mortality in COPD patients [49], as it is a better indicator of functional capacity and the need for palliative care in the future than a prognostic indicator. However, frailty has been described as a factor linked to the higher incidence of both adverse events [50] and higher mortality [51] in adults that live in a community. Similarly, several studies included in this review emphasize frailty as a risk factor of worsening outcomes, such as AECOPD [20,25,27,28,34,39], hospitalization [20,24,25,28,29,41], length of stay [20,24,39], greater difficulty in returning home [20], hospital readmission [20,23,24], MACE [28], medical costs [41], transition to a higher level of care [41], and mortality [20,24,28,29,34,43]. The authors of one of these studies [43] emphasized that vulnerability caused by frailty increased adverse outcomes in older adults, and the increment of a worsening prognosis for frailty in these patients requires regular assessment in clinical practice. As previously mentioned, frailty was a predictor of worsening health outcomes [50] and mortality [51] in older adults, as well as in older adults with COPD [20,22–25,27–29,34,38,41,43]. In fact, COPD patients who were also frail had worsening outcomes than COPD patients without frailty. Therefore, we can consider frailty an indicator of a worsening progression of COPD, and it should be detected early using a CGA in order to slow or reverse its impact.

The CGA is completed with the design of an individualized care plan, in which non-pharmacological measures are suggested to reverse or reduce present geriatric syndromes. Only one study focused on frailty intervention was found in this narrative review [21]. The physical intervention improved the functional performance of the recruited patients, leading the authors to highlight that the integration of functional rehabilitation with palliative care could improve the functional capacity of patients, along with better treatments for symptoms such as anxiety and depression, which are typically reported. These findings are consistent with some suggestions outlined in a recent review [52] that highlighted that the ability to reverse frailty without intervention was minimal but still possible through pulmonary rehabilitation (PR) programs, which also improved the prognosis of these patients. However, the mechanism by which PR reversed frailty in this patient population was not elucidated. PR programs with multidisciplinary components could reverse frailty by addressing the five components of the frailty phenotype; however, the heterogeneity of the COPD population hampers the uniformity of scheduled exercise programs, as well as the objectives. Once again, it would be necessary to individually design programs according to the patient's outcomes based on their CGA. Likewise, another review [53] highlighted the significant association between frailty and COPD that requires the early detection and treatment of frailty in order to reduce the risk of worsening health outcomes, such as increased functional impairment, disability, hospital admissions, presence of geriatric syndromes, institutionalization, and death. Similarly, a review that included 20 scientific papers that described interventions using pulmonary rehabilitation, electrical stimulation, home-based programs, geriatric rehabilitation, hospital-based exercises, physical activity, and non-standardized exercise, reported that different programs were successful when they sought a therapeutic partnership with the patient, were individualized, increased patient engagement, and improved robustness and adaptive capacity [54]. Building trust, individualizing priorities, and approaching multidimensional problems appropriately was necessary for the success of the programs. Home-based pulmonary rehabilitation was another method by which to treat frailty in these patients in order to improve functional capacity, frailty status, quality of life, and pulmonary symptoms, such as fatigue [55]. Palliative care should not be overlooked. As a recent review noted, palliative care is much more than hospice or end-of-life care [56] and could be integrated with the above-mentioned interventions and approaches to provide a holistic, comprehensive assessment and treatment, pharmacological and non-pharmacological, to manage symptoms and enhance quality of care for these patients.

Lastly, another review highlighted the need to improve the comprehension of the frailty phenotype in different chronic diseases, given its high prevalence in COPD patients. From a clinical perspective, it requires interdisciplinary cooperation in order to mitigate the impact of frailty in COPD patients [57]. In this sense, and given the particular characteristics of frail patients with COPD, not only would it be necessary to establish a "COPD–frail" phenotype but also to develop specific interventions in order to reduce the prevalence of geriatric syndromes and improve functionality, quality of life, and survival while reducing acute exacerbations and hospitalizations. Certainly, the authors were unable to determine through this narrative review which frailty tool was optimal for patients with COPD or whether it was necessary to use several scales simultaneously (nor could the authors suggest the appropriate combination of instruments) to improve diagnostic sensitivity and specificity. Once again, given the importance of frailty in the clinical course of COPD patients, the authors considered the need to emphasize the standardization of frailty screenings in both outpatient and inpatient settings. It could increase the inclusion of patients when selecting the appropriate interventions. It could improve or reverse frailty and other associated geriatric syndromes as well as the evolution of the disease (reducing AECOPD, hospital admissions, and mortality), and it could maintain or improve the quality of life in COPD patients. Though the discussion section is appropriate for any speculation concerning new hypotheses, such hypotheses need to be ratified by relevant studies that, at best, confirm them or, at worst, refute them.

The limitations of this review stem from the absence of randomized clinical trials and a considerable number of cross-sectional studies that identified associations but not causalities. The authors evaluated the published articles in the OVID and PubMed databases in order to include articles published in journals with the greatest impact. However, this decision was not unbiased, as publications in other databases that could have been of interest were not evaluated. Likewise, the authors are aware that a systematic review could have provided more evidence on the topics covered. However, the heterogeneity of the frailty assessment instruments, the different settings in which the patients were recruited, and the varied degrees of severity of disease in the selected studies led the authors to dismiss this possibility. Given the growing interest in this topic, the authors of this review are considering the development of a systematic review in a near future.

5. Conclusions

Frailty prevalence is high in COPD patients. The heterogeneity of frailty measurement tools challenges drawing broad conclusions about the results obtained with such tools. Regardless of the assessment tool used, frailty prevalence in COPD patients has been associated with a high prevalence of geriatric syndromes and worsening clinical outcomes, including mortality. It is recommended to use frailty screenings for COPD patients, regardless of the setting in which they are assessed, and to perform CGAs in order to detect associated problems and to establish individualized treatment plans in order to improve clinical outcomes in these patients.

Author Contributions: Conceptualization, F.J.T.-S., E.N., J.M.D.l.C.-d.l.H. and J.V.; methodology, F.J.T.-S., E.N. and J.M.D.l.C.-d.l.H. software, J.M.D.l.C.-d.l.H.; validation, F.J.T.-S., E.N. and J.V.; formal analysis, J.M.D.l.C.-d.l.H.; investigation, F.J.T.-S. and E.N.; resources, J.V.; data curation, C.C.-P.; writing—original draft preparation, F.J.T.-S. and E.N.; writing—review and editing, J.M.D.l.C.-d.l.H., C.C.-P. and J.V.; visualization, E.N.; supervision, F.J.T.-S. and J.V.; project administration, J.M.D.l.C.-d.l.H.; funding acquisition, E.N. All authors have read and agreed to the published version of the manuscript.

Funding: This research received no external funding.

Acknowledgments: The authors would like to thank Nadia Dolz from the Hospital Universitario de la Ribera, for translating and proofreading during the elaboration of this manuscript.

Conflicts of Interest: The authors declare no conflict of interest. The funders had no role in the design of the study; in the collection, analyses, or interpretation of data; in the writing of the manuscript; or in the decision to publish the results.

References

1. Morley, J.E.; Vellas, B.; Abellan van Kan, G.; Anker, S.D.; Bauer, J.M.; Bernabei, R.; Walston, J. Frailty consensus: A call to action. *J. Am. Med. Dir. Assoc.* **2013**, *14*, 392–397. [CrossRef] [PubMed]
2. Fried, L.P.; Tangen, C.M.; Walston, J.; Newman, A.B.; Hirsch, C.; Gottdiener, J.; Seeman, T.; Tracy, R.; Kop, W.J.; Burke, G.; et al. Frailty in older adults: Evidence for a phenotype. *J. Gerontol. A Biol. Sci. Med. Sci.* **2001**, *56*, 146–157. [CrossRef]
3. Kojima, G.; Taniguchi, Y.; Iliffe, S.; Jivraj, S.; Walters, K. Transitions between frailty states among community-dwelling older people: A systematic review and meta-analysis. *Ageing Res. Rev.* **2019**, *50*, 81–88. [CrossRef] [PubMed]
4. Dent, E.; Martin, F.C.; Bergman, H.; Woo, J.; Romero-Ortuno, R.; Walston, J.D. Management of frailty: Opportunities, challenges, and future directions. *Lancet* **2019**, *394*, 1376–1386. [CrossRef]
5. Veronese, N.; Sigeirsdottir, K.; Eiriksdottir, G.; Marques, E.A.; Chalhoub, D.; Philips, C.L.; Launer, L.J.; Maggi, S.; Gudnason, V.; Harris, T.B. Frailty and risk of cardiovascular diseases in older persons: The age, gene/environment susceptibility-Reykjavik study. *Rejuvenation Res.* **2017**, *20*, 517–524. [CrossRef]
6. Bandeen-Roche, K.; Xue, Q.L.; Ferrucci, L.; Walston, J.; Guralnik, J.M.; Chaves, P.; Zeger, S.L.; Fried, L.P. Phenotype of frailty: Characterization in the women's health and aging studies. *J. Gerontol. A Biol. Sci. Med. Sci.* **2006**, *61*, 262–266. [CrossRef]
7. Vestbo, J.; Hurd, S.S.; Agustí, A.G.; Jones, P.W.; Vogelmeier, C.; Anzueto, A.; Barnes, P.J.; Fabbri, L.M.; Martinez, F.J.; Nishimura, M.; et al. Global strategy for the diagnosis, management, and prevention of chronic obstructive pulmonary disease: GOLD executive summary. *Am. J. Respir. Crit. Care Med.* **2013**, *187*, 347–365. [CrossRef] [PubMed]
8. Collard, R.M.; Boter, H.; Schoevers, R.A.; Oude-Voshaar, R.C. Prevalence of frailty in community-dwelling older persons: A systematic review. *J. Am. Geriatr. Soc.* **2012**, *60*, 1487–1492. [CrossRef] [PubMed]
9. Singer, J.P.; Lederer, D.J.; Baldwin, M.R. Frailty in pulmonary and critical care medicine. *Ann. Am. Thorac. Soc.* **2016**, *13*, 1394–1404. [CrossRef]
10. Milne, K.M.; Kwan, J.M.; Guler, S.; Winstone, T.A.; Le, A.; Khalil, N.; Camp, P.G.; Wilcox, P.G.; Ryerson, C.J. Frailty is common and strongly associated with dyspnoea severity in fibrotic interstitial lung disease. *Respirology* **2017**, *22*, 728–734. [CrossRef]
11. Rosenwasser, Y.; Berger, I.; Loewy, Z.G. Therapeutic Approaches for Chronic Obstructive Pulmonary Disease (COPD) Exacerbations. *Pathogens* **2022**, *11*, 1513. [CrossRef] [PubMed]
12. Barnes, P.J. Cellular and molecular mechanisms of asthma and COPD. *Clin. Sci.* **2017**, *131*, 1541–1558. [CrossRef]
13. Karpman, C.; Benzo, R. Gait speed as a measure of functional status in COPD patients. *Int. J. Chron. Obstruct. Pulmon. Dis.* **2014**, *9*, 1315–1320.
14. Lahousse, L.; Ziere, G.; Verlinden, V.J.; Zillikens, M.C.; Uitterlinden, A.G.; Rivadeneira, F.; Tiemeier, H.; Joos, G.F.; Hofman, A.; Ikram, M.A.; et al. Risk of frailty in elderly with COPD: A population-based study. *J. Gerontol. A Biol. Sci. Med. Sci.* **2016**, *71*, 689–695. [CrossRef]
15. Lippi, L.; Folli, A.; Curci, C.; D'Abrosca, F.; Moalli, S.; Mezian, K.; de Sire, A.; Invernizzi, M. Osteosarcopenia in Patients with Chronic Obstructive Pulmonary Diseases: Which Pathophysiologic Implications for Rehabilitation? *Int. J. Environ. Res. Public Health* **2022**, *19*, 14314. [CrossRef] [PubMed]
16. Vázquez-Gandullo, E.; Hidalgo-Molina, A.; Montoro-Ballesteros, F.; Morales-González, M.; Muñoz-Ramírez, I.; Arnedillo-Muñoz, A. Inspiratory Muscle Training in Patients with Chronic Obstructive Pulmonary Disease (COPD) as Part of a Respiratory Rehabilitation Program Implementation of Mechanical Devices: A Systematic Review. *Int. J. Environ. Res. Public Health* **2022**, *19*, 5564. [CrossRef]
17. Marengoni, A.; Vetrano, D.L.; Manes-Gravina, E.; Bernabei, R.; Onder, G.; Palmer, K. The relationship between COPD and frailty: A systematic review and meta-analysis of observational studies. *Chest* **2018**, *154*, 21–40. [CrossRef] [PubMed]
18. Baldi, S.; Jose, P.E.; Bruschi, C.; Pinna, G.D.; Maestri, R.; Rezzani, A.; Bellinzona, E.; Fracchia, C.; Dacosto, E.; Crotti, P.; et al. The mediating role of cytokine IL-6 on the relationship of FEV1 upon 6-minute walk distance in chronic obstructive pulmonary disease. *Int. J. Chron. Obstruct. Pulmon. Dis.* **2014**, *9*, 1091–1099. [CrossRef] [PubMed]
19. Moher, D.; Liberati, A.; Tetzlaff, J.; Altman, D.G.; PRISMA Group. Preferred reporting items for systematic reviews and meta-analyses: The PRISMA Statement. *PLoS Med.* **2009**, *6*, e1000097. [CrossRef]
20. Ushida, K.; Shimizu, A.; Hori, S.; Yamamoto, Y.; Momosaki, R. Hospital Frailty Risk Score Predicts Outcomes in Chronic Obstructive Pulmonary Disease Exacerbations. *Arch. Gerontol. Geriatr.* **2022**, *100*, 104658. [CrossRef] [PubMed]
21. Neo, H.Y.; Yap, C.W.; Teo, L.M.; Lee, L.T.; Tan, H.W.; Xu, H.Y.; Hum, A.; Abisheganaden, J.A. Palliative Rehabilitation Improves Health Care Utilization and Function in Frail Older Adults with Chronic Lung Diseases. *J. Am. Med. Dir. Assoc.* **2021**, *22*, 2478–2485. [CrossRef]

22. Kennedy, C.C.; Novotny, P.J.; LeBrasseur, N.K.; Wise, R.A.; Sciurba, F.C.; Benzo, R.P. Frailty and Clinical Outcomes in Chronic Obstructive Pulmonary Disease. *Ann. Am. Thorac. Soc.* **2019**, *16*, 217–224. [CrossRef]
23. Witt, L.J.; Spacht, W.A.; Carey, K.A.; Arora, V.M.; White, S.R.; Huisingh-Scheetz, M.; Press, V.G. Weak Handgrip at Index Admission for Acute Exacerbation of COPD Predicts All-Cause 30-Day Readmission. *Front. Med.* **2021**, *8*, 611989. [CrossRef]
24. Luo, J.; Zhang, D.; Tang, W.; Dou, L.Y.; Sun, Y. Impact of Frailty on the Risk of Exacerbations and All-Cause Mortality in Elderly Patients with Stable Chronic Obstructive Pulmonary Disease. *Clin. Interv. Aging* **2021**, *16*, 593–601. [CrossRef] [PubMed]
25. Yee, N.; Locke, E.R.; Pike, K.C.; Chen, Z.; Lee, J.; Huang, J.C.; Nguyen, H.Q.; Fan, V.S. Frailty in Chronic Obstructive Pulmonary Disease and Risk of Exacerbations and Hospitalizations. *Int. J. Chron. Obstruct. Pulmon. Dis.* **2020**, *15*, 1967–1976. [CrossRef] [PubMed]
26. Bernabeu-Mora, R.; Oliveira-Sousa, S.L.; Sánchez-Martínez, M.P.; García-Vidal, J.A.; Gacto-Sánchez, M.; Medina-Mirapeix, F. Frailty transitions and associated clinical outcomes in patients with stable COPD: A longitudinal study. *PLoS ONE* **2020**, *15*, e0230116. [CrossRef] [PubMed]
27. Naval, E.; González, M.C.; Giraldós, S.; Calatayud, J.; Jornet, M.; Lluch, I.; Meseguer, M.; Cubillan, J.J.R.; Viña, J.; Tarazona-Santabalbina, F.J. Frailty Assessment in a Stable COPD Cohort: Is There a COPD-Frail Phenotype? *COPD* **2021**, *18*, 525–532. [CrossRef]
28. Hanlon, P.; Lewsey, J.; Quint, J.K.; Jani, B.D.; Nicholl, B.I.; McAllister, D.A.; Mair, F.S. Frailty in COPD: An analysis of prevalence and clinical impact using UK Biobank. *BMJ Open Respir. Res.* **2022**, *9*, e001314. [CrossRef]
29. Zhang, D.; Tang, W.; Dou, L.Y.; Luo, J.; Sun, Y. Four different frailty models predict health outcomes in older patients with stable chronic obstructive pulmonary disease. *BMC Geriatr.* **2022**, *22*, 57. [CrossRef]
30. Soni, N.; Banerjee, J.; Gunasekaran, V.; Saini, M.; Singh, V.; Kumar, P.; Chakrawarty, A.; Chatterjee, P.; Dey, A.B. Association of geriatric syndromes in older adults with chronic obstructive pulmonary disease. *Aging Med.* **2022**, *5*, 106–112. [CrossRef] [PubMed]
31. Dias, L.S.; Ferreira, A.C.G.; da Silva Junior, J.L.R.; Conte, M.B.; Rabahi, M.F. Prevalence of Frailty and Evaluation of Associated Variables among COPD Patients. *Int. J. Chron. Obstruct. Pulmon. Dis.* **2020**, *15*, 1349–1356. [CrossRef] [PubMed]
32. Gale, N.S.; Albarrati, A.M.; Munnery, M.M.; Hubbard, R.E.; Tal-Singer, R.; Cockcroft, J.R.; Shale, D.J. Frailty: A global measure of the multisystem impact of COPD. *Chron. Respir. Dis.* **2018**, *15*, 347–355. [CrossRef] [PubMed]
33. Albarrati, A.M.; Gale, N.S.; Munnery, M.M.; Reid, N.; Cockcroft, J.R.; Shale, D.J. The Timed up and Go test predicts frailty in patients with COPD. *NPJ Prim. Care Respir. Med.* **2022**, *32*, 24. [CrossRef] [PubMed]
34. Gu, J.J.; Liu, Q.; Zheng, L.J. A Frailty Assessment Tool to Predict In-Hospital Mortality in Patients with Acute Exacerbations of Chronic Obstructive Pulmonary Disease. *Int. J. Chron. Obstruct. Pulmon. Dis.* **2021**, *16*, 1093–1100. [CrossRef] [PubMed]
35. Takahashi, S.; Hirano, T.; Yasuda, K.; Donishi, T.; Suga, K.; Doi, K.; Oishi, K.; Ohata, S.; Murata, Y.; Yamaji, Y.; et al. Impact of Frailty on Hippocampal Volume in Patients with Chronic Obstructive Pulmonary Disease. *Biomedicines* **2021**, *9*, 1103. [CrossRef] [PubMed]
36. Oishi, K.; Matsunaga, K.; Harada, M.; Suizu, J.; Murakawa, K.; Chikumoto, A.; Ohteru, Y.; Matsuda, K.; Uehara, S.; Hamada, K.; et al. A New Dyspnea Evaluation System Focusing on Patients' Perceptions of Dyspnea and Their Living Disabilities: The Linkage between COPD and Frailty. *J. Clin. Med.* **2020**, *9*, 3580. [CrossRef] [PubMed]
37. Witt, L.J.; Wroblewski, K.E.; Pinto, J.M.; Wang, E.; McClintock, M.K.; Dale, W.; White, S.R.; Press, V.G.; Huisingh-Scheetz, M. Beyond the Lung: Geriatric Conditions Afflict Community-Dwelling Older Adults with Self- Reported Chronic Obstructive Pulmonary Disease. *Front. Med.* **2022**, *9*, 814606. [CrossRef] [PubMed]
38. Bernabeu-Mora, R.; García-Guillamón, G.; Valera-Novella, E.; Giménez-Giménez, L.M.; Escolar-Reina, P.; Medina-Mirapeix, F. Frailty is a predictive factor of readmission within 90 days of hospitalization for acute exacerbations of chronic obstructive pulmonary disease: A longitudinal study. *Ther. Adv. Respir. Dis.* **2017**, *11*, 383–392. [CrossRef] [PubMed]
39. Chen, P.J.; Yang, K.Y.; Perng, W.C.; Lin, K.C.; Wang, K.Y. Effect of dyspnea on frailty stages and related factors in Taiwanese men with COPD. *Int. J. Chron. Obstruct. Pulmon. Dis.* **2018**, *13*, 2463–2469. [CrossRef] [PubMed]
40. Ter Beek, L.; van der Vaart, H.; Wempe, J.B.; Krijnen, W.P.; Roodenburg, J.L.N.; van der Schans, C.P.; Jager-Wittenaar, H. Coexistence of malnutrition, frailty, physical frailty and disability in patients with COPD starting a pulmonary rehabilitation program. *Clin. Nutr.* **2020**, *39*, 2557–2563. [CrossRef] [PubMed]
41. Chin, M.; Voduc, N.; Huang, S.; Forster, A.; Mulpuru, S. Practical lessons in implementing frailty assessments for hospitalised patients with COPD. *BMJ Open Qual.* **2020**, *9*, e000782. [CrossRef] [PubMed]
42. Ierodiakonou, D.; Kampouraki, M.; Poulonirakis, I.; Papadokostakis, P.; Lintovoi, E.; Karanassos, D.; Maltezis, K.; Chorti, M.; Petrovitsos, E.; Dimopoulou, S.; et al. Determinants of frailty in primary care patients with COPD: The Greek UNLOCK study. *BMC Pulm. Med.* **2019**, *19*, 63. [CrossRef] [PubMed]
43. Patino-Hernandez, D.; Borda, M.G.; Cano-Gutiérrez, C.A.; Celis-Preciado, C.A.; Pérez-Zepeda, M.U. La fragilidad se asocia con mayor mortalidad en la enfermedad pulmonar obstructiva crónica [Frailty is associated with increased mortality in chronic obstructive pulmonary disease]. *Rev. Esp. Geriatr. Gerontol.* **2019**, *54*, 237–238. [CrossRef] [PubMed]
44. Theou, O.; Walston, J.; Rockwood, K. Operationalizing Frailty Using the Frailty Phenotype and Deficit Accumulation Approaches. *Interdiscip. Top Gerontol. Geriatr.* **2015**, *41*, 66–73.

45. Lim, Y.J.; Ng, Y.S.; Sultana, R.; Tay, E.L.; Mah, S.M.; Chan, C.H.N.; Latib, A.B.; Abu-Bakar, H.M.; Ho, J.C.Y.; Kwek, T.H.H.; et al. Frailty Assessment in Community-Dwelling Older Adults: A Comparison of 3 Diagnostic Instruments. *J. Nutr. Health Aging* **2020**, *24*, 582–590. [CrossRef]
46. Liu, S.; Yan, Q.; Jiang, Y.; Xiao, M.; Zhao, J.; Wang, Y.; Deng, R.; Wang, C.; Yang, Z. The Impact of Frailty on All-Cause Mortality in Patients with HIV Infection: A Systematic Review and Meta-Analysis. *AIDS Res. Hum. Retrovir.* **2022**, *38*, 692–699. [CrossRef]
47. Sepúlveda-Loyola, W.; Osadnik, C.; Phu, S.; Morita, A.A.; Duque, G.; Probst, V.S. Diagnosis, prevalence, and clinical impact of sarcopenia in COPD: A systematic review and meta-analysis. *J. Cachexia Sarcopenia Muscle* **2020**, *11*, 1164–1176. [CrossRef]
48. Boakye, D.; Rillmann, B.; Walter, V.; Jansen, L.; Hoffmeister, M.; Brenner, H. Impact of comorbidity and frailty on prognosis in colorectal cancer patients: A systematic review and meta-analysis. *Cancer Treat. Rev.* **2018**, *64*, 30–39. [CrossRef]
49. Antoniu, S.A.; Boiculese, L.V.; Prunoiu, V. Frailty, a Dimension of Impaired Functional Status in Advanced COPD: Utility and Clinical Applicability. *Medicina* **2021**, *57*, 474. [CrossRef]
50. Kojima, G. Frailty as a predictor of hospitalisation among community-dwelling older people: A systematic review and meta-analysis. *J. Epidemiol. Community Health* **2016**, *70*, 722–729. [CrossRef]
51. Kojima, G.; Iliffe, S.; Walters, K. Frailty index as a predictor of mortality: A systematic review and meta-analysis. *Age Ageing* **2018**, *47*, 193–200. [CrossRef]
52. Wang, Z.; Hu, X.; Dai, Q. Is it possible to reverse frailty in patients with chronic obstructive pulmonary disease? *Clinics* **2020**, *75*, e1778. [CrossRef] [PubMed]
53. Guan, C.; Niu, H. Frailty assessment in older adults with chronic obstructive respiratory diseases. *Clin. Interv. Aging* **2018**, *13*, 1513–1524. [CrossRef] [PubMed]
54. Brighton, L.J.; Evans, C.J.; Man, W.D.C.; Maddocks, M. Improving Exercise-Based Interventions for People Living with Both COPD and Frailty: A Realist Review. *Int. J. Chron. Obstruct. Pulmon. Dis.* **2020**, *15*, 841–855. [CrossRef]
55. Gephine, S.; Saey, D.; Grosbois, J.M.; Maltais, F.; Mucci, P. Home-based Pulmonary Rehabilitation is Effective in Frail COPD Patients with Chronic Respiratory Failure. *Chronic. Obstr. Pulm. Dis.* **2022**, *9*, 15–25. [CrossRef]
56. Lyer, A.S.; Sullivan, D.R.; Lindell, K.O.; Reinke, L.F. The Role of Palliative Care in COPD. *Chest* **2022**, *161*, 1250–1262.
57. Symvoulakis, E.K.; Kamekis, A.; Drakonaki, E.; Mastrodemou, S.; Ryerson, C.J.; Antoniou, K. Frailty and chronic respiratory disease: The need for a multidisciplinary care model. *Sarcoidosis Vasc. Diffuse Lung Dis.* **2021**, *38*, e2021031. [PubMed]

International Journal of
Environmental Research and Public Health

MDPI

Review

Challenges in Help-Seeking Behaviors among Rural Older People Mitigated through Family Physician-Driven Outreach: A Systematic Review

Ryuichi Ohta [1,*], Takuji Katsube [1] and Chiaki Sano [2]

1 Community Care, Unnan City Hospital, Unnan 699-1221, Japan
2 Department of Community Medicine Management, Faculty of Medicine, Shimane University, Izumo 690-0823, Japan
* Correspondence: ryuichiohta0120@gmail.com; Tel.: +81-905-0605-330

Abstract: Help-seeking behaviors (HSBs) refer to approaches taken by individuals towards their health and symptoms, and they are supported by healthcare professionals. Outreach interventions aimed at older people in rural communities can mitigate difficulties in implementing HSBs and help them remain healthy. This systematic review investigated evidence regarding family medicine-involved outreach aimed at HSBs among older individuals in rural areas. We searched three databases (PubMed, EMBASE, and Web of Science) for international and original interventional articles regarding family physicians involved in outreach to older people in rural or underserved areas between April 2000 and October 2022. The articles were analyzed and summarized based on the setting, country, health issues, and outreach outcomes. Of the 376 studies identified, four were included in this review. Our findings showed that family physician-involved outreach to rural and underserved areas improved health outcomes, including anxiety, subjective physical function, and diabetic care. The challenges of outreach interventions include the duration and continuity of outreach, the active participation of family physicians and patients in the outreach programs, and the focus of outreach participants. Although the number of studies included was small, family physician-involved outreach to rural and underserved areas was shown to improve various health outcomes.

Keywords: help-seeking; rural; family medicine; physician; outreach; older people; elderly

Citation: Ohta, R.; Katsube, T.; Sano, C. Challenges in Help-Seeking Behaviors among Rural Older People Mitigated through Family Physician-Driven Outreach: A Systematic Review. *Int. J. Environ. Res. Public Health* **2022**, *19*, 17004. https://doi.org/10.3390/ijerph192417004

Academic Editors: Francisco José Tarazona Santabalbina, Sebastià Josep Santaeugènia Gonzàlez, José Augusto García Navarro and José Viña

Received: 24 November 2022
Accepted: 15 December 2022
Published: 18 December 2022

1. Introduction

An individual's health can be affected by their approaches to health and symptoms in their everyday lives. These health-related behaviors are known as help-seeking behaviors (HSBs), which refer to concrete behaviors, including taking rest, gathering information, and consulting with relatives and healthcare professionals [1,2]. HSBs are categorized into lay and professional care. Lay care is provided by those with lay knowledge and non-professionals [1]. It involves self-management; gathering knowledge; consulting with families, relatives, and friends; buying and using over-the-counter drugs; and home remedies [1,2]. Meanwhile, professional care is provided by professionals and involves visiting primary care doctors, pharmacists, and emergency rooms in general hospitals [1,2]. According to their symptoms, effective lay and professional care are critical for people's health conditions [3,4].

HSBs may be related to subjective health conditions, including quality of life (QOL). Previous research has demonstrated that self-management as a form of lay care can be related to a high QOL [5]. Another study showed that self-medication could also be associated with high QOL [6]. Furthermore, self-management of common symptoms have been shown to improve QOL, including during the 2019 coronavirus pandemic [7]. Therefore, improvements to self-management methods and medication use may improve QOL. Notably, older people tend to experience more symptoms than younger generations and use various HSBs [8–10]. Thus, the HSBs of older people should be modified to improve their health conditions [10].

HSBs may be influenced by the environment, particularly in older people. Globally, HSBs in older people are a critical public health issue [11]. Aging causes the deterioration of physical and cognitive abilities, and older people lose accessibility to various social resources, owing to the loss of capacity to drive and difficulty in using public transportation [12,13]. Moreover, living in rural areas can affect older people's lives because of the scarcity of social resources and public transportation systems [12,13], and such conditions may prevent them from utilizing healthcare systems. Furthermore, delays in using healthcare resources may cause the progression of critical diseases, leading to morbidity and mortality.

Therefore, in the rural context, older people's HSBs should be improved for their health and the sustainability of rural healthcare systems. For sustainability, outreach by healthcare professionals to rural older people who cannot access healthcare institutions because of low accessibility and availability is an effective approach for early detection of modifiable risk factors using healthcare resources [14]. Among older people, delays in using healthcare resources in critical situations, including cardiovascular diseases and malignancy, are detrimental to their lives [15,16]. Therefore, outreach aimed at these populations in rural communities can mitigate the risks of acute diseases, helping them remain healthy. In addition, because aging is progressing, effective outreach to rural communities can reduce multimorbidity issues, healthcare usage irregularity, and the burden on rural healthcare professionals.

Family physicians who specialize in person-centered care and promoting health conditions should lead effective outreach projects as family medicine can address various health issues that occur within communities [17]. Regarding health promotion through HSBs, family physicians can collaborate with people and other healthcare professionals in communities to improve their perceptions of HSBs and concrete behaviors regarding their health [17]. Globally, there are various healthcare resources and professionals to help improve HSBs [12]. Clarifying evidence-based outreach interventions involving family physicians for improving HSBs can enhance family physician-driven outreach regarding HSBs. Therefore, this systematic review aimed to investigate current evidence regarding outreach involving family medicine aimed at HSBs in older people in rural areas.

2. Materials and Methods

This systematic review was conducted according to the PRISMA guidelines [18]. This study was registered on the PROSPERO platform with registration number 371095. In addition, we searched for interventions for HSBs related to family medicine in PubMed, Web of Science, and Embase between April 2000 and October 2022. The words used in the search were ["rural" or "remote" or "underserved"] AND ["older" or "elderly"] AND ["family physician" or "general practitioner" or "primary care"] AND "outreach" AND "community."

2.1. Study Selection

The inclusion and exclusion criteria are presented in Table 1. Original interventional articles were included in the international context, whereas conference presentations, reviews, and duplicate articles in the search results were excluded.

Table 1. Inclusion and exclusion criteria.

Criteria	Inclusion	Exclusion
Population	People > 60 to 65 years old	Other people
Setting	Rural or underserved community	Other settings
Types of study	Interventional study	Non-empirical studies (editorial, news, review, conference papers)
Interventions	Outreach including family physicians	Without outreach
Outcome	Health-related	Not health-related
Other	Abstract available Full text available in English	Abstract unavailable Full text unavailable in English

2.2. Data Extraction

The literature search, data extraction, and review were conducted by three investigators (RO, TK, and CS), and any discrepancies were resolved through discussion. The databases were searched for original studies on the health promotion through HSBs. Studies without clear descriptions of the aims, participants, or outcomes were excluded (Table 1).

Concretely, one of the investigators (R.O.) extracted the data from each original study using a purpose-designed data-extraction form. Two other investigators (T.K. and C.S.) examined the extracted data, which were categorized as follows: country, publication year, participants, purpose, research methodology, health issues, types of intervention, involved professionals, and outcomes concerning outreach.

2.3. Statistical Analysis

This study excluded statistical analysis because of the small number of included articles. However, the data from each study are presented descriptively. The quality of each study was assessed based on the best evidence medical education scale (1 to 5): grade 1 indicated that no definite conclusions could be drawn, that is, the data were not significant; grade 2 showed that the results were ambiguous, although there appeared to be a trend; grade 3 indicated that conclusions could be drawn based on the results; grade 4 indicated that the results were clear and probably very true; grade 5 indicated that the results were unequivocal [19].

3. Results

Overall, 376 studies were identified. Of these, 25 duplicate studies were excluded. After reviewing the abstracts, 333 studies were excluded for the following reasons: 69, different settings; 123, different participants; 103, no interventions; and 38, no clear health outcomes. Finally, a total of four studies were identified in the final analysis after excluding 14 articles through the assessment of eligibility (10, unoriginal articles; 4, no outreach to communities) (Figure 1). The details of the four articles are presented in Table 2. Each article was summarized in the categories of study design, participants, countries, health issues, interventions, involved professionals, and outcomes.

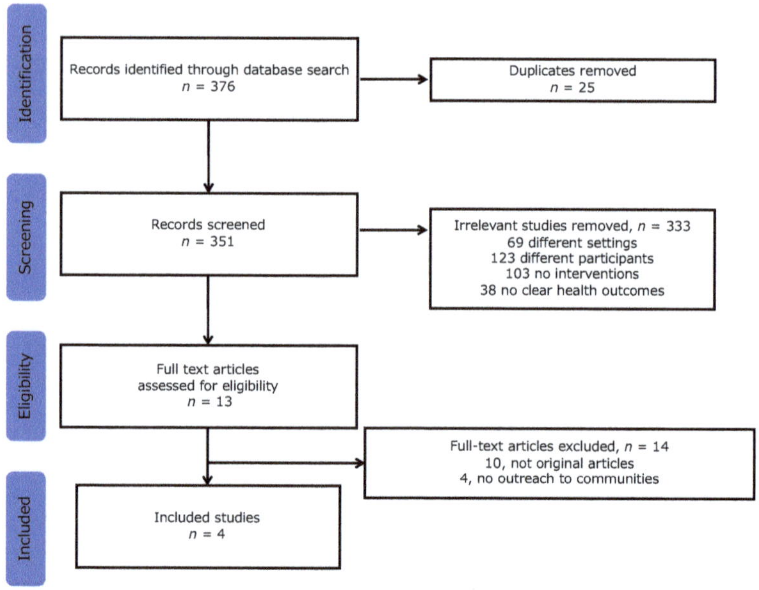

Figure 1. Study selection flow.

Table 2. Studies included in the review.

Year	Country	Purpose	Study Design	Participants	Health Issues	Interventions	Involved Professionals	Results
2003 [20]	Canada	To assess the effectiveness of a multidisciplinary diabetes outreach service	Pre–post study	Patients living in rural communities	Diabetes control	6-month interventions, home visiting educational message to patients	Family physicians Specialists Nurse educators Dieticians Pharmacists	The intervention was associated with a trend toward 10% improvement in blood pressure.
2010 [21]	Canada	To evaluate the impact of a provider initiated primary care outreach intervention to functional decline.	Randomized controlled trial	Older people in communities	Functional decline	12-month intervention comprehensive initial assessment collaborative care planning health promotion referral to community health and social support services.	Home care nurses Family physicians Patients Family	Changes in functional status and self-rated health did not significantly change.
2018 [22]	USA	To determine the effectiveness of cognitive behavioral therapy to mental conditions	Randomized controlled trial	People in underserved communities	Worry and GAD-related symptom	9-month intervention cognitive behavioral therapy with resource counseling, facilitation of communication with primary care providers about worry/anxiety, integration of religion/spirituality, person-centered skill content and delivery, and nontraditional community providers	General practitioners Healthcare providers Social worker Case manager	Moderate improvements on worry, GAD-related symptoms, anxiety, depression, sleep, trauma-related symptoms, and mental health QOL.
2020 [23]	USA	To show the effectiveness of person-centered wellness home	Randomized controlled trial	People in underserved communities	Physical functions	6-month intervention self-management resource center small group programs plus wellness coaching, as a booster intervention in older adults with chronic diseases.	Family physicians Therapists Community health worker	There was an improvement in self-reported physical functioning, not physical activity.

GAD, general anxiety disorder; QOL, quality of life.

3.1. Summary of the Study Results

In terms of the study designs, all the studies were comparative interventional studies, and two were randomized controlled trials. The participants were over 60–65 years of age. Two studies were from the United States [22,23] and two were from Canada [20,21]. In regard to the range of health issues, one study dealt with functional decline in usual life [21], one with worry and generalized anxiety disorder [22], one with diabetes control [20], and one with physical dysfunction [23]. The setting of two studies included rural communities [20,21], while the other two studies included underserved areas [22,23]. All of the study interventions included multiple professionals: All studies, with family physicians or general practitioners; three, with community workers; one each, with family, patients, social workers, therapists, specialists, dieticians, and pharmacists; and two, with nurses. Considering the outcome measurements, one study measured the QOL based on a questionnaire [22], one measured the worry and general anxiety disorder (GAD) severity with multiple questionnaires [21], one measured the satisfaction of care in diabetes [20], and one measured the perceived and objective physical functions using multiple questionnaires [23]. All of the studies demonstrated unequivocal data considering community outreach; the study grade was rated five.

3.2. Suggested limitations of the interventions

First, the involvement of family physicians in the interdisciplinary teams was limited, considering the implication on care decision-making. A previous study suggested that mutual collaboration among specialists, family physicians, and patients in chronic care can improve the objective health outcomes [20]. Second, the established primary care sufficiently supported older people, and the outreach projects may not improve the health outcomes; thus, outreach projects should be conducted in rural and underserved areas [21]. Third, the involvement of various professionals was limited. A previous study suggested that the involvement of general physicians and other healthcare professionals could improve the health outcomes of the participants [22]. Fourth, the duration of the intervention was short. Continual involvement of family physicians and multiple healthcare professionals in the healthcare of older individuals is needed for objective improvement [23].

4. Discussion

This study shows that family physician-involved outreach to rural and underserved areas can improve various health outcomes, including anxiety, subjective physical function, and diabetic care. The issues of outreach interventions are the duration and continuity of outreach, the active participation of family physicians and patients in outreach programs, and the focus of outreach participants.

For effective outreach, the focus should be specifically targeted to assessing rural community conditions. One article included in this systematic review showed no effectiveness of QOL changes among older people, based on the assessment of home care nurses [21]. As QOL is a patient-reported outcome, the patient's perception is essential for improvement [24]. If the participants were satisfied with the present conditions, the additional interventions, without considering the true needs of the participants, may not change their perceptions of their QOL, eventually leading to no improvement in research outcomes [9,25]. However, other studies of outreach focusing on rural community needs regarding the gap between primary care clinics and local people improved the participants' perceived physical functions and worries [20,22,23]. Therefore, considering the principle of family medicine, family physicians should focus on the needs of each community to improve their health conditions [26,27]. The need assessment for interventions is essential in public health for establishing effective outreach [28], and thus should include actual need assessments in communities.

The duration of outreach interventions is crucial to improve the subjective and objective outcomes in rural communities. This review reveals that some outreach involving various healthcare professionals and patients improves the subjective health conditions, including worry, subjective physical functions, and diabetic care quality [20,22,23]. However, these studies cannot change the objective health outcomes, including physical function and other chronic disease outcomes. These results could be attributed to the short intervention duration of outreach influencing changes in these outcomes. These studies may not be able to take advantage of the continuity of the care, which is crucial for improving patient health in family medicine [29,30]. Furthermore, the continuity of care improves various health outcomes in primary care [31,32]. All the included studies had less than a year of intervention. However, this duration was longer than that in other scientific studies regarding medicine. Therefore, community outreach programs should evaluate social needs and the acceptance of interventions in rural communities for effective implementation [28]. Moreover, effective acceptance of rural older people and the continuous implementation of outreach in rural communities can facilitate efficient implementation, changing the objective outcomes in healthcare [28]. Therefore, family physicians should respect the continuity of care in rural outreach programs and continue outreach to communities in collaboration with various healthcare professionals and stakeholders in order to change the objective outcomes.

The active participation of family physicians and patients in outreach programs should be promoted. Person-centered care and continuity of care are the competencies of family physicians ana are essential for effective outreach in rural contexts [26,33]. In this study, family physicians were members of outreach interdisciplinary teams [20–23]. However, the primary interventions were conducted by individuals in other specialties and healthcare professionals, including geriatricians, home care nurses, care workers, and counselors [20–23]. Each professional could approach a specific set of older individuals and patients; however, the systemic and holistic approaches in collaboration with patients, community members, and family physicians are lacking in the studies of this review. In addition, the empowerment of rural people and patients is crucial for health promotion in rural communities [34]. Without this empowerment, outreach interventions may not make sufficient changes to rural community health [35]. The reviewed studies assessed subjective and objective health rather than the perception and motivation for behavioral changes and perceptions regarding outreach. For effective outreach intervention, family physicians can create rural outreach programs in collaboration with various professionals and lay people, respecting the principles of person-centered care and the continuity of care in rural contexts [36].

This study had some limitations. First, few original studies investigated outreach programs involving family physicians and general practitioners. Many family physicians may approach this field in rural places; however, their effectiveness in rural people's health conditions may not be clarified. Therefore, future studies should use longitudinal research designs to assess outreach programs in rural contexts regarding improving health programs, which could motivate more family physicians to conduct outreach programs in their communities. Second, this systematic review excluded articles other than interventional studies to clarify the current evidence regarding rural outreach to communities. This inclusion criterion may exclude some grey studies from rural outreach by family physicians, including the term "regional" as a search term. Third, because of accessibility limitations, the review may have missed studies published in languages other than English. However, to overcome this limitation, we used search engines worldwide. As the world population is gradually aging, all countries are experiencing the issues of aging societies, in turn necessitating rural outreach from family physicians or general practitioners. Therefore, future reviews can include rural outreach research in other global contexts for other focuses, including the difficulty of implementation.

5. Conclusions

Despite the small number of included studies, this systematic review shows that family physician-involved outreach to rural and underserved areas can improve various health outcomes, including anxiety, subjective physical function, and diabetic care. The issues with outreach interventions are the duration and continuity of outreach, the active participation of family physicians and patients, and the focus of outreach participants. Therefore, future studies should use longitudinal research designs to assess outreach programs in rural contexts regarding improving health outcomes, which could motivate more family physicians to conduct outreach programs in their communities.

Author Contributions: Conceptualization, R.O.; Methodology, R.O.; Software, R.O.; Validation, R.O., T.K., and C.S.; Formal Analysis, R.O., T.K., and C.S.; Investigation, R.O.; Resources, R.O.; Data Curation, R.O., T.K., and C.S.; Writing— Original Draft Preparation, R.O., T.K., and C.S.; Writing— Review and Editing, R.O., T.K., and C.S.; Visualization, R.O., T.K., and C.S.; Supervision, C.S.; Project Administration, R.O. All authors have read and agreed to the published version of the manuscript.

Funding: This study received no external funding.

Institutional Review Board Statement: Not applicable.

Informed Consent Statement: Not applicable.

Data Availability Statement: All relevant datasets used in this study are presented in the manuscript.

Conflicts of Interest: The authors declare no conflict of interest.

References

1. Ohta, R.; Sato, M.; Ryu, Y.; Kitayuguchi, J.; Maeno, T.; Sano, C. What resources do elderly people choose for managing their symptoms? Clarification of rural older people's choices of help-seeking behaviors in Japan. *BMC Health Serv. Res.* **2021**, *21*, 640. [CrossRef] [PubMed]
2. Elliott, A.M.; McAteer, A.; Hannaford, P.C. Revisiting the symptom iceberg in today's primary care: Results from a UK population survey. *BMC Fam. Pract.* **2011**, *12*, 16. [CrossRef] [PubMed]
3. Ong, K.I.C.; Khattignavong, P.; Keomalaphet, S.; Iwagami, M.; Brey, P.; Kano, S.; Jimba, M. Health-seeking behaviours in a malaria endemic district in Lao People's Democratic Republic: A mixed methods study. *BMJ Open* **2021**, *11*, e055350. [CrossRef] [PubMed]
4. Kerrigan, D.; West, N.; Tudor, C.; Hanrahan, C.F.; Lebina, L.; Msandiwa, R.; Mmolawa, L.; Martinson, N.; Dowdy, D. Improving active case finding for tuberculosis in South Africa: Informing innovative implementation approaches in the context of the Kharitode trial through formative research. *Health Res. Policy Syst.* **2017**, *15*, 42. [CrossRef] [PubMed]
5. Ohta, R.; Sato, M.; Kitayuguchi, J.; Maeno, T.; Sano, C. The association between the self-management of mild symptoms and quality of life of elderly populations in rural communities: A cross-sectional study. *Int. J. Environ. Res. Public Health* **2021**, *18*, 8857. [CrossRef]
6. Ohta, R.; Ryu, Y.; Sano, C. Association between self-medication for mild symptoms and quality of life among older adults in Rural Japan: A cross-sectional study. *Medicina* **2022**, *58*, 701. [CrossRef]
7. Ohta, R.; Ryu, Y.; Sano, C. Improvement in quality of life through self-management of mild symptoms during the COVID-19 pandemic: A prospective cohort study. *Int. J. Environ. Res. Public Health* **2022**, *19*, 6652. [CrossRef]
8. Jiao, C.; Leng, A.; Nicholas, S.; Maitland, E.; Wang, J.; Zhao, Q.; Xu, L.; Gong, C. Multimorbidity and mental health: The role of gender among disease-causing poverty, rural, aged households in China. *Int. J. Environ. Res. Public Health* **2020**, *17*, 8855. [CrossRef]
9. Zhao, Q.; Wang, J.; Nicholas, S.; Maitland, E.; Sun, J.; Jiao, C.; Xu, L.; Leng, A. Health-related quality of life and health service use among multimorbid middle-aged and older-aged adults in China: A cross-sectional study in Shandong Province. *Int. J. Environ. Res. Public Health* **2020**, *17*, 9261. [CrossRef]
10. Hopman, P.; de Bruin, S.R.; Forjaz, M.J.; Rodriguez-Blazquez, C.; Tonnara, G.; Lemmens, L.C.; Onder, G.; Baan, C.A.; Rijken, M. Effectiveness of comprehensive care programs for patients with multiple chronic conditions or frailty: A systematic literature review. *Health Policy* **2016**, *120*, 818–832. [CrossRef]
11. Ohta, R.; Ryu, Y.; Sano, C. Older people's help-seeking behaviors in rural contexts: A systematic review. *Int. J. Environ. Res. Public Health* **2022**, *19*, 3233. [CrossRef] [PubMed]
12. Nguyen, H.; Nguyen, T.; Tran, D.; Hinton, L. "It's extremely hard but it's not a burden": A qualitative study of family caregiving for people living with dementia in Vietnam. *PLoS ONE* **2021**, *16*, e0259788. [CrossRef] [PubMed]
13. Yamaguchi, M.; Yoshida, T.; Yamada, Y.; Watanabe, Y.; Nanri, H.; Yokoyama, K.; Date, H.; Miyake, M.; Itoi, A.; Yamagata, E.; et al. Sociodemographic and physical predictors of non-participation in community based physical checkup among older neighbors: A case-control study from the Kyoto-Kameoka longitudinal study, Japan. *BMC Public Health* **2018**, *18*, 568. [CrossRef] [PubMed]
14. Ogbolu, Y.; Scrandis, D.A.; Fitzpatrick, G. Barriers and facilitators of care for diverse patients: Nurse leader perspectives and nurse manager implications. *J. Nurs. Manag.* **2018**, *26*, 3–10. [CrossRef] [PubMed]
15. Khan, S.; Ali, S.A. Exploratory study into awareness of heart disease and health care seeking behavior among Emirati women (UAE) — Cross sectional descriptive study. *BMC Womens Health* **2017**, *17*, 88. [CrossRef] [PubMed]

16. Laake, J.P.; Vulkan, D.; Quaife, S.L.; Hamilton, W.T.; Martins, T.; Waller, J.; Parmar, D.; Sasieni, P.; Duffy, S.W. Targeted encouragement of GP consultations for possible cancer symptoms: A randomised controlled trial. *Br. J. Gen. Pract.* **2021**, *71*, e339–e346. [CrossRef] [PubMed]
17. Stollenwerk, D.; Kennedy, L.B.; Hughes, L.S.; O'Connor, M. A systematic approach to understanding and implementing patient-centered care. *Fam. Med.* **2019**, *51*, 173–178. [CrossRef] [PubMed]
18. Moher, D.; Shamseer, L.; Clarke, M.; Ghersi, D.; Liberati, A.; Petticrew, M.; Shekelle, P.; Stewart, L.A.; PRISMA-P Group. Preferred reporting items for systematic review and meta-analysis protocols (PRISMA-P) 2015 statement. *Syst. Rev.* **2015**, *4*, 1. [CrossRef] [PubMed]
19. Hammick, M.; Dornan, T.; Steinert, Y. Conducting a best evidence systematic review. Part 1: From idea to data coding. BEME Guide No. 13. *BEEE Guide No 13 Med. Teach.* **2010**, *32*, 3–15. [CrossRef]
20. Majumdar, S.R.; Guirguis, L.M.; Toth, E.L.; Lewanczuk, R.Z.; Lee, T.K.; Johnson, J.A. Controlled trial of a multifaceted intervention for improving quality of care for rural patients with type 2 diabetes. *Diabetes Care* **2003**, *26*, 3061–3066. [CrossRef] [PubMed]
21. Ploeg, J.; Brazil, K.; Hutchison, B.; Kaczorowski, J.; Dalby, D.M.; Goldsmith, C.H.; Furlong, W. Effect of preventive primary care outreach on health related quality of life among older adults at risk of functional decline: Randomised controlled trial. *BMJ* **2010**, *340*, c1480. [CrossRef] [PubMed]
22. Stanley, M.A.; Wilson, N.L.; Shrestha, S.; Amspoker, A.B.; Wagener, P.; Bavineau, J.; Turner, M.; Fletcher, T.L.; Freshour, J.; Kraus-Schuman, C.; et al. Community-based outreach and treatment for underserved older adults with clinically significant worry: A randomized controlled trial. *Am. J. Geriatr. Psychiatry* **2018**, *26*, 1147–1162. [CrossRef] [PubMed]
23. Mielenz, T.J.; Tracy, M.; Jia, H.; Durbin, L.L.; Allegrante, J.P.; Arniella, G.; Sorensen, J.A. Creation of the person-centered wellness home in older adults. *Innov. Aging* **2020**, *4*, igz055. [CrossRef] [PubMed]
24. Shiroiwa, T.; Fukuda, T.; Ikeda, S.; Igarashi, A.; Noto, S.; Saito, S.; Shimozuma, K. Japanese population norms for preference-based measures: EQ-5D-3L, EQ-5D-5L, and SF-6D. *Qual. Life Res.* **2016**, *25*, 707–719. [CrossRef] [PubMed]
25. Ohta, R.; Sano, C. Associations between perception of help-seeking behaviors and quality of life among older people in rural communities: A cross-sectional study. *Int. J. Environ. Res. Public Health* **2022**, *19*, 13331. [CrossRef] [PubMed]
26. Göktaş, O. The Goktas definition of family medicine/general practice. *Aten. Primaria.* **2022**, *54*, 102468. [CrossRef] [PubMed]
27. Freeman, T.R. Principles of family medicine. In *TR, in McWhinney's Textbook of Family Medicine*; Oxford Academic: New York, NY, USA, 2016.
28. Mazloomymahmoodabad, S.; Masoudy, G.; Fallahzadeh, H.; Jalili, Z. Education based on precede-proceed on quality of life in elderly. *Glob. J. Health Sci.* **2014**, *6*, 178–184. [CrossRef] [PubMed]
29. De Maeseneer, J.M.; De Prins, L.; Gosset, C.; Heyerick, J. Provider continuity in family medicine: Does it make a difference for total health care costs? *Ann. Fam. Med.* **2003**, *1*, 144–148. [CrossRef] [PubMed]
30. Tourigny, A.; Aubin, M.; Haggerty, J.; Bonin, L.; Morin, D.; Reinharz, D.; Leduc, Y.; St-Pierre, M.; Houle, N.; Giguère, A.; et al. Patients' perceptions of the quality of care after primary care reform: Family medicine groups in Quebec. *Can. Fam. Physician* **2010**, *56*, e273–e282. [PubMed]
31. Bazemore, A.; Petterson, S.; Peterson, L.E.; Bruno, R.; Chung, Y.; Phillips, R.L. Higher primary care physician continuity is associated with lower costs and hospitalizations. *Ann. Fam. Med.* **2018**, *16*, 492–497. [CrossRef] [PubMed]
32. Tammes, P.; Purdy, S.; Salisbury, C.; MacKichan, F.; Lasserson, D.; Morris, R.W. Continuity of primary care and emergency hospital admissions among older patients in England. *Ann. Fam. Med.* **2017**, *15*, 515–522. [CrossRef]
33. Ventres, W.B. Looking within: Intentions of practice for person-centered care. *Ann. Fam. Med.* **2017**, *15*, 171–174. [CrossRef]
34. Ocloo, J.; Matthews, R. From tokenism to empowerment: Progressing patient and public involvement in healthcare improvement. *BMJ Qual. Saf.* **2016**, *25*, 626–632. [CrossRef] [PubMed]
35. Barcham, R.; Silas, E.; Irie, J. Health promotion and empowerment in Henganofi District, Papua New Guinea. *Rural Remote Health* **2016**, *16*, 3553. [CrossRef]
36. Kaneko, M.; Motomura, K.; Mori, H.; Ohta, R.; Matsuzawa, H.; Shimabukuro, A.; Matsushima, M. Gatekeeping function of primary care physicians under Japan's free-access system: A prospective open cohort study involving 14 isolated islands. *Fam. Pract.* **2019**, *36*, 452–459. [CrossRef]

International Journal of
Environmental Research and Public Health

MDPI

Review

Pharmacists' Role in Older Adults' Medication Regimen Complexity: A Systematic Review

Catharine Falch and Gilberto Alves *

CICS-UBI-Health Sciences Research Centre, University of Beira Interior, Av. Infante D. Henrique, 6200-506 Covilhã, Portugal; cathyfalch@gmail.com
* Correspondence: gilberto@fcsaude.ubi.pt; Tel.: +351-275329002

Abstract: Medication regimen complexity (MRC) may influence health outcomes, such as hospitalisation, hospital readmission and medication adherence. Pharmacists have been referred to as health professionals with the opportunity to act on MRC reduction. This study aimed to investigate pharmacists' role in studies about older adults' medication regimen complexity. A literature search was performed in PubMed, Web of Science and the Cochrane Library—CENTRAL—up to October 2019. Out of 653 potentially relevant studies, 17 articles met the inclusion criteria for this review. Most studies used the 65-item medication regimen complexity index (MRCI) to assess medication complexity. Pharmacists' role was mainly confined to data collection. It seems that pharmacists' active role in older adults' medication complexity has not been studied in depth so far. However, the few existing interventional ones suggest that, after previous training, regimen simplification is feasible.

Keywords: older adults; medication regimen complexity; pharmacist; systematic review

Citation: Falch, C.; Alves, G. Pharmacists' Role in Older Adults' Medication Regimen Complexity: A Systematic Review. *Int. J. Environ. Res. Public Health* **2021**, *18*, 8824. https://doi.org/10.3390/ijerph18168824

Academic Editors: Francisco José Tarazona Santabalbina, Sebastià Josep Santaeugènia González, José Augusto García Navarro and José Viña

Received: 5 July 2021
Accepted: 16 August 2021
Published: 21 August 2021

1. Introduction

Nowadays the world faces global ageing, often associated with a high prevalence of multimorbidities. As a natural consequence, older age frequently stands out for polypharmacy and complex medication regimens [1–3]. When considering medication regimen complexity (MRC), there is so far no clear definition for it [4,5]. However, it has already been shown that the number of drugs is only one of the relevant factors to consider, and that, in addition, dosage form, dosage frequency and administration instructions also have to be considered [6–14]. Furthermore, there is also no agreement about the reference instrument for MRC determination [4,5]. Several tools have been used, with the 65-item medication regimen complexity index (MRCI), developed by George et al. [15], being the most common, reliable and validated tool for this purpose, which has already been translated and validated to a few languages [16–18] and even standardised for older adults in primary care [19]. It is an open-ended index, with higher total MRCI scores representing more complex medication regimens.

Interest in this subject has grown because numerous studies have associated high MRC with non-adherence [4,20,21], higher hospitalisation rates [20], hospital discharge destination different than home [22] and low overall quality of life [23]. Indeed, medication management may frequently be demanding for the older population, their caregivers and healthcare professionals. In fact, older adults often present reduced manual dexterity in addition to cognitive and sensory impairment that can lead to a higher risk of medication errors and drug-related problems (DRPs) [2,3,6–8]. In order to reduce these negative aspects, it seems imperative to attempt medication regimen simplifications in many circumstances. Some research has already investigated its feasibility, with evidence suggesting that complexity can be reduced, and referring to pharmacists as healthcare professionals with a great potential to perform it in routine pharmaceutical dispensing or as part of medication reviews [2,9,11,13,14]. Indeed, pharmacists have a privileged access to the population's medication, both in community pharmacies and hospital settings, and awareness of this topic is needed, especially in the older population, for whom managing their daily medication may often represent a considerable challenge.

Up to the present date, to the best of our knowledge, there are no systematic reviews available about the role that pharmacists play in the older population's MRC and the effort made to simplify it. To address this gap, this study aimed to examine and describe pharmacists' role in studies on older adults' MRC.

2. Materials and Methods

2.1. Search Strategy

A systematic literature search was conducted in three databases (PubMed, Web of Science and the Cochrane Library—CENTRAL) from their inceptions to October 2019. The search strategy considered the PICOS elements, representing P—Population, older adults; I—Intervention, pharmacists' role in MRC; O—Outcomes of interest, any outcome related to MRC; and S—Study design, original peer-reviewed observational or experimental studies. The comprehensive search expression included the combination of keywords related to pharmacists (e.g., pharmacists, pharmaceutical services/care/intervention), older people (e.g., aged, elderly, old age, geriatric, retired, ancient) and medication regimen complexity (e.g., treatment/medicine/drug complexity) [24]. The detailed PubMed search strategy is provided in Table S1. To ensure literature saturation, references lists from included articles were screened for potential further relevant studies. The PRISMA guideline was used to perform and report items in the present review [25,26].

2.2. Study Selection

To be included in this review the article had to meet the following criteria:

- all study participants had to be aged 60 or over, since, according to the World Health Organisation (WHO), the definition of older person "varies among countries but is often associated with the age of normal retirement" (60 or 65 years) [27];
- pharmacists' role in MRC had to be mentioned in the article (for this purpose all pharmacists' actions were regarded, beginning with simple data collection and ending with pharmacist intervention). Sabater's et al. [28] definition of pharmacist intervention was considered: "pharmacist intervention is defined as the pharmacist's activity consisting in a suggested action on the patient treatment and/or an action on the patient oriented towards finding a solution for or preventing a negative clinical outcome of the pharmacotherapy";
- medication complexity had to be assessed, and for this purpose all tools were considered;
- be an original peer-reviewed observational study (i.e., cohort study, cross-sectional study, case study) or an experimental study (i.e., randomised controlled trial, quasi-experimental study);
- be written in English, Portuguese or German.

Articles were excluded if they:

- were not performed exclusively on the older population;
- did not mention any role of pharmacists in MRC, or MRC assessment was not performed;
- were qualitative studies, reviews, protocols, congress abstracts, editorials, letters, dissertations, theses, feasibility or pilot studies;
- did not fulfil the language restrictions.

Additionally, no timing or setting restrictions were applied, and only published studies were included.

2.3. Data Extraction

Literature search results were uploaded to the Covidence platform, removing duplicates using the "duplicate" function. The remaining duplicates were removed manually.

Two independent reviewers (CF and GA) analysed the studies by screening titles and abstracts to verify potential inclusion criteria correspondence. If an article potentially met the inclusion criteria or provided insufficient information in the abstract to be excluded, the full text was obtained and screened by the same investigators. Any disagreement between reviewers was solved through discussion. Data extraction was performed using a previously created data extraction form (Microsoft Word format) by a single reviewer (CF) and was independently checked afterwards by the second reviewer (GA). Reviewers were not blinded to the authors or journals when screening articles and extracting data.

The following information was collected: authors, year of publication, country, study design, participants' demographics (no. of participants, age, gender), setting, study aim, medication data (source, type of medication included, instrument to assess MRC), pharmacists' role and main outcomes.

2.4. Quality Assessment

To evaluate the risk of bias and study design, two reviewers (CF and GA) independently used the Effective Public Health Practice Project (EPHPP) quality assessment tool [29]. Any disagreement between reviewers was to be solved by discussion.

The EPHPP quality assessment tool is a generic tool that assesses six domains (a. selection bias, b. study design, c. confounders, d. blinding, e. data collection methods and f. withdrawals and dropouts); it was chosen because of its applicability in a variety of study designs [30].

3. Results

A total of 653 potentially relevant studies were identified from the databases. After title and abstract screening in addition to full-text assessment against inclusion criteria, 16 studies remained to be included in this systematic review. One study was identified from the searching reference list and added to the selection, resulting in 17 included studies (Figure 1) [10,12,22,31–44].

3.1. Study Characteristics

Characteristics of the included studies in this systematic review are shown in Tables 1 and 2. The included studies were performed in Australia (n = 7) [10,12,22,33,34,42,44], Brazil (n = 2) [31,41], Israel (n = 1) [39], Spain (n = 1) [43] and the United States (n = 6) [32,35–38,40], and were reported between 1990 and 2018. Most articles were written in English, except one that was written in Portuguese [31]. In regard to study design, most studies were cross-sectional (n = 7) [31,32,36–38,41,43] and cohort studies (n = 8) [10,12,22,33,34,39,42,44], but a quasi-experimental (n = 1) [40] and a prospective controlled trial (n = 1) [35] were also included. The majority of the studies were conducted in health care settings: hospital wards/units/clinics (n = 10) [10,12,22,33–35,38,39,43,44], primary health care units (n = 1) [41], continuing care retirement communities (n = 1) [36] and a residential aged care facility (n = 1) [42], but data were also obtained by home visits (n = 2) [31,37], in churches (n = 1) [32] or through a telephone consultation (n = 1) [40]. The total number of participants included in the 17 articles was 3652, with sample sizes ranging from 79 to 400 individuals. Participants' mean age ranged from 71.2 to 86.8 years, with females representing between 39.5 and 79.8% of participants.

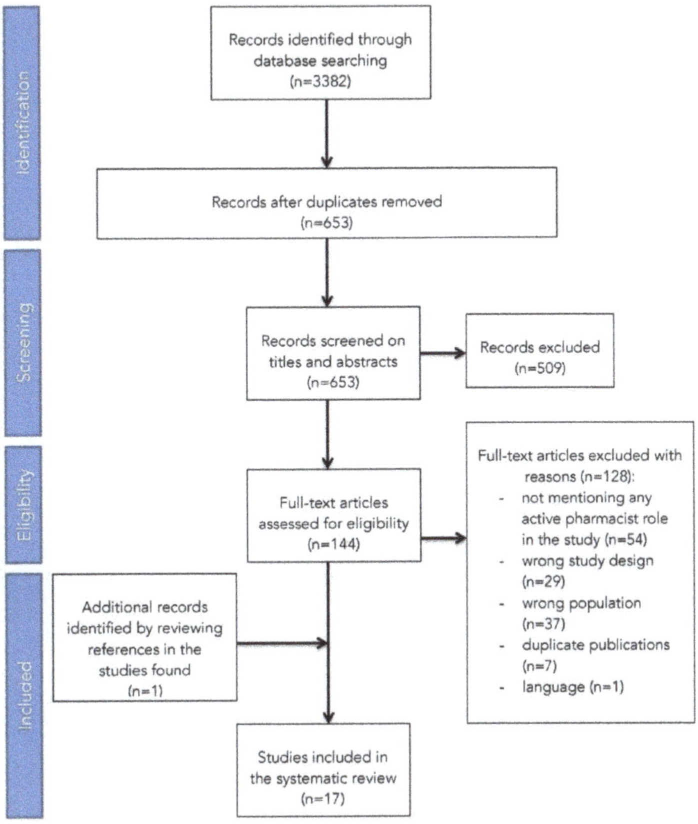

Figure 1. Study selection process using the PRISMA flowchart.

Table 1. Description of the included studies: country, study design, participants' demographics, setting and study aim.

Author, Year	Country	Study Design	Participants' Demographics			Setting	Study Aim
			No. of Participants (n)	Mean Age, Years (SD) [Range]	Female		
Acurcio et al. [31], 2009	Brazil	Epidemiological cross-sectional	377	72.4 (61–102)	69.2%	Home visits.	To examine factors associated with therapeutic regimen complexity of drug prescriptions for elderly people.
Bazargan et al. [32], 2017	USA	Cross-sectional study	400	73.5 (7) (65–94)	65%	Sixteen predominantly African-American churches in SPA6 of Los Angeles County.	To examine the association between adherence to drug regimens and a spectrum of medication-related factors, including polypharmacy, medication regimen complexity, use of PIMs, knowledge about their therapeutic purpose and instructions of proper medication use.
Chang et al. [33], 2017	Australia	Retrospective cohort study	100	82 (9.15)	60%	General medical units of a tertiary care hospital.	To assess the changes in the MRCI before and after hospitalisation. To examine the prevalence of prescribing PIMs at the time of hospital discharge, using the 2015 Beers Criteria.
Elliot [34], 2012	Australia	Cohort study	205	81.3 (8.0)	58%	Two acute general medicine wards and two subacute aged care wards at a major metropolitan public hospital.	To explore the feasibility of incorporating medication regimen simplifications into routine clinical pharmacists' care for older hospital inpatients, and to identify barriers to regimen simplification.
Elliot et al. [10], 2011	Australia	Cohort study	186 Acute wards: 115 Subacute wards: 71	Acute wards: 79 (77–80) Subacute wards: 81 (80–83)	Acute wards: 59% Subacute wards: 55%	Two acute general medicine wards and two subacute aged care wards at a large public hospital.	To investigate the impact of hospitalisation on the complexity of older patients' medication regimens, and to determine whether discharge medication regimens could be simplified.
Elliot et al. [12], 2013	Australia	Cohort study	391 Pre-intervention: 186 Intervention: 205	Pre-intervention: 79.7 (8.2) Intervention: 81.3 (8.0)	57.8% Pre-intervention: 57.5% Intervention: 58.%	Two acute general medicine wards and two subacute aged care wards at a major metropolitan public hospital.	To investigate the impact of pharmacists' medication reviews, together with an educational intervention targeting inpatient clinical pharmacists and junior medical officers on the increase in medication regimen complexity during hospitalisation.
Kroenke et al. [35], 1990	USA	Prospective controlled trial	79 Intervention: 38 Control: 41	Intervention: 72.3 Control: 71.4	40.5% Intervention: 39.5% Control: 41.5%	Internal Medicine Clinic at Brooke Army Medical Center.	To determine the effectiveness of specific feedback to prescribing physicians in reducing polypharmacy in elderly outpatients.
Lakey et al. [36], 2009	USA	Cross-sectional	109	85.9 (5.1) (73–98)	79.8%	Continuing care retirement community in Seattle.	To assess older adults' current use of, knowledge of and preferences for medication management tools and supports.
Lindquist et al. [37], 2014	USA	Cross-sectional	200	79.6 (6.4) (70–100)	58%	Home visits after discharge from Northwestern Memorial Hospital.	To determine whether seniors consolidate their home medications or if there is evidence of unnecessary regimen complexity.

Table 1. *Cont.*

Author, Year	Country	Study Design	Participants' Demographics			Setting	Study Aim
			No. of Participants (n)	Mean Age, Years (SD) [Range]	Female		
Linnebur et al. [38], 2014	USA	Retrospective cross-sectional	200 CA: 100 CO: 100	CA: 74.3 (7.4) CO: 79.7 (6.1)	78.5% CA: 76% CO: 81%	Ambulatory clinics at the University of CA and the University of CO.	To evaluate the entire medication regimen of older adults with depression, and determine potential targets to simplify the regimen and improve adherence.
Mansur et al. [39], 2012	Israel	Cohort study	212	81.1 (7.3) (66–103)	61.8%	Acute Geriatric Ward at the Beilinson Hospital, Rabin Medical Center.	To test the convergent, discriminant and predictive validity of the MRCI in older hospitalised patients with varying functional and cognitive levels.
Moczygemba et al. [40], 2012	USA	Quasi-experimental	120 Intervention: 60 Control: 60	Intervention: 71.2 (7.5) Control: 73.9 (8.0)	60% Intervention: 48.3% Control: 71.7%	Telephone consultation.	To determine the impact of telephone MTM on MHRPs, medication adherence and total drug costs for Medicare Part D participants.
Pinto et al. [41], 2016	Brazil	Cross-sectional	227	71.4	70.9%	Two PHUs in the municipality of Belo Horizonte.	To evaluate the level of understanding of pharmacotherapy and the associated factors amongst older people in two PHUs.
Pouranayatihosseinab et al. [42], 2018	Australia	Retrospective observational study	285	85.5 (7.7)	68%	Residential ACFs.	To investigate the impact of RMMRs on simplifying medication regimen complexity in Australian ACF residents using the MRCI.
Sevilla-Sánchez et al. [43], 2017	Spain	Prospective cross-sectional study	235	86.80 (5.37)	65.50%	AGU in a second-level hospital.	To determine the prevalence of PIMs among patients with advanced chronic conditions and palliative care needs, and to analyse the associated risk factors and resulting clinical consequences.
Wimmer et al. [44], 2014	Australia	Prospective cohort	163 Readmitted: 99 Not readmitted: 64	Readmitted: 84.9 (6.2) Not readmitted: 85.6 (6.74)	72.4% Readmitted: 68.7% Not readmitted: 78.1%	GEM unit of a public hospital in Adelaide.	To investigate the association between discharge medication regimen complexity and unplanned re-hospitalisation over 12 months.
Wimmer et al. [22], 2014	Australia	Prospective cohort	163 DD home: 87 DD NCS: 76	85.2 (6.4) (71–101) DD home: 84.6 (6.9) DD NCS: 85.8 (5.8)	72.4% DD home: 68.7% DD NCS: 77.6%	GEM unit at the Queen Elizabeth Hospital.	To investigate the association between polypharmacy and medication regimen complexity with hospital discharge destination among older people.

ACFs, aged care facilities; AGU, acute care geriatric unit; CA, California San Diego; CO, Colorado Anschutz Medical Campus; DD, discharge destination; GEM, geriatric evaluation and management; MHRP, medication- and health-related problems; MRCI, medication regimen complexity index; MTM, medication therapy management; NCS, non-community setting; PHUs, primary health care units; RMMRs, residential medication management reviews; and SPA6, Service Planning Area 6.

Table 2. Description of the included studies: medication data, pharmacists' role and main outcomes.

Author, Year	Medication Data			Pharmacists' Role in MRC	Main Outcomes
	Source	Type of Medication Included	Instrument to Assess MRC		
Acurcio et al. [31], 2009	Medical prescription	Prescription medication	MCI	Household data collection, after previous training.	MCI ranged from 1 to 24 (mean = 6.1; median = 5.0). Nr. of drugs prescribed (> 2), less schooling, worse perception of health and lower benefit payment associated positively with greater complexity ($p < 0.05$). An association was observed between RC and failure to use some drugs in the preceding 15 days ($p = 0.34$).
Bazargan et al. [32], 2017	The brown bag method	Prescription medication	MRCI	Evaluation of any medication duplication; application of the Beers Criteria to document the number of PIM use; and comparison of subjects perceived purpose of each prescription drug with all therapeutic indications used in clinical practice.	The mean value of the MRCI was 15.1 (SD = 8.2; minimum = 2.5; and maximum = 55.5). Of the participants, 70% (278) engaged in PIM use and used at least one medication that was classified as "Avoid" (27%) and "Use Conditionally" (43%) through the Beers Criteria. Participants with increased knowledge about the therapeutic purpose of the dosage regimen were almost seven times more likely to adhere to their medication.
Chang et al. [33], 2017	Hospital electronic medical record system	Prescription medication	MRCI	Medication history record.	The mean MRCI increased from 28.70 at the time of admission to 32.46 at discharge ($p = 0.007$). Hospitalisation resulted in a statistically significant reduction in the prevalence of the use of PIMs.
Elliot [34], 2012	Hospital medical record	Long-term, short-term and "when required" medication	Not mentioned	Pharmacists were encouraged to minimise RC and discuss simplifications with hospital doctors and patients. Afterwards, they were asked to indicate if they reviewed the patient's RC (and why not) and whether regimen changes were considered or attempted. Changes had to be recorded and whether they were successfully implemented (and if not why).	Pharmacists reviewed medication RC for 173/205 (84.4%) patients and identified 149 potential changes to reduce RC for 79/173 (45.7%) reviewed patients. Ninety-four (63.1%) changes were successfully implemented in 54/205 (26.3%) patients.
Elliot et al. [10], 2011	Pre-admission: medication history on admission within the patient's hospital medical record Discharge: discharge prescription	Long-term, short-term and "when required" medication	MRCI	Review of discharge medication regimens and identification of any potential change that could make the regimen simpler.	MRCI scores increased by 22% (18 to 22; $p < 0.0001$) for regularly scheduled long-term medications and 32% (21 to 27; $p < 0.001$) for all medications. Ninety simplifications to regularly long-term medications were proposed and 84 (93%) were rated as feasible and likely to have the same or similar outcome.

Table 2. *Cont.*

Author, Year	Medication Data			Pharmacists' Role in MRC	Main Outcomes
	Source	Type of Medication Included	Instrument to Assess MRC		
Elliot et al. [12], 2013	Pre-admission: hospital medical record Discharge: discharge prescription	Long-term, short-term and "when required" medication	MRCI	Clinical pharmacists were encouraged to review RC and make recommendations to hospital medical officers to simplify medication regimens when clinically appropriate.	MRCI score for all medications: pre-intervention patients = 20.7 (SD = 12.5); intervention patients = 21.7 (SD = 11.6). MRCI score for all regularly scheduled long-term medication: pre-intervention patients = 18.2 (SD = 11.2); intervention patients = 19.1 (SD = 10.3). The mean increase in MRCI score between admission and discharge was significantly smaller in the intervention patients than in the usual care patients (2.5 vs. 4.0; $p = 0.02$, adjusted difference 1.6; 95% CI 0.3, 2.9).
Kroenke et al. [35], 1990	Patients were asked to bring all prescription bottles to the interview, where medications were used regularly and the dosage schedules were confirmed	Prescription medication	A complexity score was calculated by summing the different dosage intervals, weighted for frequency	A clinical pharmacist interviewed patients to determine the precise drug regimen. Investigating physicians discussed and agreed upon recommendations that might reduce polypharmacy.	All four indices of polypharmacy improved in the intervention group (7.2 vs. 6.6; $p = 0.007$). Physicians complied 100% with recommendations to simplify a dosage schedule, 62% to substitute a new drug for the old one and only 40% with recommendations to stop a medication ($p = 0.04$).
Lakey et al. [36], 2009	Participants were asked to show containers for all medication taken in the week before	Prescription medication, non-prescription medication and herbal products	The frequency of dosing for each medication was summed to calculate a complexity score	The investigating pharmacist performed data collection.	Medication complexity score: only for prescription drugs = 5.0 ± 3.8; total = 8.3 ± 4.4.
Lindquist et al. [37], 2014	Subjects were asked to show how they take their medication on a typical day and all medications were registered and compared to discharge instructions	Medication subjects take a day	The number of times medications were taken in a 24 h period for each subject was calculated	A pharmacist acted as one of two coders, determining the fewest number of times a day that a patient could take the regimen.	Home medication regimens could be simplified for 85 (42.5%) subjects.
Linnebur et al. [38], 2014	Electronic health record	Depression medications, other prescription medications and over-the-counter (OTC) medications	MRCI/patient-level MRCI (pMRCI)	A clinical pharmacist coded the pMRCI using an electronic data capture tool that calculated three subscores.	Individual pMRCI scores average: 17.62 (CA) and 19.36 (CO). Dosing frequency contributed to 57–58% of the MRCI score, with patients facing an average of 7–8 unique dosing frequencies in their regimen.

Table 2. *Cont.*

Author, Year	Medication Data			Pharmacists' Role in MRC	Main Outcomes
	Source	Type of Medication Included	Instrument to Assess MRC		
Mansur et al. [39], 2012	Patient interview on admission. Retrospectively, patients' medical files were reviewed to calculate their MRCI score for discharge medication regimens	Long-term discharge medication	MRCI	The pharmacist interviewed the patients to collect clinical and demographical data as well as patients' medication on admission.	Mean (SD) MRCI at discharge: 30.27 (13.95). The MRCI score was strongly correlated with the number of medications (r = 0.94; $p < 0.001$) and the number of daily doses (r = 0.87; $p < 0.001$), and increased as the number of medications taken ≥ 3 times a day increased (27.35; 34.45 and 43.00 for none, 1 and 2 drugs; $p < 0.001$).
Moczygemba et al. [40], 2012	Electronic medical records and prescription claims	Prescription medication	MRCI	Pharmacists reviewed the patient's electronic medical records to identify potential MHRPs. Furtherly, in the intervention group, an MTM telephone consultation was made and recommendations were given to the patient to resolve MHRPs.	MRCI intervention group: 21.5 ± 7.8. MRCI control group: 22.8 ± 6.9.
Pinto et al. [41], 2016	Prescribed medication that each individual had in his hand at the time of the interview	Prescription medication	MRCI	Resident pharmacists and academics studying pharmacy were part of the team carrying out data collection.	Mean value of the rate of complexity: 22.7 (DP = 10.9; CV = 48.0%), with a minimum of 4.0 and a maximum of 65.5.
Pouranayatihosseinabad et al. [42], 2018	RMMR report	Prescription medication	MRCI	Accredited pharmacists compiled RMMR reports, which included pharmacists' findings and recommendations.	The median MRCI at baseline was 25.5 (19.0–32.5). The main contributing factor to the MRCI score was dosing frequency. Pharmacists made 764 recommendations, of which GPs accepted 74.5%. There were no significant differences in the MRCI scores after pharmacists' recommendations ($p = 0.53$) or after GPs' acceptance of these recommendations ($p = 0.07$) compared to baseline.
Sevilla-Sánchez et al. [43], 2017	Not mentioned	Routine chronic medication	MRCI	A multidisciplinary team consisting of a pharmacist and two physicians determined the PIMs.	MRCI (mean; SD) = 38.00 (16.54). Of the population, 88.50% had at least one STOPP criterion and 97.40% had some criterion according to the MAI criteria. The identified risk factors for the existence of PIMs were: insomnia, anxiety–depressive disorders, falls, pain, excessive polypharmacy and therapeutic complexity.
Wimmer et al. [44], 2014	Hospital separation summary recorded in the OACIS	All prescription and non-prescription medications, nutritional supplements, health products, dermatological preparations and short-term medications were considered	MRCI	The same pharmacist researcher extracted all medication data.	Mean MRCI for: - discharged patients: 27.86 (SD = 11.63) - readmitted: 28.01 (SD = 12.48) - not readmitted: 27.62 (SD = 10.26). The MRCI was not significantly different in patients who were readmitted and not readmitted over 12 months (mean difference = −0.39; 95% CI = −4.09 to 3.30).

Table 2. *Cont.*

Author, Year	Medication Data			Pharmacists' Role in MRC	Main Outcomes
	Source	Type of Medication Included	Instrument to Assess MRC		
Wimmer et al. [22], 2014	Hospital separation summary recorded in OACIS	Prescription, non-prescription and CAMs	MRCI	The same pharmacist researcher extracted all medication data.	Patients discharged directly to home: mean MRCI = 26.1; SD 9.7. Patients discharged to NCSs: mean MRCI = 29.9; SD 9.7. High medication RC (MRCI > 35) inversely associated with discharge directly to home (RR 0.39; 95% CI 0.20–0.73). Polypharmacy (≥ 9 medications) not significantly associated with discharge directly to home (RR 0.97; 95% CI 0.53–1.58).

CAMs, complementary and alternative medications; GP, general practitioner; MCI, medication complexity index; MRCI, medication regimen complexity index; NCS, non-community setting; OACIS, Open Architecture Clinical Information System; PIMs, potentially inappropriate medications; RC, regimen complexity; and RMMR, residential medication management review.

3.2. Study Quality

Concerning quality assessment, using the EPHPP global rating decision tool, only two studies were rated as being of strong quality [33,34], eight of moderate [10,12,22,34,35,39,40,42] and seven of weak quality [31,32,36–38,41,43] (Table S2).

3.3. Medication Regimen Complexity Assessment

To assess MRC, six studies considered only prescription medication [31–33,35,40,42], while four included both prescription and non-prescription medication [22,36,38,44]. Other studies did not mention the prescription/non-prescription status of medication, considering instead long-term, short-term and "when required" medication (n = 3) [10,12,34], or only long-term medication (n = 1) [39]. Additionally, one study included all medication that subjects take a day, without any other mention [37] and another one included patients' routine chronic medication [43].

In regard to the instrument used to assess MRC, the majority of studies (n = 11) [10,12,22,32,33,39–44] applied the 65-item MRCI, which is an open-ended tool, where the final score is the result of the sum of three sections (dosage forms, dosage frequency and additional instructions). Besides this, one study used the medication complexity index (MCI) [31], two studies calculated a complexity score by summing the different dosage intervals, weighted for frequency [35,36], one study calculated the number of times that medications were taken in a 24 h period for each subject [37] and another study did not make reference to the instrument used to determine regimen complexity [34]. Additionally, one study [38] used the patient-level MRCI (pMRCI), which is the sum of three MRCI sub-scores for: prescription disease state medications, prescription for other non-disease medications and over-the-counter (OTC) medications [45,46].

3.4. Pharmacists' Role on Older People's Medication Regimen Complexity and Main Outcomes When Intervention Is Performed

In most studies pharmacists' role was limited to data collection (n = 8) [22,31,33,35,36,39,41,44]. In two studies, pharmacists acted as coders [37,38] and in two other they contributed to data analysis [32,43]. Four studies referred specifically to pharmacists' actions on MRC [10,12,34,42]: in one, pharmacists only determined regimen simplification potential [10], in two [12,34], simplifications were implemented, and at last, in the fourth study [42], the impact of pharmacists' residential medication management reviews (RMMRs) on the MRCI were retrospectively analysed. Furthermore, another study reported pharmacists' intervention on medication- and health-related problems (MHRPs), with MRC being one of the health-related covariates that could be changed [40].

Concerning the main outcomes when pharmacist intervention is performed, Elliot [34] states the proportion of identified and implemented regimen simplifications, and the reasons for non-implementation, as endpoints, while Elliot et al. [12] consider the change in MRCI score between hospital admission and discharge as main outcome measures. Moreover, Pouranayatihosseinabad et al. [42] present their outcomes as MRCI score change (at baseline, after pharmacists' recommendation and after general practitioners (GPs) acceptance of those recommendations) and number and type of pharmacists' recommendations and further GPs' acceptance. Finally, Moczygemba et al. [40] present their study results as clinical (change in MHRPs and medication adherence) and economic outcomes (Part D drug costs).

Lack of time [12,34], non-acceptance of recommendations by the physicians and patients [12,34,35] as well as medication prescribed by another physician [35] were pointed out as the main reasons for non-implementation of regimen simplifications.

4. Discussion

Three recent systematic reviews focused on MRC: Wimmer et al. [20] reviewed the association of clinical outcomes with MRC in older people, Pantuzza et al. [4] investigated the association between MRC and pharmacotherapy adherence and Alves-Conceição et al. [47] identified health outcomes related to MRC measured by MRCI.

This work is the first systematic review to explore pharmacists' role in studies on MRC of the older population to the best of our knowledge.

4.1. Medication Regimen Complexity Assessment

At first, it is essential to mention the heterogeneity of instruments used to assess MRC. Several instruments were used in the different studies, including the medication complexity index (MCI) [31], which failed to show satisfactory reliability with complex regimens, and did not demonstrate any significant correlation with outcomes such as medication adherence [15,48]. However, most studies already use the MRCI, which already shows good evidence of classifying complexity better than a simple medication count [6], discriminating between regimens with an equal number of medications, resulting in higher complexity scores for regimens with fewer drugs [15] and being a better overall predictor of all-cause mortality [49] and discharge destination [22] than polypharmacy. Additionally, in a few studies, the MRCI has been regarded as beneficial in targeting patients who may benefit from additional services such as domiciliary reviews and medication therapy management (MTM) services [14,15,45]. These strengths of the MRCI over other instruments should be taken into account in future investigations, especially regarding the importance of using a universal tool for MRC determination.

A greater consensus should also be achieved about the type of medication included to determine regimen complexity, which varied from prescription to non-prescription; long-term, short-term and "when required" as well as routine chronic medication. Even concerning the MRCI, there is no uniformity in the medications to be included. Although the instrument was initially developed and validated only for prescribed medications [15], several studies already indicate that prescription and non-prescription medications contribute to regimen complexity and should be considered [14,20,22,44–46]. However, even in that case, there is still no harmony in the practical applicability of the instrument: some authors [22,44] use the original MRCI, while Linnebur et al. [38] use the pMRCI. This aspect may be relevant to set high and low complexity scores, which has not yet been achieved despite some research in that area [9].

4.2. Measured Outcomes

Regarding the overall measured outcomes, in pharmacy practice research, the Economic, Clinical and Humanistic Outcomes (ECHO) model should be followed, with clinically meaningful outcomes being the most desirable [50,51]. However, in the present review none of the included studies present their results entirely according to this recommendation. Despite that, most of the reported results were related to the type of regimen simplification and its feasibility, reasons for non-implementation, change in the MRCI, the effect of recommendations as well as knowledge and preference of patients, which are endpoints whose relation to better patient outcomes are unknown [51]. Collection and further publication of relevant outcomes should be considered in future research.

4.3. Study Setting

In contrast with what was expected, most of the included studies were conducted in hospitals or clinics, but none in community pharmacies. This can reflect different factors: on the one hand there may exist an underreporting of provided pharmaceutical services, while on the other hand, it is also possible that still little attention has been given to this subject, even though several studies already state that the MRCI may be a valuable tool to prioritise patients who could take advantage of medication reviews or drug therapy management services [14,15,45]. At that time, MRC determination tools can be included, side by side with those that identify potentially inappropriate medications (PIMs), such as Beers [52] and STOPP criteria [53], as starting points for medication reduction, which are already an onset for regimen simplification. The frequently polymedicated older population may benefit most from this proximity, as the study findings show that overcomplexity is frequent among seniors [37] and that regimen complexities are higher in older adults with worse socio-economic and health conditions [31]. Additionally, insufficient pharmacotherapy understanding was high, especially among older adults with low levels of education and dependency on medication use [41]. These findings reinforce the need for pharmacists' intervention regarding older peoples' medication.

4.4. Pharmacists' Role

Only four [12,34,40,42] of the 17 included studies mention pharmacists' intervention: two studies focused directly on regimen complexity simplification while the other focused on MHRPS, with MRC being one of the variables. Elliot [34] demonstrated that a clinical pharmacist's simplification of older inpatients' medication is feasible when previous training about simplification is provided. In addition, Elliot et al. [12] concluded that after an educational intervention, a pharmacist-led medication review reduced the impact of hospitalisation on the complexity of older patients' medication regimens. Furthermore, Moczygemba et al. [40] obtained results that show that a telephone MTM telephone program from a pharmacist reduced MHRPs. Finally, Pouranayatihosseinabad et al. [42] concluded that pharmacists could use the MRCI to identify older adults with complex medication regimens, but they failed to show significant benefits of RMMRs in reducing MRC. However, other of the included studies refer to pharmacists' potential role in MRC: Elliot et al. [8] concluded that "most regimens had potential to be simplified by a clinical pharmacist review"; Lakey et al. [36] mentioned that "Educational strategies are needed to increase awareness of the pharmacist's role in facilitating medication management and the option of simplifying complex regimens" (p. 1011); Lindquist et al. [37] stated that "health care professionals need to be aware of how patients are taking their medications. ... another option would be to partner with pharmacists in reducing medication regimen complexity" (p.96); and Linnebur et al. [38] indicated that "our results suggest a need for pharmacist review of the patient's entire medication regimen ... to assess and reduce complexity to a manageable level for the patient if possible" (p. 1545).

One aspect that has to be mentioned under this topic is that the included studies were performed in many countries, where factors like national policies and culture may influence the recognition of pharmacists as a trusted profession in the community and for other health care providers, and therefore may be responsible for the differences observed in pharmacists' roles [54]. This fact may explain why Australia and the USA were the most representative countries in this review, with three of the four studies mentioning pharmacists' intervention being performed in Australia [12,34,42].

Given all the above, the present review highlights that the pharmacist's active role in improving MRC in the older population has been minimal. Nevertheless, the little evidence where pharmacists had an active role showed that medication regimen simplifications are feasible and emphasise the pharmacist's role to achieve them. However, it is also essential to bear in mind that several studies point out that previous educational sessions for pharmacists are necessary to raise awareness of this topic and give them the skills and practice to minimise regimen complexities [12,34,36]. Even so, a vital opponent to achieve regimen simplification in daily practice seems to be the lack of time of healthcare professionals [12,34]. Having this in mind, pharmacists can, however, take regimen simplifications into account in a more general way whenever they perform OTC advice in their daily routine, and more carefully when performing medication review services.

At last, one of the biggest challenges seems to be multidisciplinary collaboration. Among the findings of the studies, non-acceptance of recommendations by the prescribers is mentioned as one of the most common reasons for noncompliance with suggested regimen changes. As difficulties in the relationship between pharmacists/physicians are well known, we think it is also imperative to sensitise physicians to this subject and make clear that the ultimate goal of this collaboration is health gain, including optimising patients' health care.

Based on the present review findings, it seems that, until now, pharmacists have not played a relevant role in older people's MRC. For this reason, future high-quality research should focus on this subject, and in particular should include community pharmacists' interventions and the resulting possible benefits, not only for patients, e.g., in terms of safety, clinical outcomes and quality of life, but also for the healthcare system, in particular in terms of cost reductions.

4.5. Limitations

Although the search was conducted in three major databases, it is always possible that some studies have not been included. Scanning the reference list of the included studies only added one study to the selected ones, indicating that selection bias was minimal. The keywords and synonyms used in the search strategy may have been too restrictive, which may have led to the possible loss of some papers. Despite language bias being frequently reported, only one study has been rejected based on language in our review. Publication bias may also have occurred because only published full papers were included, leading to possibly missing relevant information. Moreover, the included studies were heterogeneous in study design, setting, data collection method, pharmacists' role and outcomes, which made the comparison difficult and meta-analysis impossible. Finally, it should be noted that this review took under consideration studies performed until October 2019, that is, in a pre-COVID-19 pandemic period. During the COVID-19 pandemic, pharmacists faced new approaches and had to adapt their routine procedures, therefore their role on older adults' MRC may have been different, and it may be reviewed and eventually compared to the pre-pandemic period.

5. Conclusions

Old age is often synonymous with multiple comorbidities and consequently polypharmacy and complex medication regimens. As the latter has been associated with several negative outcomes, particularly in the older population, an effort should be made to reduce MRC whenever possible. Pharmacists may play a relevant role at this point after previous training, which has, however, been underexplored. There is almost no research on pharmacists' intervention on older people's MRC; that which does exist is of moderate quality. This aspect leaves an open door for future high-quality evidence investigations on pharmacists' interventions and their relation to better outcomes. Therefore, pharmacists should be provided with the necessary skills, either during graduation or in post-graduate education and training programs, and encouraged to assess the possibility of simplifying the medication regimen in their daily routine or even on a service-based remuneration model.

Supplementary Materials: The following are available online at https://www.mdpi.com/article/10.3390/ijerph18168824/s1, Table S1: Pubmed search strategy, Table S2: EPHPP quality assessment tool rating for individual studies.

Author Contributions: Conceptualisation, C.F. and G.A.; methodology, C.F. and G.A.; formal analysis, C.F. and G.A.; writing—original draft preparation, C.F.; writing—review and editing, G.A.; visualisation and supervision, G.A. All authors have read and agreed to the published version of the manuscript.

Funding: This research received no external funding.

Institutional Review Board Statement: Not applicable.

Informed Consent Statement: Not applicable.

Data Availability Statement: Not applicable.

Conflicts of Interest: The authors declare no conflict of interest.

References

1. World Health Organization. *World Report on Ageing and Health*; World Health Organization Press: Geneva, Switzerland, 2015.
2. Advinha, A.M.; De Oliveira-Martins, S.; Mateus, V.; Pajote, S.G.; Lopes, M.J. Medication regimen complexity in institutionalized elderly people in an aging society. *Int. J. Clin. Pharm.* **2014**, *36*, 750–756. [CrossRef] [PubMed]
3. Silva, C.; Ramalho, C.; Luz, I.; Monteiro, J.; Fresco, P. Drug-related problems in institutionalized, polymedicated elderly patients: Opportunities for pharmacist intervention. *Int. J. Clin. Pharm.* **2015**, *37*, 327–334. [CrossRef] [PubMed]
4. Pantuzza, L.L.N.; Ceccato, M.D.G.B.; Silveira, M.R.; Junqueira, L.M.R.; Reis, A. Association between medication regimen complexity and pharmacotherapy adherence: A systematic review. *Eur. J. Clin. Pharmacol.* **2017**, *73*, 1475–1489. [CrossRef] [PubMed]
5. Paquin, A.M.; Zimmerman, K.M.; Kostas, T.R.; Pelletier, L.; Hwang, A.; Simone, M.; Skarf, L.M.; Rudolph, J.L. Complexity perplexity: A systematic review to describe the measurement of medication regimen complexity. *Expert Opin. Drug Saf.* **2013**, *12*, 829–840. [CrossRef]
6. Advinha, A.M.; Lopes, M.; De Oliveira-Martins, S. Assessment of the elderly's functional ability to manage their medication: A systematic literature review. *Int. J. Clin. Pharm.* **2016**, *39*, 1–15. [CrossRef]
7. Wimmer, B.C.; Johnell, K.; Fastbom, J.; Wiese, M.; Bell, J.S. Factors associated with medication regimen complexity in older people: A cross-sectional population-based study. *Eur. J. Clin. Pharmacol.* **2015**, *71*, 1099–1108. [CrossRef]
8. Herson, M.; Bell, J.; Tan, E.; Emery, T.; Robson, L.; Wimmer, B. Factors associated with medication regimen complexity in residents of long-term care facilities. *Eur. Geriatr. Med.* **2015**, *6*, 561–564. [CrossRef]
9. Ferreira, J.M.; Galato, D.; Melo, A.C. Medication regimen complexity in adults and the elderly in a primary healthcare setting: Determination of high and low complexities. *Pharm. Pr.* **2015**, *13*, 659. [CrossRef]
10. Elliott, R.A.; O'Callaghan, C.J. Impact of Hospitalisation on the Complexity of Older Patients' Medication Regimens and Potential for Regimen Simplification. *J. Pharm. Pr. Res.* **2011**, *41*, 21–25. [CrossRef]
11. Witticke, D.; Seidling, H.M.; Lohmann, K.; Send, A.F.J.; Haefeli, W.E. Opportunities to Reduce Medication Regimen Complexity: A retrospective analysis of patients discharged from a university hospital in Germany. *Drug Saf.* **2012**, *36*, 31–41. [CrossRef] [PubMed]
12. Elliott, R.A.; O'Callaghan, C.; Paul, E.; George, J. Impact of an intervention to reduce medication regimen complexity for older hospital inpatients. *Int. J. Clin. Pharm.* **2013**, *35*, 217–224. [CrossRef]
13. Stange, D.; Kriston, L.; Von-Wolff, A.; Baehr, M.; Dartsch, D.C. Reducing Cardiovascular Medication Complexity in a German University Hospital: Effects of a Structured Pharmaceutical Management Intervention on Adherence. *J. Manag. Care Pharm.* **2013**, *19*, 396–407. [CrossRef] [PubMed]
14. Rettig, S.M.; Wood, Y.; Hirsch, J.D. Medication regimen complexity in patients with uncontrolled hypertension and/or diabetes. *J. Am. Pharm. Assoc.* **2013**, *53*, 427–431. [CrossRef]
15. George, J.; Phun, Y.-T.; Bailey, M.J.; Kong, D.C.; Stewart, K. Development and Validation of the Medication Regimen Complexity Index. *Ann. Pharmacother.* **2004**, *38*, 1369–1376. [CrossRef]
16. Melchiors, A.C.; Correr, C.J.; Fernandez-Llimos, F. Tradução e validação para o português do Medication Regimen Complexity Index. *Arq. Bras. Cardiol.* **2007**, *89*, 210–218. [CrossRef]
17. Stange, D.; Kriston, L.; Langebrake, C.; Cameron, L.K.; Wollacott, J.D.; Baehr, M.; Dartsch, D.C. Development and psychometric evaluation of the German version of the Medication Regimen Complexity Index (MRCI-D). *J. Eval. Clin. Pr.* **2011**, *18*, 515–522. [CrossRef]
18. De La Fuente, J.S.; Díaz, A.S.; Cañamares, I.; Ramila, E.; Izquierdo-Garcia, E.; Esteban, C.; Escobar-Rodríguez, I. Cross-cultural Adaptation and Validation of the Medication Regimen Complexity Index Adapted to Spanish. *Ann. Pharmacother.* **2016**, *50*, 918–925. [CrossRef]
19. Pantuzza, L.L.N.; Ceccato, M.D.G.B.; Silveira, M.R.; Pinto, I.V.; Reis, A. Validation and standardization of the Brazilian version of the Medication Regimen Complexity Index for older adults in primary care. *Geriatr. Gerontol. Int.* **2018**, *18*, 853–859. [CrossRef] [PubMed]
20. Wimmer, B.C.; Cross, A.J.; Jokanovic, N.; Wiese, M.; George, J.; Johnell, K.; Diug, B.; Bell, J.S. Clinical Outcomes Associated with Medication Regimen Complexity in Older People: A Systematic Review. *J. Am. Geriatr. Soc.* **2016**, *65*, 747–753. [CrossRef] [PubMed]
21. De Vries, S.T.; Keers, J.C.; Visser, R.; De Zeeuw, D.; Haaijer-Ruskamp, F.M.; Voorham, J.; Denig, P. Medication beliefs, treatment complexity, and non-adherence to different drug classes in patients with type 2 diabetes. *J. Psychosom. Res.* **2014**, *76*, 134–138. [CrossRef] [PubMed]
22. Wimmer, B.C.; Dent, E.; Visvanathan, R.; Wiese, M.; Johnell, K.; Chapman, I.; Bell, J.S. Polypharmacy and Medication Regimen Complexity as Factors Associated with Hospital Discharge Destination Among Older People: A Prospective Cohort Study. *Drugs Aging* **2014**, *31*, 623–630. [CrossRef]
23. Fröhlich, S.E.; Zaccolo, A.V.; Da Silva, S.L.C.; Mengue, S.S. Association between drug prescribing and quality of life in primary care. *Pharm. World Sci.* **2010**, *32*, 744–751. [CrossRef] [PubMed]
24. Center for Reviews and Dissemination. *Systematic Reviews: CRD's Guidance for Undertaking Reviews in Health Care*; CRD University of York: York, UK, 2009.

25. Moher, D.; Liberati, A.; Tetzlaff, J.; Altman, D.G. The PRISMA Group Preferred Reporting Items for Systematic Reviews and Meta-Analyses: The PRISMA Statement. *PLoS Med.* **2009**, *6*, e1000097. [CrossRef]
26. Liberati, A.; Altman, D.G.; Tetzlaff, J.; Mulrow, C.; Gøtzsche, P.C.; Ioannidis, J.P.A.; Clarke, M.; Devereaux, P.; Kleijnen, J.; Moher, D. The PRISMA statement for reporting systematic reviews and meta-analyses of studies that evaluate healthcare interventions: Explanation and elaboration. *BMJ* **2009**, *339*, b2700. [CrossRef] [PubMed]
27. World Health Organization Centre for Health Development. *A Glossary of Terms for Community Health Care and Services for Older Persons*; WHO Centre for Health Development: Kobe, Japan, 2004; p. 42. Available online: https://apps.who.int/iris/handle/10665/68896 (accessed on 12 November 2019).
28. Sabater, D.; Fernandez-Llimos, F.; Parras, M.; Faus, M.J. Tipos de intervenciones farmacéuticas en seguimiento farmacoterapéutico. Types of pharmacist intervention in pharmacotherapy follow-up. *Seguim. Farmacoter.* **2005**, *3*, 90–97.
29. Effective Public Health Practice Project (EPHPP). Quality Assessment Tool for Quantitative Studies. Available online: http://www.ephpp.ca/tools.html (accessed on 20 October 2019).
30. Armijo-Olivo, S.; Stiles, C.R.; Hagen, N.A.; Biondo, P.D.; Cummings, G.G. Assessment of study quality for systematic reviews: A comparison of the Cochrane Collaboration Risk of Bias Tool and the Effective Public Health Practice Project Quality Assessment Tool: Methodological research. *J. Evaluation Clin. Pr.* **2010**, *18*, 12–18. [CrossRef]
31. Acurcio, F.A.; Silva, A.L.; Ribeiro, A.Q.; Rocha, N.P.; Silveira, M.R.; Klein, C.H.; Rozenfeld, S. Complexidade do regime terapêutico prescrito para idosos. *Rev. Assoc. Med. Bras.* **2009**, *55*, 468–474. [CrossRef]
32. Bazargan, M.; Smith, J.; Yazdanshenas, H.; Movassaghi, M.; Martins, D.; Orum, G. Non-adherence to medication regimens among older African-American adults. *BMC Geriatr.* **2017**, *17*, 163. [CrossRef]
33. Chang, W.T.; Kowalski, S.R.; Sorich, W.; Alderman, C.P. Medication regimen complexity and prevalence of potentially inappropriate medicines in older patients after hospitalisation. *Int. J. Clin. Pharm.* **2017**, *39*, 867–873. [CrossRef]
34. Elliott, R.A. Reducing medication regimen complexity for older patients prior to discharge from hospital: Feasibility and barriers. *J. Clin. Pharm. Ther.* **2012**, *37*, 637–642. [CrossRef]
35. Kroenke, K.; Pinholt, E.M. Reducing polypharmacy in the elderly. a controlled trial of physician feedback. *J. Am. Geriatr. Soc.* **1990**, *38*, 31–36. [CrossRef]
36. Lakey, S.L.; Gray, S.L.; Borson, S. Assessment of Older Adults' Knowledge of and Preferences for Medication Management Tools and Support Systems. *Ann. Pharmacother.* **2009**, *43*, 1011–1019. [CrossRef] [PubMed]
37. Lindquist, L.A.; Lindquist, L.M.; Zickuhr, L.; Friesema, E.; Wolf, M.S. Unnecessary complexity of home medication regimens among seniors. *Patient Educ. Couns.* **2014**, *96*, 93–97. [CrossRef] [PubMed]
38. Linnebur, S.A.; Griend, J.P.V.; Metz, K.R.; Hosokawa, P.W.; Hirsch, J.D.; Libby, A.M. Patient-level Medication Regimen Complexity in Older Adults with Depression. *Clin. Ther.* **2014**, *36*, 1538–1546.e1. [CrossRef]
39. Mansur, N.; Weiss, A.; Beloosesky, Y. Looking Beyond Polypharmacy: Quantification of Medication Regimen Complexity in the Elderly. *Am. J. Geriatr. Pharmacother.* **2012**, *10*, 223–229. [CrossRef]
40. Moczygemba, L.R.; Barner, J.C.; Gabrillo, E.R. Outcomes of a Medicare Part D telephone medication therapy management program. *J. Am. Pharm. Assoc.* **2012**, *52*, e144–e152. [CrossRef] [PubMed]
41. Pinto, I.V.L.; Reis, A.; Almeida-Brasil, C.C.; Da Silveira, M.R.; Lima, M.G.; Ceccato, M.D.G.B. Avaliação da compreensão da farmacoterapia entre idosos atendidos na Atenção Primária à Saúde de Belo Horizonte, MG, Brasil. *Cien. Saude Colet.* **2016**, *21*, 3469–3481. [CrossRef] [PubMed]
42. Pouranayatihosseinabad, M.; Zaidi, S.T.R.; Peterson, G.; Nishtala, P.S.; Hannan, P.; Castelino, R. The impact of residential medication management reviews (RMMRs) on medication regimen complexity. *Postgrad. Med.* **2018**, *130*, 575–579. [CrossRef]
43. Sevilla-Sánchez, D.; Molist-Brunet, N.; Amblàs-Novellas, J.; Espaulella-Panicot, J.; Codina-Jané, C. Potentially inappropriate medication at hospital admission in patients with palliative care needs. *Int. J. Clin. Pharm.* **2017**, *39*, 1018–1030. [CrossRef]
44. Wimmer, B.C.; Dent, E.; Bell, J.S.; Wiese, M.D.; Chapman, I.; Johnell, K.; Visvanathan, R. Medication Regimen Complexity and Unplanned Hospital Readmissions in Older People. *Ann. Pharmacother.* **2014**, *48*, 1120–1128. [CrossRef]
45. Hirsch, J.D.; Metz, K.R.; Hosokawa, P.W.; Libby, A. Validation of a Patient-Level Medication Regimen Complexity Index as a Possible Tool to Identify Patients for Medication Therapy Management Intervention. *Pharmacother. J. Hum. Pharmacol. Drug Ther.* **2014**, *34*, 826–835. [CrossRef]
46. Libby, A.M.; Fish, D.N.; Hosokawa, P.W.; Linnebur, S.A.; Metz, K.R.; Nair, K.V.; Saseen, J.; Griend, J.P.V.; Vu, S.P.; Hirsch, J.D. Patient-Level Medication Regimen Complexity Across Populations With Chronic Disease. *Clin. Ther.* **2013**, *35*, 385–398. [CrossRef]
47. Alves-Conceição, V.; Rocha, K.S.S.; Silva, F.V.N.; Silva, R.O.S.; Da Silva, D.T.; De Lyra, D.P., Jr. Medication Regimen Complexity Measured by MRCI: A Systematic Review to Identify Health Outcomes. *Ann. Pharmacother.* **2018**, *52*, 1117–1134. [CrossRef]
48. Conn, V.S.; Taylor, S.G.; Kelley, S. Medication Regimen Complexity and Adherence Among Older Adults. *Image J. Nurs. Sch.* **1991**, *23*, 231–236. [CrossRef]
49. Wimmer, B.C.; Bell, J.S.; Fastbom, J.; Wiese, M.; Johnell, K. Medication Regimen Complexity and Polypharmacy as Factors Associated With All-Cause Mortality in Older People: A population-based cohort study. *Ann. Pharmacother.* **2015**, *50*, 89–95. [CrossRef]
50. Kozma, C.M.; Reeder, C.E.; Schultz, R.M. Economic, clinical, and humanistic outcomes: A planning model for pharmacoeconomic research. *Clin. Ther.* **1993**, *15*, 1121–1132.

51. Charrois, T.L.; Durec, T.; Tsuyuki, R.T. Systematic Reviews of Pharmacy Practice Research: Methodologic Issues in Searching, Evaluating, Interpreting, and Disseminating Results. *Ann. Pharmacother.* **2009**, *43*, 118–122. [CrossRef] [PubMed]
52. American Geriatrics Society. American Geriatrics Society 2019 updated AGS Beers Criteria®for potentially inappropriate medication use in older adults. *J. Am. Geriatr. Soc.* **2019**, *67*, 674–694. [CrossRef] [PubMed]
53. O'Mahony, D.; O'Sullivan, D.; Byrne, S.; O'Connor, M.N.; Ryan, C.; Gallagher, P. STOPP/START criteria for potentially inappropriate prescribing in older people: Version 2. *Age Ageing* **2014**, *44*, 213–218. [CrossRef] [PubMed]
54. Anderson, S. The state of the world's pharmacy: A portrait of the pharmacy profession. *J. Interprofessional Care* **2002**, *16*, 391–404. [CrossRef]

MDPI
St. Alban-Anlage 66
4052 Basel
Switzerland
www.mdpi.com

International Journal of Environmental Research and Public Health Editorial Office
E-mail: ijerph@mdpi.com
www.mdpi.com/journal/ijerph

www.ingramcontent.com/pod-product-compliance
Lightning Source LLC
Chambersburg PA
CBHW061519210926
51850CB00008B/1978
9783725809813